PHASEFOCUS

2

Quantitative Phase Imaging of Cells and Tissues

About the Author

Gabriel Popescu is Assistant Professor at the University of Illinois at Urbana-Champaign in the Department of Electrical and Computer Engineering, Beckman Institute for Advanced Science & Technology. He is the editor of *Nanobiophotonics*.

Quantitative Phase Imaging of Cells and Tissues

Gabriel Popescu

Mc
Graw
Hill

New York   Chicago   San Francisco
Lisbon   London   Madrid   Mexico City
Milan   New Delhi   San Juan
Seoul   Singapore   Sydney   Toronto

The McGraw·Hill Companies

Cataloging-in-Publication Data is on file with the Library of Congress

McGraw-Hill books are available at special quantity discounts to use as premiums and sales promotions, or for use in corporate training programs. To contact a representative please e-mail us at bulksales@mcgraw-hill.com.

Quantitative Phase Imaging of Cells and Tissues

1 2 3 4 5 6 7 8 9 0   CTPS/CTPS   1 9 8 7 6 5 4 3 2 1

ISBN 978-0-07-166342-7
MHID 0-07-166342-8

The pages within this book were printed on acid-free paper.

Sponsoring Editor
Michael Penn

Proofreader
Upendra Prasad, Glyph International

Acquisitions Coordinator
Michael Mulcahy

Indexer
Valerie Haynes Perry

Editorial Supervisor
David E. Fogarty

Production Supervisor
Richard C. Ruzycka

Project Manager
Ranjit Kaur, Glyph International

Composition
Glyph International

Copy Editor
Erica Orloff

Art Director, Cover
Jeff Weeks

To Catherine, Sophia, Sorin, and my mother

"The microscope image is the interference effect of a diffraction phenomenon"

Ernst Abbe, 1873.

Contents

Foreword

This book is an authoritative presentation of a relatively new and rapidly expanding subject that is of particular interest to biologists, chemists, physicists, physicians, and engineers. The techniques discussed have been developed in recent years and they provide information, for example, about the dynamics of both the thickness and the refractive index fluctuations of specimens on the nanoscale.

The text, written by a leading expert in the field, is one of the first, if not the first, comprehensive presentation of the subject. It is clearly written and beautifully illustrated. It is a joy to read and to learn from it.

EMIL WOLF
Wilson Professor of Optical Physics
University of Rochester
Rochester, New York

Preface

This book is intended to give insight into the emerging field of *quantitative phase imaging (QPI)* as it applies to biomedicine. It is an invitation extended to researchers at various stages in their careers to explore this new and exciting field of study. I devoted particular effort toward providing enough introductory material to make the book self-contained, as much as possible. Thus, the structure of the book progresses from basic to advanced concepts as follows.

The motivation and key concepts behind QPI are presented in the *Introduction*, with particular attention to clarifying ideas such as "nanoscale" and "three-dimensional", which often generate confusion. *Chapter 2 (Groundwork)* reviews the basics of light propagation in a vacuum and inhomogeneous media (scattering), with emphasis on Fourier optics. Coherence properties of optical fields are described in *Chapter 3*. Very basic properties of images (e.g., resolution, contrast, contrast-to-noise ratio) are reviewed in *Chapter 4*. Light microscopy concepts, from Abbe's image description to Zernike's phase contrast are discussed in *Chapter 5*, while the main developments in holography are reviewed in *Chapter 6*. The remaining chapters, 7-15 are devoted specifically to various aspects of QPI.

Except for *Chapter 7*, which is dedicated to point-scanning QPI methods and includes a thorough introduction to low-coherence interferometry, all chapters deal with full-field QPI methods. *Chapter 8* presents the main ideas behind full-field QPI and specifies the main figures of merit in QPI: acquisition rate, transverse resolution, temporal phase stability, and spatial phase uniformity. The following four chapters, 9-12, describe four QPI approaches that, by default, excel in one of these figures of merit: off-axis (high acquisition rates), phase-shifting (high transverse resolution), common-path (high temporal stability), and white light (high spatial uniformity).

Chapter 13 presents *Fourier transform light scattering*, essentially establishing the equivalence between QPI and light scattering measurements. The last two chapters, 14 and 15, are devoted to recent developments, both in methods and applications, which currently appear to be very promising. Finally, the book contains three appendices on complex analytic signals (A), Fourier transforms (B), and interesting QPI images (C).

It is my belief that QPI will continue to grow at an accelerated pace and become a dominant field in biomedical optics in the years to come. It is hoped that this book will contribute to this process by providing a logical structure of this new field and a condensed summary of the current research.

GABRIEL POPESCU
Urbana, Illinois

Aknowledgments

In the past decade, I have learned immensely from the work of my peers in the field of Quantitative Phase Imaging. In writing this book, it has been an extremely challenging task to go through thousands of publications and select the main developments to be discussed. While the book cites a combined number of 794 references, it is possible that some important work was overlooked, for which I apologize in advance.

I learned a great deal from my thesis advisor, Aristide Dogariu, and postdoctoral mentor, the late Michael Feld. I am especially thankful to Ramachandra Dasari for giving me a old microscope from Ali Javan's laboratory, with which I obtained my first quantitative phase images (one makes the cover of this book).

I am indebted to Emil Wolf for generous encouragements and for reviewing Chapter 3 of this book. I am grateful to the current and previous colleagues with whom I have been privileged to work in the past ten years: Huafeng Ding, Zhuo Wang, Mustafa Mir, and Ru Wang (Quantitative Light Imaging Laboratory, Department of Electrical and Computer Engineering, University of Illinois at Urbana-Champaign); YoungKeun Park, Niyom Lue, Shahrooz Amin, Lauren Deflores, Seungeun Oh, Christopher Fang-Yen, Wonshik Choi, Kamran Badizadegan, and Ramachandra Dasari (Spectroscopy Laboratory, Massachusetts Institute of Technology); Takahiro Ikeda and Hidenao Iwai (Hamamatsu Photonics KK).

I would like to acknowledge contributions from my collaborators: Catherine Best-Popescu (College of Medicine, University of Illinois at Urbana-Champaign), Martha Gillette (Department of Cell and Developmental Biology, University of Illinois at Urbana-Champaign), Alex Levine (Department of Chemistry and Biochemistry, University of California at Los Angeles), Scott Carney (Department of Electrical and Computer Engineering, University of Illinois at Urbana-Champaign), Dan Marks (Department of Electrical and Computer Engineering, University of Illinois at Urbana-Champaign; current address Duke University), Krishna Tangella (Department of Pathology, Provena Covenant Medical Center), Supriya Prasanth (Department of Cell and

Developmental Biology, University of Illinois at Urbana-Champaign), Marni Boppart (Department of Kinesiology and Community Health, Beckman Institute for Advanced Science & Technology, University of Illinois at Urbana-Champaign), Stephen A. Boppart (Department of Electrical and Computer Engineering, Beckman Institute for Advanced Science & Technology, University of Illinois at Urbana-Champaign), Subra Suresh (Massachusetts Institute of Technology), Michael Laposata (Division of Laboratory Medicine and Clinical Laboratories, Vanderbilt University Medical Center), Carlo Brugnara (Department of Laboratory Medicine, Children's Hospital Boston).

This book would not be possible without the generous support from the National Science Foundation (NSF). In particular, my proposal entitled *Quantitative phase imaging of cells and tissues* was funded as CAREER Award (CBET 08-46660) and allowed new progress in my laboratory in both research and education. The Network for Computational Nanoscience (NCN), an NSF-funded center, also supported undergraduate student activities toward the development of new education material. I am grateful to NCN's Umberto Ravioli and Nahil Sobh for their enthusiasm toward such scholarly activities. Joe Leigh has been the main student supported by NCN to assist with preparing the material in a presentable form. I am extremely grateful to Julie McCartney for assisting me throughout the process.

I am indebted to the McGraw-Hill team, especially Michael Penn, David Fogarty, and Richard Ruzycka. Special thanks to Glyph International for carefully editing, illustrating, and typesetting the book.

GABRIEL POPESCU

CHAPTER 1

Introduction

1.1 Light Microscopy

The light microscope is one of the significant inventions in the history of humankind that, along with the telescope, played a central role in the Scientific Revolution, which started around 1600.[1] Despite its long history, of more than four centuries, the microscopy field has continuously expanded, covering a growing number of methods.[2] Much of the effort in the field has been devoted to improving two main properties of the microscopic image: *resolution* and *contrast.*

In terms of resolution, Abbe showed in 1873 that the ultimate, theoretical limit for far-field imaging is, essentially, half the wavelength of light.[3] Since then, researchers mainly have worked on approaching this limit (by aberration correction, etc.) rather than exceeding it. The first superresolution optical imaging occurred much more recently, by employing evanescent rather than propagating waves.[4] Remarkably, in the past decade or so, several methods have been developed to exceed the diffraction barrier in *far-field* fluorescence microscopy. Thus, techniques such as STED (stimulated emission depletion),[5-7] (f)PALM (fluorescence photoactivation localization microscopy),[8-14] STORM (stochastic optical reconstruction microscopy),[15-17] and structured illumination[18] represent a departure from the diffraction-limited imaging formulated by Abbe.[3]

There are two main kinds of contrast: *endogenous* and *exogenous.* The endogenous (intrinsic) contrast is generated by revealing the structures as they appear naturally. This challenge is typically addressed via *optical* solutions, i.e., by exploiting the phenomenon of light-matter interaction. On the other hand, exogenous contrast is produced by attaching a contrast agent (e.g., stain, fluorescent dye) to the structure of interest. A remarkable development in fluorescence microscopy is the technology based on green fluorescent protein (GFP).[19] In this case, live cells are genetically modified to express GFP, a protein purified from jellyfish, which essentially converts the exogenous into endogenous contrast generation. Deep-tissue imaging is also a contrast problem, which is addressed via techniques such as confocal microscopy, nonlinear microscopy, and optical coherence tomography.

1.2 Quantitative-Phase Imaging (QPI)

The great obstacle in generating intrinsic contrast from optically thin specimens (including live cells) is that, generally, they do not absorb or scatter light significantly, i.e., they are transparent, or *phase objects*. In his theory, Abbe described image formation as an interference phenomenon,[3] opening the door for formulating the problem of contrast precisely as in interferometry. Exploiting this concept, in the 1930s Zernike developed phase-contrast microscopy (PCM), in which the contrast of the interferogram generated by the scattered and unscattered light (i.e., the contrast of the image) is enhanced by shifting their relative phase by a quarter wavelength and further matching their relative power.[20,21] PCM represents a major advance in intrinsic contrast imaging, as it reveals inner details of transparent structures without staining or tagging. However, the resulting phase-contrast image is an *intensity* distribution, in which the *phase* information is coupled nonlinearly and cannot be retrieved *quantitatively*.

In the 1940s, Gabor understood the significance of the phase information and proposed *holography* as an approach to exploit it for imaging purposes.[22] It became clear that knowing both the amplitude and phase of the field allows imaging to be treated as transmission of information, akin to radio communication.[23]

QPI essentially combines the pioneering ideas of Abbe, Zernike, and Gabor (Fig. 1.1). The resulting image is a map of pathlength shifts associated with the specimen. Of course, this image contains information about both the local thickness and refractive index of the structure, which makes the decoupling of their respective contributions

Ernst Abbe   Frits Zernike   Dennis Gabor
(1840-1905)   (1888-1966)   (1900-1979)

FIGURE 1.1   Pioneers of coherent light microscopy.

challenging. At the same time, recent work shows that QPI provides a powerful means to study dynamics associated with both thickness and refractive index fluctuations. Remarkably, the quantitative-phase map associated with a live cell reports on the cell's dry mass density, i.e., its non-aqueous content.[24,25] Thus, QPI has the ability to quantify cell growth with *femtogram* sensitivity and without contact.[26]

1.3 QPI and Multimodal Investigation

From the knowledge of spatially resolved phase distribution, $\phi(x, y)$, various other visualization modalities can be easily obtained by simple numerical calculations. Figure 1.2 shows how, by taking the 1D gradient of a quantitative-phase image, an image similar to that of differential interference contrast (DIC) microscopy is obtained (Fig. 1.2b). In QPI, we have the additional flexibility of numerically removing the "shadow artifact" associated with DIC images. This artifact is due to the first-order derivative-changing sign across an edge and can be easily eliminated by taking the modulus of the gradient (Fig. 1.2c). Further, the Laplacian of the phase image reveals fine details (high-frequency content) from the specimen (Fig. 1.2d).

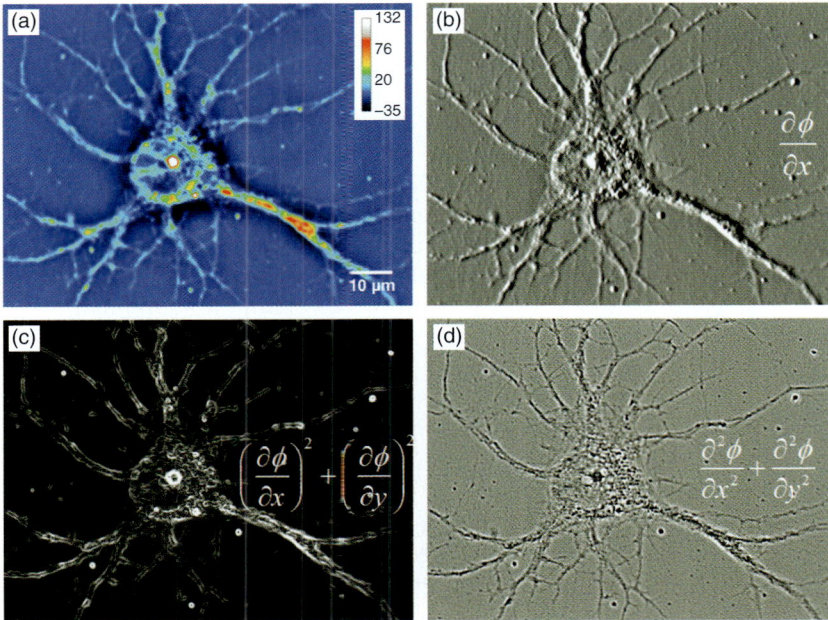

FIGURE 1.2 (a) QPI of a neuron (colorbar shows pathlength in nanometer). (b) Synthetic DIC image obtained by taking a 1D derivative of a. (c) Image obtained by taking the modulus squared of the gradient associated with a. (d) Image obtained by taking the Laplacian of a.

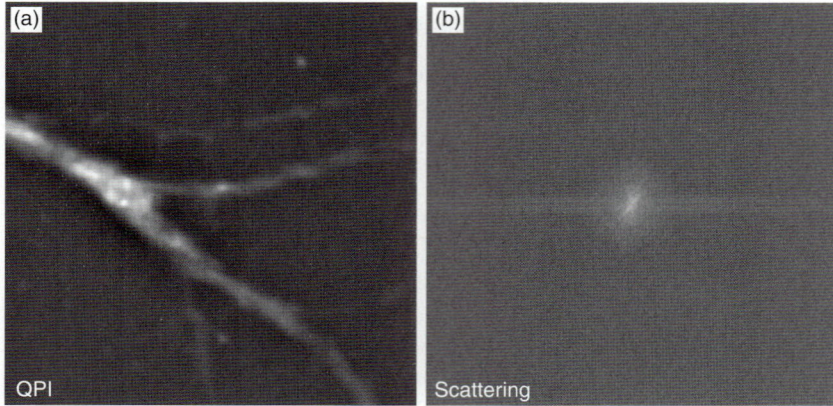

FIGURE **1.3**   Fourier transform light scattering. (*a*) QPI of dendrites; (*b*) scattering map from dendrites in *a*.

Perhaps one of the most striking features of QPI is that it can generate *light-scattering* data with extreme sensitivity. This happens because full knowledge of the complex (i.e., amplitude and phase) field at a given plane (the image plane) allows us to infer the field distribution at any other plane, including in the far zone. In other words, the image and scattering fields are simply Fourier transforms of each other; this relationship does not hold in intensity. Figure 1.3 shows the QPI of a dendritic structure (lower-right corner of Fig. 1.2*a*) and its corresponding scattering map obtained by taking the Fourier transform of the image field. This approach, called Fourier transform light scattering (FTLS) is much more sensitive than common, goniometer-based angular scattering because the measurement takes place at the image plane, where the optical field is most uniform. As a result, FTLS can render with ease scattering properties of minute subcellular structures which is unprecedented.

1.4   Nanoscale and Three-Dimensional Imaging

Sometimes, some confusion emerges regarding QPI, especially in the context of *nanoscale* and *3D imaging*. Here we briefly address these two issues.

First, it is clear that QPI provides sensitivity to spatial and temporal pathlength changes down to the *nanoscale*. This has been exploited, for example, in studies of red blood cell fluctuations and topography of nanostructures. Figure 1.4 illustrates the nanoscale sensitivity to pathlength changes caused by transport along a neuron dendrite. However, this *sensitivity* should not be referred to as axial *resolution*. Nanometer resolution, or resolving power, would describe the ability

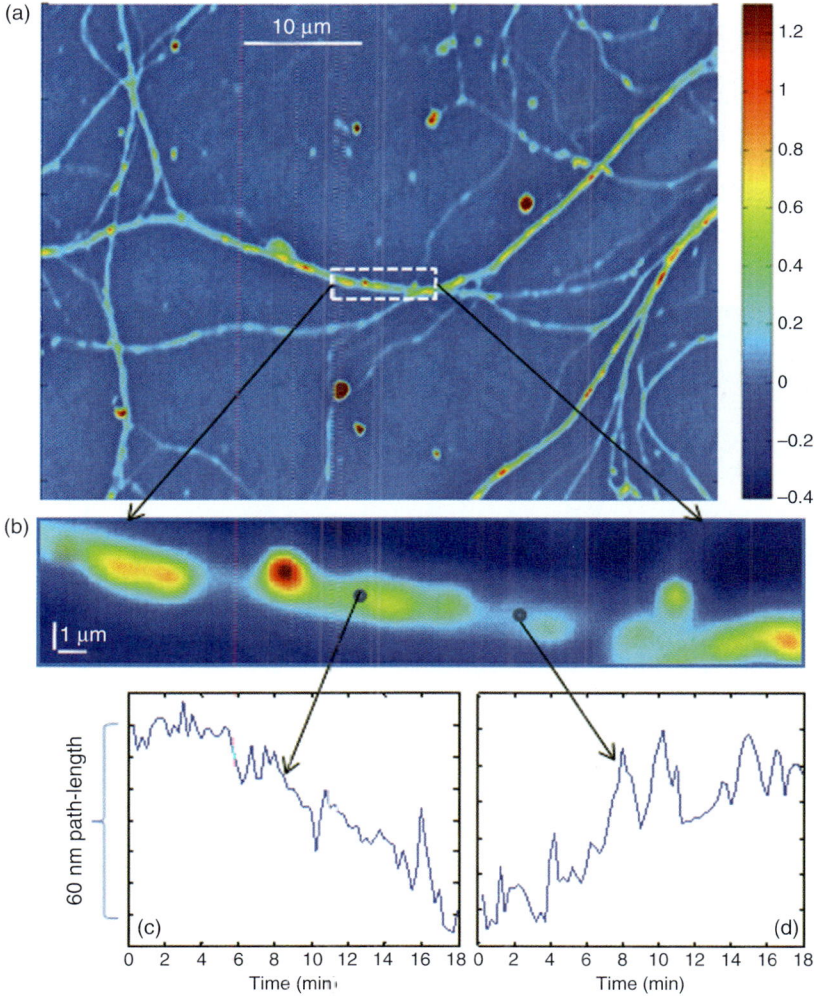

FIGURE 1.4   (a) QPI image of live neuronal processes. (b) The magnified region indicated in a. (c and d) Nanoscale temporal sensitivity to pathlength fluctuations associated with the respective points in b.

of QPI to resolve two objects separated axially by 1 nm. Of course, this is impossible, as it would violate the uncertainty principle.

Second, QPI images are occasionally represented as surface plots of the form in Fig. 1.5c. Perhaps because these plots contain three axes, (ϕ, x, y), sometimes QPI is erroneously referred to as *3D imaging*. Note that 3D imaging, or *tomography*, means resolving a physical property of the object (in this case refractive index, n) in all three dimensions. Rrepresenting these tomographic data requires four axes

(a)

(b)

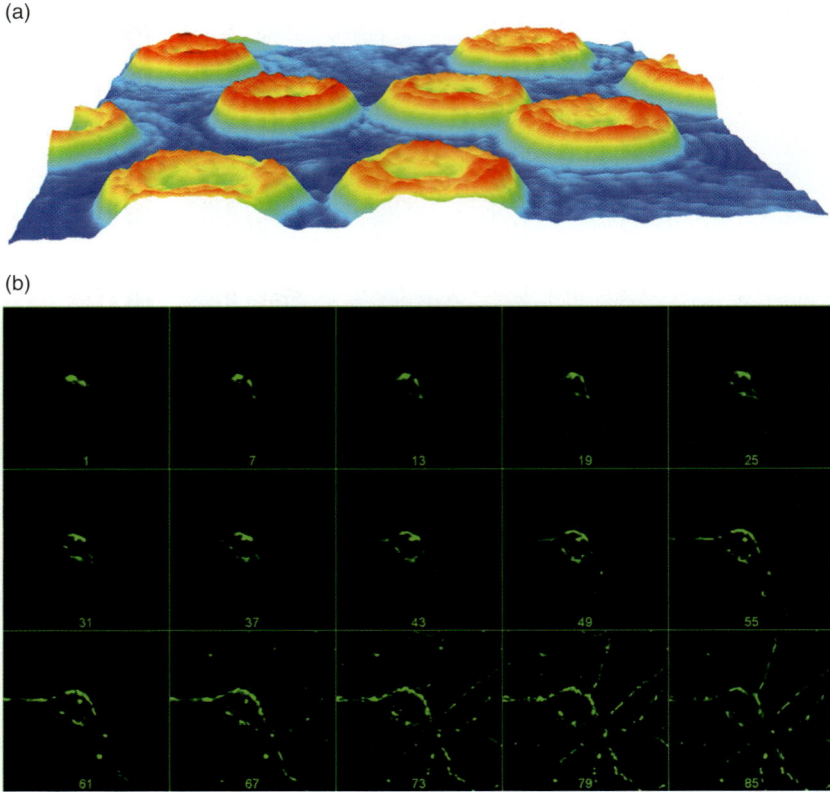

FIGURE 1.5   (a) Surface plot representation of quantitative-phase image, $\phi\,(x, y)$. (b) Montage representation of a tomogram, $n\,(x, y, z)$. Images 1 to 85 represent different slices along z.

(n, x, y, z), which means that in a plot only certain sections or projections of the data can be represented (Fig. 1.5b). While QPI can be used to perform tomography (see Chap. 14), this operation requires acquisition of images vs. an additional dimension, e.g., wavelength, sample rotation angle, or sample axial position.

References

1. R. P. Carlisle, *Scientific American Inventions and Discoveries: All the Milestones in Ingenuity—from the Discovery of Fire to the Invention of the Microwave Oven* (John Wiley & Sons, Hoboken, N, 2004).
2. Milestones in light microscopy, *Nat Cell Biol*, 11, 1165–1165 (2009).
3. E. Abbe, Beiträge zur Theorie des Mikroskops und der mikroskopischen Wahrnehmung, *Arch. Mikrosk. Anat.*, 9, 431 (1873).
4. E. Betzig, J. K. Trautman, T. D. Harris, J. S. Weiner and R. L. Kostelak, Breaking the diffraction barrier-optical microscopy on a nanometric scale, *Science*, 251, 1468–1470 (1991).

5. R. Schmidt, C. A. Wurm, S. Jakobs, J. Engelhardt, A. Egner and S. W. Hell, Spherical nanosized focal spot unravels the interior of cells, *Nature Methods*, 5, 539–544 (2008).
6. J. Folling, M. Bossi, H. Bock, R. Medda, C. A. Wurm, B. Hein, S. Jakobs, C. Eggeling and S. W. Hell, Fluorescence nanoscopy by ground-state depletion and single-molecule return, *Nature Methods*, 5, 943–945 (2008).
7. K. I. Willig, R. R. Kellner, R. Medda, B. Hein, S. Jakobs and S. W. Hell, Nanoscale resolution in GFP-based microscopy, *Nature Methods*, 3, 721–723 (2006).
8. H. Shroff, C. G. Galbraith, J. A. Galbraith and E. Betzig, Live-cell photoactivated localization microscopy of nanoscale adhesion dynamics, *Nature Methods*, 5, 417–423 (2008).
9. S. Manley, J. M. Gillette, G. H. Patterson, H. Shroff, H. F. Hess, E. Betzig and J. Lippincott-Schwartz, High-density mapping of single-molecule trajectories with photoactivated localization microscopy, *Nature Methods*, 5, 155–157 (2008).
10. H. Shroff, C. G. Galbraith, J. A. Galbraith, H. White, J. Gillette, S. Olenych, M. W. Davidson and E. Betzig, Dual-color superresolution imaging of genetically expressed probes within individual adhesion complexes, *Proceedings of the National Academy of Sciences of the United States of America*, 104, 20308–20313 (2007).
11. E. Betzig, G. H. Patterson, R. Sougrat, O. W. Lindwasser, S. Olenych, J. S. Bonifacino, M. W. Davidson, J. Lippincott-Schwartz and H. F. Hess, Imaging intracellular fluorescent proteins at nanometer resolution, *Science*, 313, 1642–1645 (2006).
12. T. J. Gould, M. S. Gunewardene, M. V. Gudheti, V. V. Verkhusha, S. R. Yin, J. A. Gosse and S. T. Hess, Nanoscale imaging of molecular positions and anisotropies, *Nature Methods*, 5, 1027–1030 (2008).
13. S. T. Hess, T. P. K. Girirajan and M. D. Mason, Ultra-high resolution imaging by fluorescence photoactivation localization microscopy, *Biophysical Journal*, 91, 4258–4272 (2006).
14. M. F. Juette, T. J. Gould, M. D. Lessard, M. J. Mlodzianoski, B. S. Nagpure, B. T. Bennett, S. T. Hess and J. Bewersdorf, Three-dimensional sub-100 nm resolution fluorescence microscopy of thick samples, *Nature Methods*, 5, 527–529 (2008).
15. B. Huang, W. Q. Wang, M. Bates and X. W. Zhuang, Three-dimensional super-resolution imaging by stochastic optical reconstruction microscopy, *Science*, 319, 810–813 (2008).
16. M. Bates, B. Huang, G. T. Dempsey and X. W. Zhuang, Multicolor super-resolution imaging with photo-switchable fluorescent probes, *Science*, 317, 1749–1753 (2007).
17. M. J. Rust, M. Bates and X. W. Zhuang, Sub-diffraction-limit imaging by stochastic optical reconstruction microscopy (STORM), *Nature Methods*, 3, 793–795 (2006).
18. M. G. L. Gustafsson, Nonlinear structured-illumination microscopy: Wide-field fluorescence imaging with theoretically unlimited resolution, *Proceedings of the National Academy of Sciences of the United States of America*, 102, 13081–13086 (2005).
19. R. Y. Tsien, The green fluorescent protein, *Annual Review of Biochemistry*, 67, 509–544 (1998).
20. F. Zernike, Phase contrast, a new method for the microscopic observation of transparent objects, Part 2, *Physica*, 9, 974–986 (1942).
21. F. Zernike, Phase contrast, a new method for the microscopic observation of transparent objects, Part 1, *Physica*, 9, 686–698 (1942).
22. D. Gabor, A new microscopic principle, *Nature*, 161, 777 (1948).
23. D. Gabor, theory of communicaton, *J. Inst. Electr. Eng.*, 93, 329 (1946).
24. H. G. Davies and M. H. Wilkins, Interference microscopy and mass determination, *Nature*, 169, 541 (1952).
25. R. Barer, Interference microscopy and mass determination, *Nature*, 169, 366–367 (1952).
26. G. Popescu, Y. Park, N. Lue, C. Best-Popescu, L. Deflores, R. R. Dasari, M. S. Feld and K. Badizadegan, Optical imaging of cell mass and growth dynamics, *Am J Physiol Cell Physiol*, 295, C538–544 (2008).

CHAPTER 2

Groundwork

2.1 Light Propagation in Free Space

We will now review the propagation of light in free space and the principles of Fourier optics. These concepts will be used as a foundation for light microscopy. Let us first review the well-known solutions of the wave propagation in free space (or vacuum). The Helmholtz equation that describes the propagation of field emitted at frequency ω by a source s has the form

$$\nabla^2 U(\mathbf{r},\omega) + k_0^2 U(\mathbf{r},\omega) = s(\mathbf{r}) \tag{2.1}$$

In Eq. (2.1), U is the scalar field, which is a function of position $\mathbf{r}$ and frequency ω, k_0 is the vacuum wave number (or propagation constant), $k_0 = \omega/c$. In order to solve Eq. (2.1), one needs to specify whether the propagation takes place in 1D, 2D, or 3D, and take advantage of any symmetries of the problem (e.g., spherical or cylindrical symmetry in 3D). Then, the *fundamental equation* associated with Eq. (2.1), which provides the Green's function of the problem, is obtained by replacing the source term with an impulse function, i.e., Dirac delta function. The fundamental equation is then solved in the frequency domain. Below, we derive the well-known solutions of *plane* and *spherical waves*, for the 1D and 3D propagation, respectively.

2.1.1 1D Propagation: Plane Waves

For the 1D propagation (see Fig. 2.1), the fundamental equation is

$$\frac{\partial^2 g(x,\omega)}{\partial x^2} + k_0^2 g(x,\omega) = \delta(x) \tag{2.2}$$

where g is Green's function, and $\delta(x)$ is the 1D delta-function, which in 1D describes a planar source of infinite size placed at the origin.

Taking the Fourier transform with respect to x of Eq. (2.2) gives an algebraic equation

$$-k_x^2 g(k_x,\omega) + k_0^2 g(k_x,\omega) = 1 \tag{2.3}$$

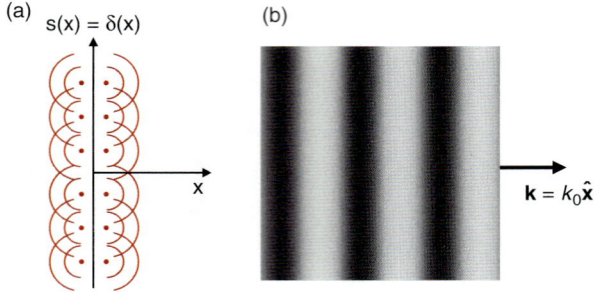

Figure 2.1   One-dimensional propagation of the field from a planar source s(x): (*a*) source; (*b*) real part of the field propagating along x; $\mathbf{k} = k_0\hat{\mathbf{x}}$ is the wave vector associate with the plane wave.

where we used the differentiation theorem of Fourier transforms, $(d/dx) \leftrightarrow ik_x$ (see App. B for details). Thus, the frequency domain solution is simply

$$g(k_x, \omega) = \frac{1}{k_0^2 - k_x^2}$$

$$= \frac{1}{2k_0}\left[\frac{1}{k_0 - k_x} + \frac{1}{k_0 + k_x}\right] \qquad (2.4)$$

In order to find *Green's function*, we Fourier transform back $g(k_x, \omega)$ to the spatial domain. In taking the inverse Fourier transform of Eq. (2.4), we note that $1/(k_x - k_0)$ is the function $1/k_x$, shifted by k_0. Thus, we invoke the inverse Fourier transform of $1/k_x$ and the shift theorem (see App. B)

$$\frac{1}{k_x} \leftrightarrow i\,\mathrm{sign}(x)$$

$$\frac{1}{k_0 \pm k_x} \leftrightarrow -e^{\mp ik_0 x} \cdot i\,\mathrm{sign}(x) \qquad (2.5)$$

By combining Eqs. (2.4) and (2.5), we can readily obtain

$$g(x, \omega) = \frac{-i}{2k_0}(e^{ik_0 x} + e^{-ik_0 x}) \qquad (2.6)$$

Equation (2.6) gives the solution to the wave propagation, as emitted by the infinite planar source at the origin. Note that ignoring the prefactor

$-i = \exp(-i\pi / 2)$, which is just phase shift, the solution through the 1D space, $x \in (-\infty, \infty)$, is (see Fig. 2.1*b*)

$$g(x,\omega) = \frac{1}{k_0} \cdot \cos(k_0 x) \qquad (2.7)$$

Function $g(x,\omega)$ consists of a superposition of two counter-propagating waves, of wave numbers k_0 and $-k_0$. If instead we prefer to work with the *complex analytic signal* associated with $g(x,\omega)$, we suppress the negative frequency component (i.e., $k_x = -k_0$), as described in App. A. In this case, the complex analytic solution is (we keep the same notation, g)

$$g(x,\omega) = \frac{1}{2k_0} \cdot e^{ik_0 \cdot x} \qquad (2.8)$$

Thus, we arrived at the well-known, complex representation of the *plane wave* solution of the wave equation, and established that it can be regarded as the complex analytic signal associated with the real field in Eq. (2.7). Note that since the wave equation contains second-order derivatives in space, both $e^{ik_0 x}$, $e^{-ik_0 x}$ and their linear combinations are all valid solutions. Green's function in Eq. (2.7), which we derived by solving the *fundamental equation* [Eq. (2.2)], is such a linear combination. Thus, the amplitude of the plane wave is constant, but its phase is linearly increasing with the propagation distance (Fig. 2.2*a* and *b*). If the 1D source has a certain distribution along x, say $s(x)$, the field emitted by such a source will be the convolution between $s(x)$ and $g(x)$.

Finally, we note that, if the propagation direction is not parallel to one of the axes, the plane wave equation takes the general form

$$g(\mathbf{r},\omega) = \frac{i}{2k_0} e^{ik_0 \hat{\mathbf{k}} \cdot \mathbf{r}} \qquad (2.9)$$

where $\hat{k}$ is the unit vector defining the direction of propagation, $\hat{\mathbf{k}} = \mathbf{k}/k_0$. Thus, the phase delay at a point described by position vector $\mathbf{r}$ is $\phi = k_\| r$, where $k_\|$ is the component of $\mathbf{k}$ parallel to $\mathbf{r}$ (Fig. 2.2*c* and *d*).

The plane wave Green's function allows for an arbitrary field consisting of a distribution of propagation directions to be represented in terms of plane waves. As we will see later, this decomposition is very helpful in many problems, especially imaging.

2.1.2   3D Propagation: Spherical Waves

Let us obtain the Green's function associated with the vacuum propagation, or the response of free-space to a monochromatic point source (i.e., impulse response). The fundamental equation becomes

$$\nabla^2 g(\mathbf{r},\omega) + k_0^2 g(\mathbf{r},\omega) = \delta^{(3)}(\mathbf{r}) \qquad (2.10)$$

(a)

$|g(x, \omega)|$

$1/k_0$

x

(b)

$\arg [g(x, \omega)]$

k_0

x

(c)

y

$k_0 \cdot \hat{\mathbf{k}}$

x

(d)

y

$\mathbf{k}_y$ $\mathbf{k}$

$\mathbf{k}_x$ $\mathbf{k}_\perp$ $\mathbf{k}$

$\mathbf{r}$ $\mathbf{k}_\parallel$

x

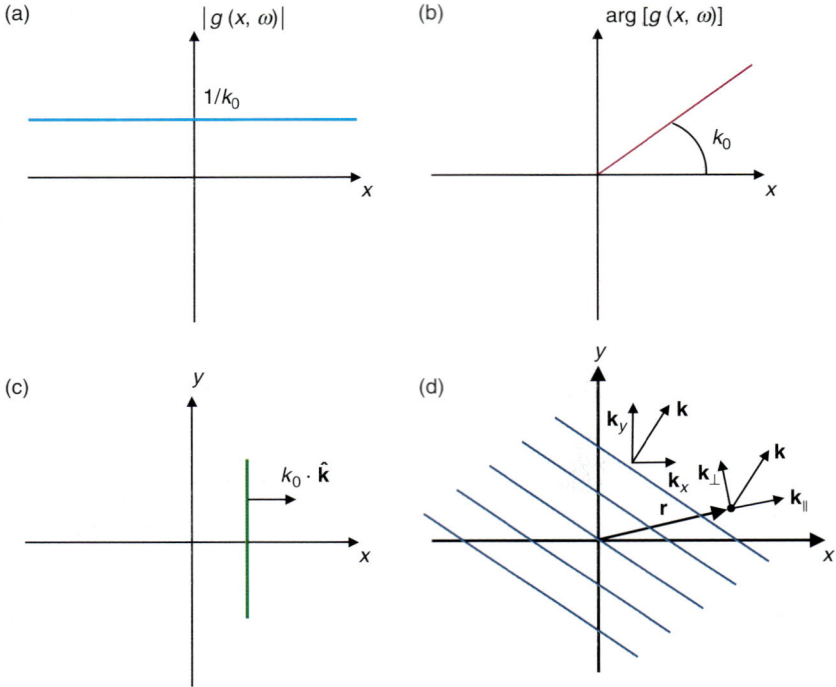

FIGURE 2.2   Plane wave propagation.

where $\delta^{(3)}$ represents a 3D delta function. We Fourier transform Eq. (2.10) with respect to variable $\mathbf{r}$ and use the relationship $\bar{\nabla} \leftrightarrow i\mathbf{k}$ (see App. B), which gives

$$-k^2 g(\mathbf{k}, \omega) + k_0^2 g(\mathbf{k}, \omega) = 1 \qquad (2.11)$$

where $k^2 = \mathbf{k} \cdot \mathbf{k}$. Equation 2.11 readily yields the solution in the $(\mathbf{k}, \omega)$ representation

$$g(\mathbf{k}, \omega) = \frac{1}{k_0^2 - k^2} \qquad (2.12)$$

where g depends only on the modulus of $\mathbf{k}$ and not its orientation, i.e., as the propagation in vacuum is isotropic. Note that Eq. (2.12) looks very similar to its analog in 1D [Eq. (2.4)] except that the x component of the wave vector, k_x, is now replaced by the modulus of the wave vector, k. Like in the 1D case, in order to obtain the spatial domain solution, $g(\mathbf{r}, \omega)$, we now take the inverse Fourier transform Eq. (2.12) back to spatial domain. For propagation in isotropic media such as free space, the problem is spherically symmetric and the Fourier transform of Eq. (2.12) can be written in spherical coordinates as a 1D integral (see App. B)

$$g(r,\omega) = \int_0^\infty \frac{1}{k_0^2 - k^2} \frac{\sin(kr)}{kr} \cdot k^2 dk \qquad (2.13)$$

Expanding the first factor under the integrand and expressing $\sin(kr)$ in exponential form,

$$g(r,\omega) = \frac{1}{2r} \int_0^\infty \left(\frac{1}{k_0 - k} - \frac{1}{k_0 + k} \right) \frac{e^{ikr} - e^{-ikr}}{2i} dk$$

$$= \frac{1}{4ir} \int_{-\infty}^\infty \left(\frac{1}{k_0 - k} - \frac{1}{k_0 + k} \right) \cdot e^{ikr} dk \qquad (2.14)$$

Note that the integral in Eq. (2.14) is a 1D Fourier transform of a function similar to that in Eq. (2.4). Thus, we obtain

$$g(r,\omega) = \frac{1}{r} \cdot \cos k_0 r \qquad (2.15)$$

where we ignored irrelevant constant factors.

This solution describes the spherically symmetric wave propagation from a point source, i.e., a *spherical wave* (see Fig. 2.3). The *complex analytic signal* associated with this solution has the well known form (up to an unimportant constant, $1/2i$)

$$g(r,\omega) = \frac{e^{ik_0 r}}{r} \qquad (2.16)$$

Knowing the Green's function associated with free space (Eq. 2.16), i.e., the response to a point source, we can easily

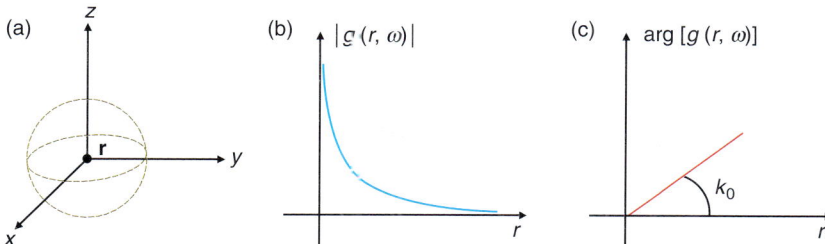

FIGURE 2.3   (a) Propagation of spherical waves; (b) amplitude vs. r; (c) phase vs. r.

calculate the field emitted by an arbitrary source $s(\mathbf{r},\omega)$ via a 3D convolution integral,

$$U(\mathbf{r},\omega) = \int_V s(\mathbf{r}',\omega) \cdot \frac{e^{ik_0|\mathbf{r}-\mathbf{r}'|}}{|\mathbf{r}-\mathbf{r}'|} d^3\mathbf{r}' \tag{2.17}$$

where the integral is performed over the volume of the source.

Equation (2.17) is the essence of the *Huygens' principle* (seventeenth century), which establishes that, upon propagation, points set in oscillation by the field become new (secondary) sources, and the emerging field is the summation of spherical wavelets emitted by all these secondary sources. Thus, each point reached by the field becomes a secondary source, which emits a new spherical wavelet and so on. It is clear that, generally, the integral in Eq. (2.17) is difficult to evaluate. In the following section, we study the propagation of fields at large distances from the source, i.e., in the *far zone* (we will use the phrase "far zone" instead of "far field" to avoid confusions with the actual complex *field*). Note that the source term, *s*, can be due to both *primary* (self-emitting) and *secondary* (scattering) sources, as described below in Sections 2.5–2.9.

2.2 Fresnel Approximation of Wave Propagation

Next we describe some useful approximations of the spherical wave, which makes the integral in Eq. (2.17) more tractable. When the propagation distance along one axis, say z, is far greater than along the other two axes (Fig. 2.4), the spherical wave can be first approximated by

$$\frac{e^{ik_0 r}}{r} \propto \frac{e^{ik_0 r}}{z} \tag{2.18}$$

where we used the fact that the amplitude attenuation is a slow function of x and y, when $z^2 > x^2 + y^2$, such that $1/\sqrt{(x^2 + y^2 + z^2)} \sim 1/z$.

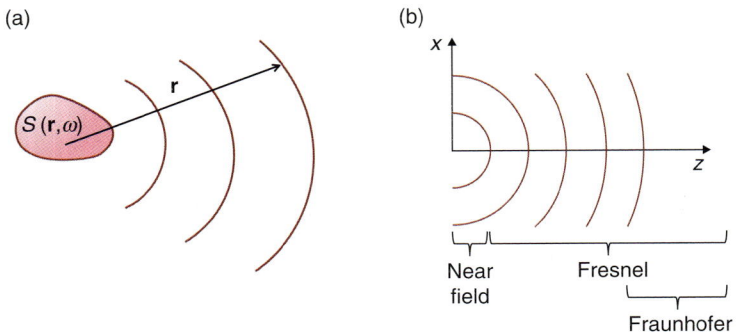

FIGURE 2.4   (*a*) Propagation from an arbitrary source; (*b*) illustration of the validity for the Fresnel and Fraunhofer approximations.

However, the phase term is significantly more sensitive to x and y variations, such that the next-order approximation is needed.

Expanding the radial distance in Taylor series, we obtain

$$r = \sqrt{x^2 + y^2 + z^2}$$

$$= z\sqrt{1 + \frac{x^2 + y^2}{z^2}} \qquad (2.19)$$

$$\simeq z\left[1 + \frac{x^2 + y^2}{2z^2}\right]$$

The spherical wavelet is now approximated by

$$g(\mathbf{r}, \omega) \simeq \frac{e^{ik_0 z}}{z} \cdot e^{ik_0 \frac{x^2 + y^2}{2z}} \qquad (2.20)$$

Equation (2.20) represents the *Fresnel approximation* of the wave propagation; the region of distance z where this approximation holds is called the *Fresnel zone*. Note that the transverse, x-y dependence comes in the form of a quadratic (parabolic) phase term.

For a given planar field distribution at $z = 0$, $U(u, v)$, we can calculate the resulting propagated field at distance z, $U(x, y, z)$ by simply convolving U with the Fresnel wavelet (Fig. 2.5), $e^{ik_0[(x^2 + y^2)/2z]}$

$$U_1(x, y, z) = \frac{e^{ik_0 z}}{z} \cdot \int_{-\infty}^{\infty}\int_{-\infty}^{\infty} U(u, v) \cdot e^{\frac{ik_0}{2z}\left[(x-u)^2 + (y-v)^2\right]} du\, dv$$

$$\propto U \circledv\, e^{ik_0\left(\frac{x^2 + y^2}{2z}\right)} \qquad (2.21)$$

Equation (2.21) represents the so called *Fresnel diffraction equation*, which essentially explains the field propagation using Huygens' concept of secondary point sources, except that now each of those secondary point sources emits Fresnel wavelets (parabolic wavefronts) rather than spherical waves ("circled v" denotes the convolution operation).

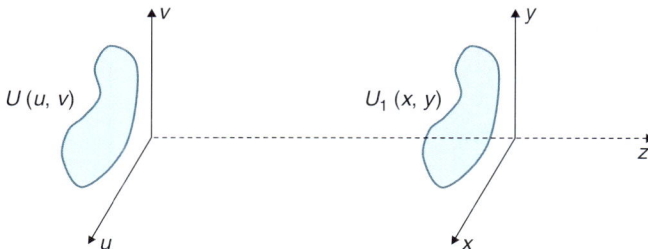

Figure 2.5 Fresnel propagation of field U at distance z.

Interestingly, although a drastic approximation, the Fresnel equation captures the essence of many important phenomena in microscopy. Note that the prefactor $e^{ik_0 z}/z$ is typically neglected, as it contains no information about the x-y field dependence, which is the only one ultimately relevant to 2D imaging.

2.3   Fourier Transform Properties of Free Space

A further approximation, called the *Fraunhofer approximation*, can be made if the observation plane is even farther away. Thus, for $z \gg k(u^2 + v^2)$, the quadratic phase terms can be ignored in Eq. (2.21), such that

$$\frac{ik_0}{2z}\left[(x-u)^2 + (y-u)^2\right] \simeq \frac{ik_0}{2z}(-2xu - 2yv) \tag{2.22}$$

The field distribution in the *Fraunhofer region* is obtained from Eq. (2.21) as

$$U_1(x,y,z) = \frac{e^{ik_0 z}}{z} \int_{-\infty}^{\infty}\int_{-\infty}^{\infty} U(u,v) \cdot e^{-i\frac{2\pi}{\lambda z}(ux+vy)}\,du\,dv \tag{2.23}$$

From Fig. 2.6, we note that, for a particular direction of propagation θ with respect to the optical axis, $\theta \simeq x/z = k_x/k_z \simeq k_x/k_0$. An analogous equation can be written for k_y. Thus, we can rewrite the Fourier transform in Eq. (2.23) as (ignoring the z-dependent prefactor)

$$U_1(k_x,k_y) = \int_{-\infty}^{\infty}\int_{-\infty}^{\infty} U(u,v) \cdot e^{-i(k_x u + k_y v)}\,du\,dv \tag{2.24a}$$

$$k_x = \frac{k_0 x}{z} = \frac{2\pi x}{\lambda z} \tag{2.24b}$$

$$k_y = \frac{k_0 y}{z} = \frac{2\pi y}{\lambda z} \tag{2.24c}$$

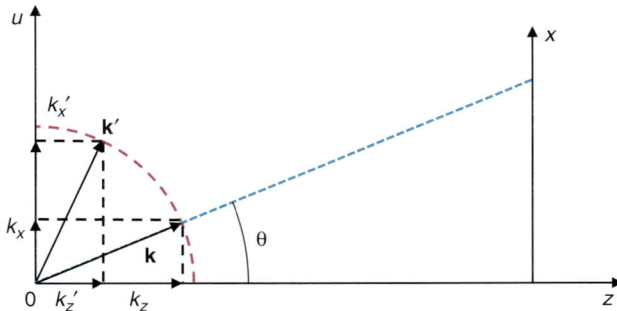

FIGURE 2.6   Fraunhofer propagation.

Equation (2.24) establishes that upon long propagation distances, the free space performs the Fourier transform of a given (input) field.

Note that the spatial frequency k_x of the input field $U(u,v)$ is associated with the propagation of a plane wave along the direction of the wave vector $\mathbf{k} = \mathbf{k}_x + \mathbf{k}_z$. Thus, we can think of the Fraunhofer regime as the situation where different plane waves (propagating at different angles) are generated by different spatial frequencies of the input field. Another way to put it is that, if distance z is large enough, the propagation angles corresponding to different spatial frequencies do not mix. This remarkable property allows us to solve many problems of practical interest with extreme ease, by invoking various Fourier transform pairs and their properties, some of which are described in App. B.

Example 2.1   Diffraction by a sinusoidal grating
Consider a plane wave incident on a one-dimensional amplitude grating of transmission $t(u) = 1 + \cos(2\pi u / \Lambda)$, with Λ the period of the grating (Fig. 2.7).

In the far zone, we expect the following diffraction pattern

$$U(x) = \int_{-\infty}^{\infty} \left[1 + \cos\left(2\pi u / \Lambda\right)\right] e^{-\frac{i2\pi}{\lambda z}xu} du \tag{2.25}$$

or, changing notations

$$U(k) = \int \left[1 + \cos(k_1 \cdot u)\right] \cdot e^{-iku} du$$

$$k_1 = \frac{2\pi}{\Lambda} \tag{2.26}$$

$$k = \frac{2\pi x}{\lambda z}$$

Using that the Fourier transform of a cosine function is a sum of two delta functions, we finally obtain:

$$U(k) = \frac{1}{2}\delta(k - k_1) + \frac{1}{2}\delta(k + k_1) + \delta(0) \tag{2.27}$$

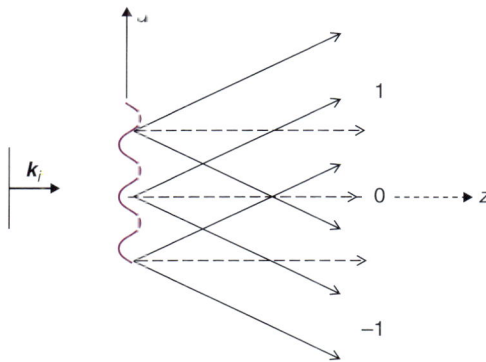

FIGURE 2.7   Diffraction by a sinusoidal grating; the diffraction orders are indicated.

Equation (2.27) shows that the diffraction by a sinusoidal grating generates two distinct off-axis plane waves in the far zone and another along the optical axis (called DC, or zeroth order). While this result is well known, it has deeper implications of practical use in solving *inverse problems*. Thus, consider a 3D object, whose spatial distribution of refractive index can be expressed as a Fourier transform, i.e., the object can be thought of as a superposition of sinusoidal gratings of various periods. Under certain conditions (object is weakly scattering, such that the frequencies do not mix), measuring the angularly scattered light from the object reveals the entire structure of the object. This solution of the inverse problem relies on each angular component reporting on a unique spatial frequency (sinusoidal) associated with the object. We will come back to this point repeatedly throughout the book.

2.4 Fourier Transform Properties of Lenses

Here we show that lenses have the capability to perform Fourier transforms, much like free space, and with the added benefit of eliminating the need for large distances of propagation.

Let us consider the biconvex lens in Fig. 2.8. We would like to determine the effect that the lens has on an incident plane wave. This effect can be incorporated via a transmission function of the form

$$t(x,y) = e^{i\phi(x,y)} \tag{2.28}$$

The transmission function in Eq. 2.28 indicates that the lens only acts on the phase and not the amplitude of the field (we ignore losses due to reflections). The problem reduces to evaluating the phase delay produced by the lens as a function of the off-axis distance, or the polar coordinate $r = \sqrt{x^2 + y^2}$ (Fig. 2.8b)

$$\phi(r) = \phi_{glass}(r) + \phi_{air}(r)$$

$$= nk_0 b(r) + k_0[b_0 - b(r)] \tag{2.29}$$

$$= k_0 b_0 + (n-1)k_0 \cdot b(r)$$

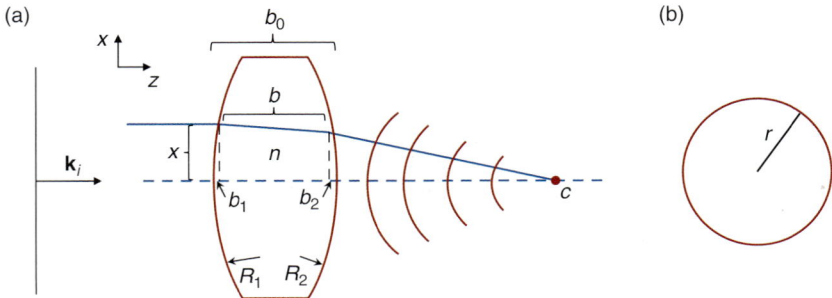

FIGURE 2.8 (a) Phase transformation by a thin convergent lens; (b) polar coordinate r.

In Eq. (2.29), ϕ_{glass} and ϕ_{air} are the phase shifts due to the glass and air portions, respectively, k_c is the wave number in air, $k_0 = 2\pi/\lambda$, b_0 is the thickness along the optical axis, i.e., at $r = 0$ (maximum thickness), $b(r)$ is the thickness at distance r off axis, and n is the refractive index of the glass.

The local thickness, $b(r)$, can be expressed as

$$b(r) = b_0 - b_1(r) - b_2(r) \tag{2.30}$$

where b_1 and b_2 are the segments shown in Fig. 2.8, which can be calculated using simple geometry, as follows (Fig. 2.9).

For small angles, ABC becomes a right triangle, where the following identity applies (the *perpendicular theorem*)

$$|AD|^2 = 2|BD| \cdot |DC| \tag{2.31}$$

Since $|AD| = r$, $|BD| = b_1$, and $|DC| \approx R_1$, we finally obtain

$$b_1(r) = \frac{r^2}{2R_1} \tag{2.32}$$

where a symmetric relationship can be obtained for $b_2(r)$. It follows that the thickness, $b(r)$, can be expressed from Eq. (2.30),

$$b(r) = b_0 - \frac{r^2}{2}\left(\frac{1}{R_1} - \frac{1}{R_2}\right) \tag{2.33}$$

With this, the phase distribution in Eq. (2.29) becomes

$$\phi(r) = \phi_C - \frac{k_0 r^2}{2}(n-1)\left(\frac{1}{R_1} - \frac{1}{R_2}\right) \tag{2.34}$$

where $\phi_0 = k_0 \cdot b_0$.

Note that in Eqs. (2.33) and (2.34), we used the geometrical optics convention whereby surfaces with centers to the left (right) are considered of negative (positive) radius; in our case $R_1 > 0$ and $R_2 < 0$.

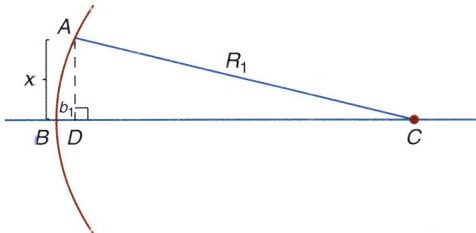

FIGURE 2.9 Geometry of the small angle propagation through the lens.

We recognize that the focal distance associated with a thin lens is given by the "lens makers equation"

$$\frac{1}{f} = (n-1)\left(\frac{1}{R_1} - \frac{1}{R_2}\right) \tag{2.35}$$

Such that Eq. (2.34) becomes

$$\phi(r) = \phi_0 - \frac{k_0 r^2}{2f} \tag{2.36}$$

Finally, the lens transmission function, which establishes how the plane wave field at a plane right before the lens is transmitted at a plane right after the lens, has the form

$$t(r) = e^{iknb_0} \cdot e^{-i\frac{k_0 r^2}{2f}} \tag{2.37}$$

As expected, Eq. (2.37) shows that the effect of the lens is to transform a plane wave into a parabolic wavefront. The negative sign $[-i(k_0 r^2/2f)]$ conventionally denotes a convergent field, while the positive sign marks a divergent field.

By comparing Eqs. (2.20) and (2.37), it is clear that the effect of propagation through free space is qualitatively similar to transmission through a thin *divergent* lens (Fig. 2.10).

Now let us consider the field propagation through a combination of free space and convergent lens (Fig. 2.11).

The problem of deriving an expression for the output field $U_4(x_4, y_4)$ as a function of input field $U_1(x_1, y_1)$ can be broken down into a Fresnel propagation over distance d_1, followed by a transformation by the lens of focal distance f, and, finally, a propagation over distance d_2.

In Sec. 2.2 [Eq. (2.21)] we found that the Fresnel propagation can be described as a convolution with the quadratic phase function (Fresnel wavelet). Thus the propagation can be written symbolically as

$$U_2(x_2, y_2) = U_1(x_1, y_1) \circledcirc e^{i\frac{k_0(x_1^2 + y_1^2)}{2d_1}}$$

$$U_3(x_3, y_3) = U_2(x_2, y_2) \cdot e^{-i\frac{k_0(x_2^2 + y_2^2)}{2f}} \tag{2.38}$$

$$U_4(x_4, x_4) = U_3(x_3, y_3) \circledcirc e^{i\frac{k_0(x_3^2 + y_3^2)}{2d_2}}$$

Carrying out these calculations is straightforward but somewhat tedious. However, a great simplification arises in the special case where

$$d_1 = d_2 = f \tag{2.39}$$

(a)

$$e^{\frac{ikr^2}{2z}}$$

Point source

z

(b)   Divergent lens

$$e^{\frac{ikr^2}{2f}}$$

$\mathbf{k}_i$ Virtual
              focal point

z

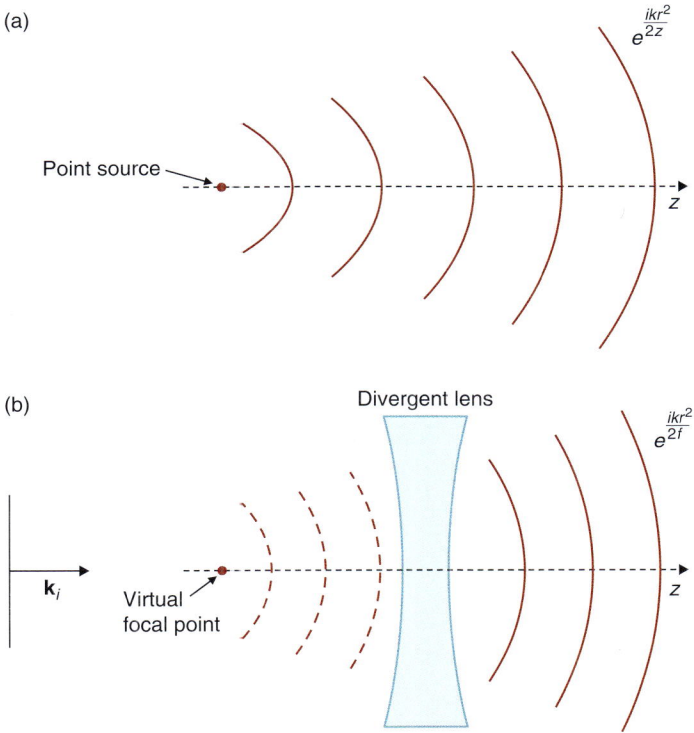

FIGURE 2.10   Propagation in free space (a) and through a divergent lens (b).

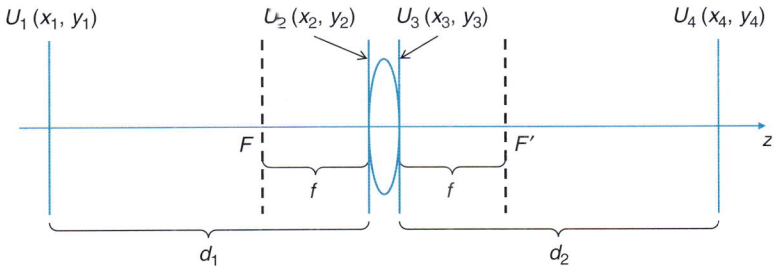

$U_1(x_1, y_1)$ $U_2(x_2, y_2)$ $U_3(x_3, y_3)$ $U_4(x_4, y_4)$

F F' z

f f

d_1 d_2

FIGURE 2.11   Propagation through free space and convergent lens. The fields at the focal planes F and F' are Fourier transforms on each other.

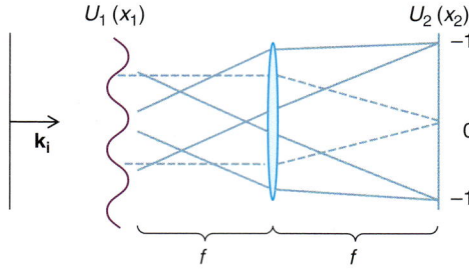

FIGURE 2.12   Fourier transform of the field diffracted by a sinusoidal grating.

Thus, if the input field is at the front focal plane of the lens and the output field is observed at the back focal plane, then the two fields are related via an exact Fourier transform

$$U_4(k_{x4}, k_{y4}) = \int\limits_{-\infty}^{\infty} \int\limits_{-\infty}^{\infty} U_1(x_1, y_1) \cdot e^{-i(k_{x4} \cdot x_1 + k_{y4} \cdot y_1)} dx_1 dy_1 \qquad (2.40a)$$

$$k_{x4} = \frac{2\pi x_4}{\lambda f} \qquad (2.40b)$$

$$k_{y4} = \frac{2\pi y_4}{\lambda f} \qquad (2.40c)$$

where we ignored trivial factors preceding the integral in Eq. (2.40a). This result is significant, as it establishes a simple yet powerful way to compute analog Fourier transforms, virtually instantaneously, using light.

Example 2.2   Sinusoidal transmission grating
Let us revisit Example 2.1, where we studied the diffraction by a sinusoidal grating. Now the grating is placed in the focal plane object of the lens and the observation is in the focal plane image (see Fig. 2.12).
    The Fourier transform of the 1D grating transmission function is

$$U(x_2) = \int\limits_{-\infty}^{\infty} \left[1 + \cos\left(2\pi x_1 \Big/ \Lambda \right) \right] \cdot e^{-i\frac{2\pi x_1 x_2}{\lambda f}} dx_1$$

$$= \frac{1}{2}\delta\left(\frac{2\pi x_2}{\lambda z} - \frac{2\pi}{\Lambda} \right) + \frac{1}{2}\delta\left(\frac{2\pi x_2}{\lambda z} + \frac{2\pi}{\Lambda} \right) + \delta(0) \qquad (2.41)$$

Thus, a thin lens can generate the diffraction pattern of a grating just as the free space. This simple picture is extremely useful in understanding the coherent image formation in a light microscope, as first described by Abbe in 1873. This theory will be presented in more detail in Chap. 5.

2.5   (The First-Order) Born Approximation of Light Scattering in Inhomogeneous Media

In many situations of biomedical interest, light interacts with *inhomogeneous media,* a process generally referred to as *scattering*. If the wavelength of the field is unchanged in the process, then we deal with *elastic light scattering*.

The general goal in elastic light scattering experiments is to infer information about the refractive index distribution in the 3D space, $n(\mathbf{r})$, from measurements on the scattered light, i.e., to solve the *scattering inverse problem*. In the following, we show that this problem can be solved analytically if we assume *weakly scattering media*. This is the *first-order Born approximation*, or simply the Born approximation. Thus, we derive an expression for the far-zone scattered field generated by a weakly scattering medium illuminated with a plane wave (Fig. 2.13).

Let us recall the Helmholtz equation

$$\nabla^2 U(\mathbf{r},\omega)+\beta^2(\mathbf{r},\omega)U(\mathbf{r},\omega)=0 \tag{2.42a}$$

$$\beta(\mathbf{r},\omega)=n(\mathbf{r},\omega)k_0 \tag{2.42b}$$

$$k_0=\omega/c \tag{2.42c}$$

where β is the (inhomogeneous, i.e., $\mathbf{r}$-dependent) propagation constant.

Equation (2.42a) can be rearranged to show the inhomogeneous term on the right hand side. Thus, the scalar field satisfies

$$\nabla^2 U(\mathbf{r},\omega)+k_0^2 U(\mathbf{r},\omega)=-4\pi F(\mathbf{r},\omega)\cdot U(\mathbf{r},\omega)$$

$$F(\mathbf{r},\omega)=\frac{1}{4\pi}k_0^2\left[n^2(\mathbf{r},\omega)-1\right] \tag{2.43}$$

Function $F(\mathbf{r},\omega)$ is called the *scattering potential* associated with the medium. Equation (2.43) is analog to Eq. 2.1. encountered earlier and explicitly shows the inhomogeneous portion of the refractive index as a source of secondary (scattered) light.

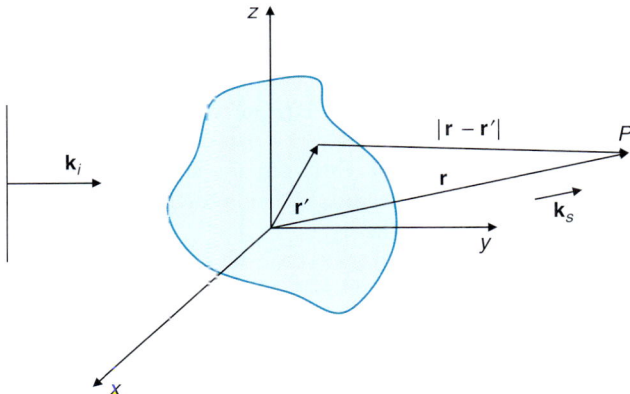

FIGURE 2.13 Light scattering by an inhomogeneous medium.

The fundamental equation that yields Green's function, $g(\mathbf{r},\omega)$, has the form

$$\nabla^2 g(\mathbf{r},\omega) + k^2_{\,0} g(\mathbf{r},\omega) = -\delta^{(3)}(\mathbf{r}) \qquad (2.44)$$

Again, as in Sec. 2.1., we solve the equation by Fourier transforming it with respect to spatial variable and arrive at the well-known spherical wave solution [Eq. (2.16)]

$$g(\mathbf{r},\omega) = \frac{e^{ik_0 r}}{r} \qquad (2.45)$$

Thus, ignoring the constant prefactors, the solution for the scattered field is a convolution between the source term, i.e., $F(\mathbf{r},\omega) \cdot U(\mathbf{r},\omega)$, and the Green function, $g(\mathbf{r},\omega)$,

$$U(\mathbf{r},\omega) = \int F(\mathbf{r}',\omega) \cdot U(\mathbf{r}',\omega) \cdot \frac{e^{ik_0|\mathbf{r}-\mathbf{r}'|}}{|\mathbf{r}-\mathbf{r}'|} d^3\mathbf{r}' \qquad (2.46)$$

The integral in Eq. (2.46) can be simplified if we assume that the measurements are performed in the *far zone*, i.e., $r' \ll r$ (Fig. 2.13). Thus, we can invoke the following approximation, which is equivalent to the Fraunhofer approximation [Eq. (2.22)]

$$|\mathbf{r}-\mathbf{r}'| = \sqrt{r^2 + r'^2 - 2\mathbf{r}\mathbf{r}'}$$

$$\simeq r - \frac{\mathbf{r}}{r} \cdot \mathbf{r}' \qquad (2.47)$$

$$\simeq r - \frac{\mathbf{k}_s}{k_0} \cdot \mathbf{r}'$$

In Eq. (2.47), $\mathbf{r} \cdot \mathbf{r}'$ is the scalar product of vectors $\mathbf{r}$ and $\mathbf{r}'$ and $\mathbf{k}_s/k_0$ is the unit vector associated with the direction of propagation. With this so-called *far-zone* approximation, Eq. (2.47) can be rewritten as

$$U(\mathbf{r},\omega) = \frac{e^{ik_0 r}}{r} \int F(\mathbf{r}',\omega) \cdot U(\mathbf{r}',\omega) \cdot e^{-i\mathbf{k}_s \cdot \mathbf{r}'} d^3\mathbf{r}' \qquad (2.48)$$

Equation (2.48) indicates that, far from the scattering medium, the field behaves as a spherical wave, $e^{ik_0 r}/r$, which is perturbed by the *scattering amplitude*, defined as

$$f(\mathbf{k}_s,\omega) = \int F(\mathbf{r}',\omega) \cdot U(\mathbf{r}',\omega) \cdot e^{-i\mathbf{k}_s \cdot \mathbf{r}'} d^3\mathbf{r}' \qquad (2.49)$$

In order to obtain a tractable expression for the integral in Eq. (2.49), we assume the scattering is *weak*, which allows us to approximate $U(\mathbf{r}',\omega)$. The (first-order) *Born approximation* assumes that the field inside the scattering volume is constant and equal to the incident field, assumed to be a plain wave

$$U_i(\mathbf{r'},\omega) = e^{i\mathbf{k}_i\mathbf{r'}} \tag{2.50}$$

With this approximation plugged into Eq. (2.49), we finally obtain for the scattering amplitude

$$f(\mathbf{k}_s,\omega) = \int_v F(\mathbf{r'},\omega) \cdot e^{-i(\mathbf{k}_s-\mathbf{k}_i)\mathbf{r'}} d^3r' \tag{2.51}$$

Note that the integral on the right hand side is a 3D Fourier transform. Thus, within the first Born approximation, measurements of the field scattered at a given angle gives access to the Fourier component $\mathbf{q} = \mathbf{k}_s - \mathbf{k}_i$ of the scattering potential F,

$$\hat{f}(\mathbf{q},\omega) = \int_v F(\mathbf{r'},\omega) \cdot e^{-i\mathbf{q}\mathbf{r'}} d^3r' \tag{2.52}$$

The physical meaning of $\mathbf{q}$ is that of the difference between the scattered and incident wave vectors, sometime called *scattering wave vector* (Fig. 2.14) and in quantum mechanics referred to as *momentum transfer*.

From the geometry in Fig. 2.14 it can be seen that $q = 2k_0 \cdot \sin\theta/2$, with θ the scattering angle. The remarkable feature of Eq. (2.52) is that, due to the reversibility of the Fourier integral, it can be inverted to provide the scattering potential,

$$F(\mathbf{r'},\omega) = \int_{V_q} U(\mathbf{q},\omega) \cdot e^{i\mathbf{q}\mathbf{r'}} d^3q \tag{2.53}$$

where V_q is the 3Dq-domain of integration

Equation (2.53) establishes the solution to the *inverse scattering problem*, i.e., it provides a way of retrieving information about the medium under investigation via angular scattering measurements. Equivalently, measuring U at a multitude of scattering angles allows the reconstruction of F from its Fourier components. For *far-zone measurements* at a fixed distance R, the scattering amplitude f and the scattered field U differ only by a constant $e^{i k R}/R$, and, thus, can be used interchangeably.

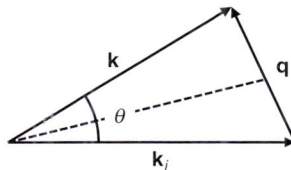

FIGURE 2.14   Momentum transfer.

Note that in order to retrieve the scattering potential $F(\mathbf{r}', \omega)$ experimentally, two essential conditions must be met:

1. The measurement has to provide the complex scattered field (i.e., amplitude *and* phase).

2. The scattered field has to be measured over an infinite range of spatial frequencies q [i.e., the limits of integration in Eq. (2.53) are $-\infty$ to ∞].

On the first issue, we note that great progress has been made recently in terms of measuring phase information, some of which is within the theme of this book. Nevertheless, most measurements are intensity-based. Thus, it is important to realize that if one only has access to the intensity of the scattered light, $|U(q, \omega)|^2$, then the auto-correlation of $F(\mathbf{r}', \omega)$ and *not F* itself is retrieved,

$$\int_{V_q} |U(\mathbf{q}, \omega)|^2 \cdot e^{i\mathbf{q}\mathbf{r}'} d^3\mathbf{q} = F(\mathbf{r}', \omega) \otimes F(\mathbf{r}', \omega) \qquad (2.54)$$

The result in Eq. (2.54) is simply the *correlation theorem* applied to 3D Fourier transforms (see App. B).

Second, clearly, we have only experimental access to a limited *frequency range*, or *bandwidth*. Therefore, the spatial frequency coverage (or range of momentum transfer) is intrinsically limited. Specifically, for a given incident wave vector $\mathbf{k}_i$, with $k_i = k_0$, the highest possible $\mathbf{q}$ is obtained for backscattering, $|\mathbf{k}_b - \mathbf{k}_i| = 2k_0$ (Fig. 2.15a).

Similarly for an incident wave vector in the opposite direction, $-\mathbf{k}_i$, the maximum momentum transfer is also $q_b = 2k_0$ (Fig. 2.15b). Altogether, the maximum frequency coverage is $4k_0$, as illustrated in Fig. 2.16.

As we rotate the incident wave vector from $\mathbf{k}_i$ to $-\mathbf{k}_i$, the respective backscattering wave vector rotates from $\mathbf{k}_b$ to $-\mathbf{k}_b$, such that the tip of $\mathbf{q}$ describes a sphere of radius $2k_0$. This is known as the *Ewald sphere, or Ewald limiting sphere*.

Let us study the effect of this bandwidth limitation in the best case scenario of the entire Ewald's sphere coverage. The measured (i.e., truncated in frequency) field, $\underline{U}(\mathbf{q}, \omega)$, can be expressed as

$$\underline{U}(\mathbf{q}, \omega) = \begin{vmatrix} U(\mathbf{q}, \omega), 0 \le q \le 2k_0 \\ 0, \text{ rest} \end{vmatrix} \qquad (2.55)$$

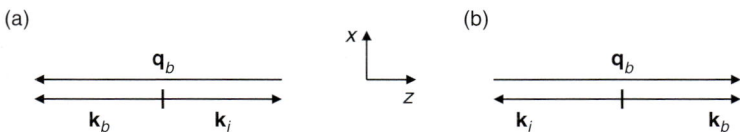

FIGURE 2.15 Momentum transfer for backscattering configuration for $\mathbf{k}_i \| \mathbf{z}$ (*a*) and $\mathbf{k}_i \| \mathbf{-z}$ (*b*).

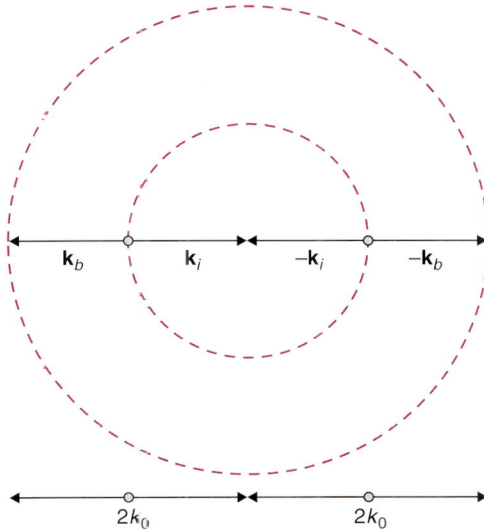

FIGURE 2.16   Ewald scattering sphere.

Using the definition of a rectangular function in 3D, or the "ball" function, defined as $\Pi[q/(4k_0)] = \begin{vmatrix} 1, \text{ if } \sqrt{q_x^2 + q_y^2 + q_y^2} \le 2k_0 \\ 0, \text{ rest} \end{vmatrix}$, we can rewrite Eq. (2.55) as (see also App. B)

$$\underline{U}(\mathbf{q}, \omega) = U(\mathbf{q}, \omega) \cdot \Pi\left(\frac{q}{4k_0}\right) \tag{2.56}$$

Thus, the scattering potential retrieved by measuring the scattered field $\underline{U}$ can be obtained via the 3D Fourier transform of Eq. (2.56)

$$\underline{F}(\mathbf{r}', \omega) = F(\mathbf{r}', \omega) \otimes_{xyz} \tilde{\Pi}(r') \tag{2.57}$$

In Eq. (2.57), $\tilde{\Pi}$ is the Fourier transform of the ball function Π and has the form (see App. B)

$$\tilde{\Pi}(r') = \frac{\sin(2k_0 r') - 2k_0 r' \cos(2k_0 r')}{(2k_0 r')^3} \cdot (2k_0)^3 \tag{2.58}$$

Clearly, even in the best case scenario, i.e., full Ewald sphere coverage, the reconstructed object $F(\mathbf{r}', \omega)$ is a "smooth" version of the original object, where the smoothing function is $\Pi(r')$, a radially symmetric function of radial coordinate r'. Practically, covering the entire Ewald sphere requires illuminating the object from all directions and measuring the scattered complex field over the entire solid angle for each illumination direction. This is, of course, a challenging task,

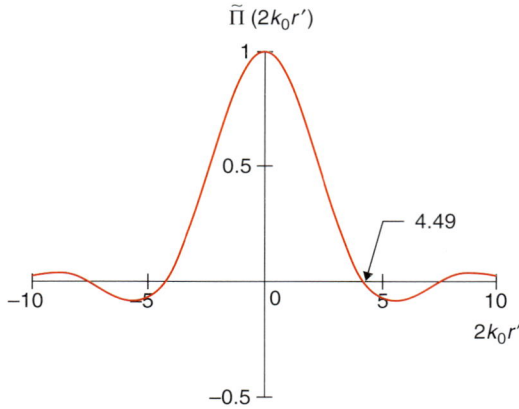

FIGURE **2.17**   Profile through function $\tilde{\Pi}(r')$ in Eq. (2.58).

rarely achieved in practice. Instead, fewer measurements are performed, at the expense of degrading resolution.

Figure 2.17 depicts the 1D profile of the 3D function $\tilde{\Pi}(r')$. As can be seen in Fig. 2.17, the first root of function $\tilde{\Pi}$ is at $2k_0r'_0 = 4.49$, or $r'_0 = 0.36\lambda$. Therefore, according to Rayleigh's criterion, the best achievable resolution in reconstructing the 3D object is better than Abbe's limit of $0.61\,\lambda$ calculated for 2D imaging. Physically, the higher resolving power obtained in 3D can be attributed to the illumination that covers the entire Ewald sphere.

2.6 Scattering by Single Particles

Here, we introduce the main concepts and definitions used in the context of light scattering by particles. By *particle*, we mean a region in space that is characterized by a dielectric permeability $\varepsilon = n^2 \in \mathbb{C}$, which is different from that of the surrounding medium. The scattering geometry is illustrated in Fig. 2.18.

The field scattered in the far zone has the general form of a *perturbed* spherical wave,

$$\mathbf{U}_s(\mathbf{r}) = \mathbf{U}_i \cdot \frac{e^{ikr}}{r} \cdot f(\mathbf{k}_s, \mathbf{k}_i) \tag{2.59}$$

where $\mathbf{U}_s(\mathbf{r})$ is the scattered *vector* field at position $\mathbf{r}$, $r = |\mathbf{r}|$, and $f(\mathbf{k}_s, \mathbf{k}_i)$ defines the *scattering amplitude*. The function f is physically similar to that encountered above, when discussing the Born approximation, [Eq. (2.52)] except that here it also includes *polarization* information, in addition to the amplitude and phase.

The *differential cross section* associated with the particle is defined as

$$\sigma_d(\mathbf{k}_s, \mathbf{k}_i) = \lim_{r \to \infty} r^2 \left| \frac{\mathbf{S}_s}{\mathbf{S}_i} \right| \tag{2.60}$$

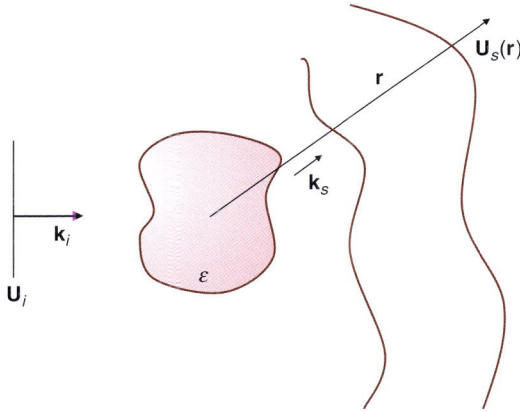

Figure **2.18** Light scattering by a single particle.

where $\mathbf{S}_s$ and $\mathbf{S}_i$ are the Poynting vectors along the scattered and initial direction, respectively, with moduli $|\mathbf{S}_{i,s}| = (1/2\eta)|U_{i,s}|^2$. ($\eta = 377$ ohms is the vacuum impedence).

Note that, following from Eq. (2.60), σ_d is only defined in the far zone. It follows immediately that the differential cross section equals the modulus squared of the scattering amplitude,

$$\sigma_d(\mathbf{k}_s, \mathbf{k}_i) = |f(\mathbf{k}_s, \mathbf{k}_i)|^2 \qquad (2.61)$$

The unit for σ_d is $[\sigma_d] = m^2/srad$.

One particular case is obtained for backscattering, when $\mathbf{k}_s = -\mathbf{k}_i$,

$$\sigma_b = \sigma_d(-\mathbf{k}_i, \mathbf{k}_i) \qquad (2.62)$$

where σ_b is referred to as the *backscattering cross section*.

The normalized version of σ_d describes the so-called *phase function*,

$$p(\mathbf{k}_s, \mathbf{k}_i) = 4\pi \frac{\sigma_d(\mathbf{k}_s, \mathbf{k}_i)}{\int_{4\pi} \sigma_d(\mathbf{k}_s, \mathbf{k}_i)d\Omega} \qquad (2.63)$$

The phase function p defines the angular probability density function associated with the scattered light (note that the phrase "phase function" was borrowed from nuclear physics and does not refer to the phase of the field). The integral in the denominator of Eq. (2.63) defines the *scattering cross section* as

$$\sigma_s = \int_{4\pi} \sigma_d(\mathbf{k}_s, \mathbf{k}_i)d\Omega \qquad (2.64)$$

Thus, the unit of the scattering cross section is $[\sigma_s] = m^2$. In general, if the particle also absorbs light, we can define an analogous *absorption cross section*, such that the attenuation due to the combined effect is governed by a *total cross section*,

$$\sigma = \sigma_a + \sigma_s \qquad (2.65)$$

For particles of arbitrary shapes, sizes, and refractive indices, deriving expression for the scattering cross sections, σ_d and σ_s, is very difficult. However, if simplifying assumptions can be made, the problem becomes tractable, as described in the next section.

2.7 Particles Under the Born Approximation

When the refractive index of a particle is only slightly different from that of the surrounding medium, its scattering properties can be derived analytically within the framework of the Born approximation described earlier (Sec. 2.5). Thus the scalar scattering amplitude from such a particle is the Fourier transform of the scattering potential of the particle [Eq. (2.52)]

$$f(\mathbf{q}, \omega) = \int F_p(\mathbf{r}') \cdot e^{i\mathbf{q}\mathbf{r}'} d^3\mathbf{r}' \qquad (2.66)$$

In Eq. (2.66), F_p is the scattering potential of the particle. We can conclude that the scattering amplitude and the scattering potential form a Fourier pair,

$$f(\mathbf{q}, \omega) \rightarrow F_p(\mathbf{r}', \omega) \qquad (2.67)$$

The rule of thumb for this scattering regime to apply is that the total phase shift accumulation through the particle is small, say, smaller than 1 *rad*. For a particle of diameter d and refractive index n in air, this condition is $(n-1)k_0 d < 1$. Under these conditions, the problem becomes easily tractable for arbitrarily shaped particles. For intricate shapes, the 3D Fourier transform in Eq. (2.66) can be at least solved numerically via fast Fourier transform (FFT) algorithms. For some regular shapes, we can find the scattered field in analytic form, as described below.

2.7.1 Spherical Particles

For a spherical particle, the scattering potential has the form of the *ball* function, which was introduced earlier to describe the Ewald sphere (see also App. B)

$$F_p(\mathbf{r}') = \Pi\left(\frac{r'}{2r}\right) \cdot F_0 \qquad (2.68a)$$

$$F_0 = \frac{1}{4\pi} k_0^2 (n^2 - 1) \qquad (2.68b)$$

Equations (2.68a and b) establish that the particle is spherical in shape, of radius r, and is characterized by a constant scattering potential F_0 inside the domain and zero outside.

Thus, plugging Eqs. (2.68) into Eq. (2.66) (see also App. B), we obtain the scattering amplitude distribution

$$f(\mathbf{q},\omega) \propto (n^2 - 1)k_0^2 \cdot r^3 \cdot \frac{\sin(qr) - qr \cdot \cos(qr)}{(qr)^3}$$

$$\mathbf{q} = \mathbf{k}_s - \mathbf{k}_i$$

(2.69)

We encountered a function of the same form as on the right hand side of Eq. (2.69) (i.e., the Fourier transform of the *ball* function) when estimating the resolution of structures determined by angular light scattering [Eq. (2.58]. Note, however, the two functions operate in the conjugate domains, i.e., in deriving Eq. (2.58) we used the ball function to describe the frequency support, while here the ball function defines the scattering potential in the spatial domain.

The differential cross section is [from Eq. (2.61)]

$$\sigma_d(\mathbf{q},\omega) = |f(\mathbf{q},\omega)|^2$$

$$\propto (n^2 - 1)^2 V^2 k_0^4 \left[\frac{\sin(qr) - qr \cdot \cos(qr)}{(qr)^3} \right]^2$$

(2.70)

where $V = 4\pi r^3/3$ is the volume of the particle.

Equation (2.70) establishes the differential cross section associated with a spherical particle under the Born approximation. Sometimes this scattering regime is referred to as Rayleigh-Gans, and the particle for which this formula holds as *Rayleigh-Gans particles*. Figure 2.19 illustrates the angular scattering according to Eq. (2.70). The scattering angle θ enters explicitly Eq. (2.70), by expressing the modulus of the momentum transfer as $q = 2k_0 \sin(\theta/2)$.

A very interesting particular case is obtained when $qr \to 0$. Note that this asymptotic case may happen both when the particle is very small, $r \to 0$, and also when the measurement is performed at very small angles, i.e., $q \to 0$. If we expand around the origin the right hand side of Eq. 2.70, we obtain

$$f(x) = \left(\frac{\sin x - x \cdot \cos x}{x^3} \right)^2$$

$$\approx \left[x - x^3/6 - x\left(1 - \frac{x^2}{2}\right) \right]^2 \Big/ x^6$$

(2.71)

$$= \frac{1}{6}$$

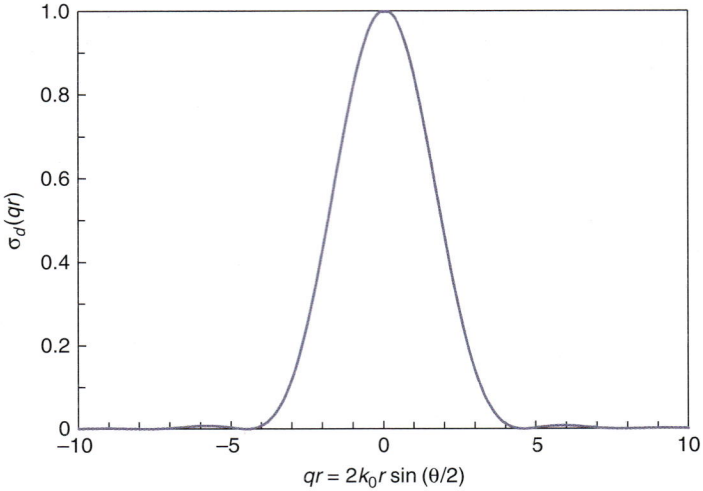

FIGURE 2.19   Angular scattering (differential cross section) for a Rayleigh-Gans particle.

Thus, remarkably, measurements at small scattering angles, can reveal the volume of the particle,

$$\left.\sigma_d(\mathbf{q},\omega)\right|_{q\to 0} \propto (n^2-1)^2 V^2 k_0^4 \qquad (2.72)$$

This result is the basis for many flow cytometry instruments, where the volume of cells is estimated via measurements of *forward scattering*. The cell structure, i.e., higher-frequency components are retrieved through larger angle, or *side scattering* measurements.

On the other hand, Eq. (2.72) applies equally well where the particle is very small, commonly referred to as the Rayleigh regime (or Rayleigh particle). Thus, the scattering cross section, σ_s, for Rayleigh particles, which is just the integral over the solid angle of σ_d, $\sigma_s = 4\pi\sigma_d$ has the same form, up to some constant prefactors

$$\sigma_s(\omega) \propto (n-1)^2 k_0^4 V^2 \qquad (2.73)$$

In this case, the fact that σ_d is independent of angles, indicates that the Rayleigh scattering is *isotropic, which is* a well known result. Still, the scattering cross section of Rayleigh particles is characterized by strong dependence on the particle radius, r ($V = 4\pi r^3/3$) and wavelength ($k_0 = 2\pi/\lambda$),

$$\sigma_s \propto r^6$$
$$\sigma_s \propto \lambda^{-4} \qquad (2.74)$$

One implication of the strong dependence on wavelength is that the nanoparticles in the atmosphere have scattering cross sections that are 16 times larger for a wavelength $\lambda_b = 400\ nm$ (blue) than for $\lambda_r = 800\ nm$ (red). This explains why the clear sky looks bluish due to the scattered light, and the sun itself looks reddish, due to the remaining, unscattered portion of the initial white light spectrum.

2.7.2 Cubical Particles

For a cubical particle (Fig. 2.20), the scattering potential can be expressed as a product of three 1D rectangular functions, along each direction,

$$F(x,y,z) = \Pi\left(\frac{x}{2a}\right) \cdot \Pi\left(\frac{y}{2a}\right) \cdot \Pi\left(\frac{z}{2a}\right) \cdot F_0$$

$$(2.75)$$

$$F_0 = \frac{1}{4\pi} k_0^2 (n^2 - 1)$$

The scattering amplitude in this case is (using the 1D Fourier transform of a rectangular function, see App. B),

$$f(q_x, q_y, q_z) = (n^2 - 1)k_0^2 V \operatorname{sinc}(q_x a) \cdot \operatorname{sinc}(q_y a) \cdot \operatorname{sinc}(q_z a) \qquad (2.76)$$

Note that, according to Fig. 20, the incident wave vector is parallel to the y-axis, such that $(q_x, q_y, q_z) = (k_{sx}, k_{sy} - k_i, k_{sz})$. It can be seen that the differential cross section $\sigma_d = |f|^2$ has the same V^2 and k_0^4 leading dependence as for spherical particles. As the size of the particle decreases, $a \rightarrow 0$, we recover the Rayleigh regime. In essence, this result is due to the fact that, for particles much smaller

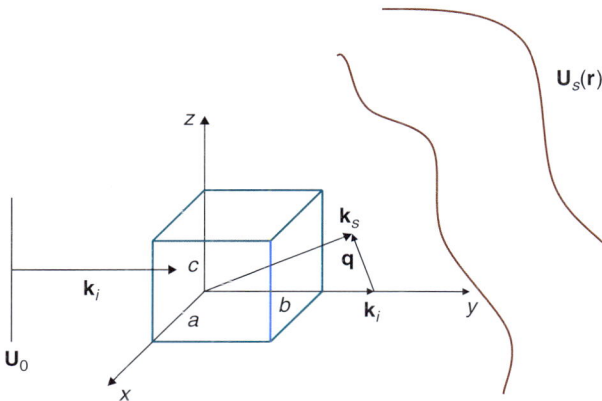

FIGURE 2.20 Scattering by a cubical particle; $q = k_s - k_i$.

than wavelength, the details about the particle shape do not affect the far-zone scattering. In other words, information about the particle shape at scales much below the wavelength is contained in the evanescent near field and not in the propagating, scattered field.

2.7.3   Cylindrical Particles

For a cylindrical particle of radius a and length b (Fig. 2.21), the scattering potential can be written as a product between a 2D (disk function) and a 1D rectangular function

$$F(x,y,z) = \Pi\left(\frac{\sqrt{x^2+y^2}}{2a}\right) \cdot \Pi\left(\frac{z}{2b}\right) \cdot F_0$$

$$(2.77)$$

$$F_0 = \frac{k_0^2}{4\pi}(n^2-1)$$

The 3D Fourier transform of F yields the scattering amplitude (using the 3D Fourier transform in cylindrical coordinates, see App. B)

$$f(q_x,q_y,q_z) = F_0\pi a^2 b \cdot \frac{J_1\left(\sqrt{q_x^2+q_y^2}\cdot a\right)}{2\sqrt{q_x^2+q_y^2}} \cdot \text{sinc}(q_z \cdot b)$$

$$(2.78)$$

where J_1 is the Bessel function of first order and kind. As before, σ_d and σ_s can be easily obtained from $|f|^2$. Note that, according to Fig. 2.21, the incident wave vector is parallel to the y-axis, such that $(q_x,q_y,q_z)=(k_{sx},k_{sy}-k_i,k_s)$.

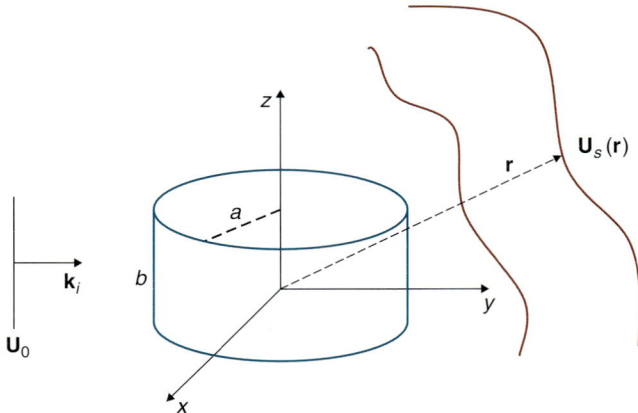

FIGURE 2.21   Scattering by a cylindrical particle.

2.8   Scattering from Ensembles of Particles within the Born Approximation

Generally, studying biological structures with light entails measuring scattering signals from an ensemble. Here we consider the situation where the scattering experiment is performed over an ensemble of particles randomly distributed in space, as illustrated in Fig. 2.22.

If we assume that the ensemble is made of identical particles of scattering potential $F_0(\mathbf{r})$, then the scattering potential of the system can be expressed by a sum of δ-functions in the 3D space, which describe the discrete positions, convolved with $F_0(\mathbf{r})$,

$$F(\mathbf{r}) = F_0(\mathbf{r}) \circledv \sum_j \delta(\mathbf{r} - \mathbf{r}_j) \tag{2.79}$$

Equation (2.79) establishes the distribution of the scattering potential, where each particle is positioned at r_j. Note that, in order for the Born approximation to apply, the particle distribution must be sparse. The scattering amplitude is simply the 3D Fourier transform of $F(\mathbf{r})$,

$$f(\mathbf{q}) = \int F_0(\mathbf{r}) \circledv \sum_j \delta(\mathbf{r} - \mathbf{r}_j) e^{i\mathbf{q}\mathbf{r}} d^3\mathbf{r}$$
$$= f_0(\mathbf{q}) \cdot \sum_j e^{i\mathbf{q}\mathbf{r}_j} \tag{2.80}$$

where, as before, the scattering wave vector is $\mathbf{q} = \mathbf{k}_s - \mathbf{k}_i$, and we used the shift theorem, i.e., $\delta(\mathbf{r} - \mathbf{r}_j) \rightarrow e^{i\mathbf{q}\mathbf{r}_j}$ (see App. B).

Therefore, the scattering amplitude of the ensemble is the scattering amplitude of a single particle, $f_0(\mathbf{q})$, multiplied (*modulated*) by the so-called *structure function*, defined as

$$S(\mathbf{q}) = \sum_j e^{i\mathbf{q}\mathbf{r}_j} \tag{2.81}$$

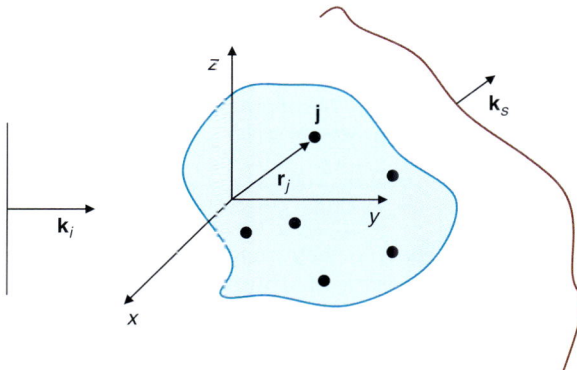

FIGURE 2.22   Scattering by an ensemble of particles.

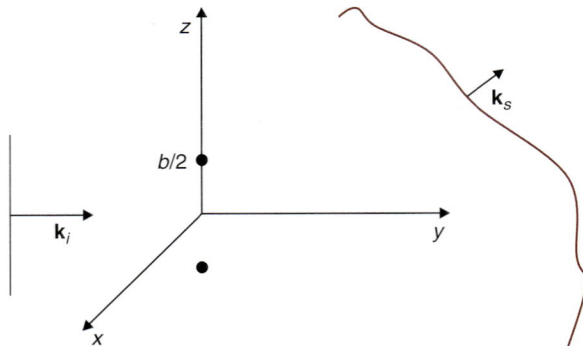

Figure 2.23 Scattering by two particles separated by a distance b.

Now we can express the scattering amplitude as a product,

$$f(\mathbf{q}) = f_0(\mathbf{q}) \cdot S(\mathbf{q}) \tag{2.82}$$

In Eq. (2.82), $f_0(\mathbf{q})$ is sometimes referred to as the *form function*. The physical meaning of the form and structure functions becomes apparent if we note that the size of the particle is smaller than the inter-particle distance. It follows that $f_0(q)$ is a *broader* function than $S(q)$, i.e., $f_0(q)$ is the envelope (*form*) of f and $S(q)$ is its rapidly varying component (*structure*).

Example 2.3 Scattering from two spherical particles (radius a) separated by a distance b (Fig. 2.23).

According to Eqs. (2.81) and (2.82), the far-zone scattering amplitude is easily obtained as

$$\begin{aligned} f(\mathbf{q}) &= f_0(\mathbf{q}) \cdot \cos\!\left(q_z \cdot \frac{b}{2}\right) \\ &= (n^2 - 1)k_0^2 V \cdot \frac{\sin(qr) - qr \cdot \cos(qr)}{(qr)^3} \cdot \cos\!\left(q_z \cdot \frac{b}{2}\right) \end{aligned} \tag{2.83}$$

Note that, according to Fig. 2.23, the incident wave vector is parallel to the y-axis, such that $(q_x, q_y, q_z) = (k_{sx}, k_{sy} - k_i, k_s)$. In Eq. (2.83), we wrote explicitly the form factor of a Rayleigh-Gans particle of volume V and refractive index n. The structure function is a simple cosine in this case. This approach is the basis for extracting crystal structures form x-ray scattering measurements.

2.9   Mie Scattering

In 1908, Mie provided the full electromagnetic solutions of Maxwell's equations for a *spherical particle* of arbitrary size and refractive index. The scattering cross section has the form

$$\sigma_s = \pi a^2 \frac{2}{\alpha^2} \sum_{n=1}^{\infty} (2n+1)\left(\left|a_n\right|^2 + \left|b_n\right|^2\right) \tag{2.84}$$

where a_n and b_n are functions of $k = k_0 a$, $\beta = k_0 na/n_0$, with a the radius of the particle, k_0 the wave number in the medium, n_0 the refractive index of the medium, and n the refractive index of the particle. Equation (2.84) shows that the Mie solution is expressed in terms of an infinite series, which can only be evaluated numerically. Although today common personal computers can evaluate σ_s very fast, we note that as the particle increases in size, the summation converges more slowly, because a higher number of terms contribute significantly. Physically, as a increases, standing waves (modes) with higher number of maxima and minima "fit" inside the sphere. Although restricted to spherical particles, Mie theory is sometimes used for modeling tissue scattering.

Further Reading

1. J. D. Jackson. *Classical electrodynamics*. (Wiley, New York, 1999).
2. L. D. Landau, E. M. Lifshits, and L. P. Pitaevskii. *Electrodynamics of continuous media*. (Pergamon, Oxford [Oxfordshire]; New York, 1984).
3. R. P. Feynman, R. B. Leighton, and M. L. Sands. *The Feynman lectures on physics*. (Addison-Wesley Pub. Co., Reading, Mass., 1963).
4. M. Born and E. Wolf. *Principles of optics: Electromagnetic theory of propagation, interference and diffraction of light*. (Cambridge University Press, Cambridge; New York, 1999).
5. B. E. A. Saleh and M. C. Teich. *Fundamentals of photonics*. (Wiley, New York, 1991).
6. M. Bass, V. N. Mahajan, and Optical Society of America. *Handbook of optics*. (McGraw-Hill, New York, 2010).
7. J. D. Gaskill. *Linear systems, Fourier transforms, and optics*. (Wiley, New York, 1978).
8. J. W. Goodman. *Introduction to Fourier optics*. (McGraw-Hill, New York, 1996).
9. H. C. van de Hulst. *Light scattering by small particles* (Dover Publications, New York, 1981).
10. C. F. Bohren and D. R. Huffman. *Absorption and scattering of light by small particles*. (Wiley, New York, 1983).
11. L. Tsang, J. A. Kong, and K.-H. Ding. *Scattering of electromagnetic waves. Theories and applications*. (Wiley, New York, 2000).
12. A. Ishimaru. *Electromagnetic wave propagation, radiation, and scattering*. (Prentice Hall, Englewood Cliffs, NJ., 1991).

CHAPTER 3

Spatiotemporal Field Correlations

3.1 Spatiotemporal Correlation Function: Coherence Volume

All optical fields encountered in practice fluctuate randomly in both time and space and are, therefore, subject to a *statistical* description.[1] These field fluctuations depend on both the emission process (primary sources) and propagation media (secondary sources). *Optical coherence* is a manifestation of the *field statistical similarities* in space and time and coherence theory is the discipline that mathematically describes this statistics.[2] A deterministic field distribution in both time and space is the *monochromatic plane wave*, which, of course, is only a mathematical construct, impossible to obtain in practice due to the uncertainty principle.

The formalism presented below for describing the field correlations is mathematically similar to that used for mechanical fluctuations, for example, in the case of vibrating membranes. The analogy between the two different types of fluctuations and their mathematical description in terms of spatiotemporal correlations has been recently emphasized.[3]

Perhaps a starting point in understanding the physical meaning of a *statistical optical field* is to ask the question: what is the *effective (average) temporal* sinusoid, i.e., $\left\langle e^{-i\omega t}\right\rangle_\omega$, for a broadband field? Similarly, what is the *average spatial* sinusoid, i.e., $\left\langle e^{i\mathbf{k}\cdot\mathbf{r}}\right\rangle_\mathbf{k}$. Here we use the sign convention, whereby a monochromatic plane wave is described by $e^{-i(\omega t - \mathbf{k}\cdot\mathbf{r})}$. These two averages can be performed using the *probability densities* associated with the temporal and spatial frequencies, $S(\omega)$ and $P(k)$, which are normalized to satisfy $\int S(\omega)d\omega = 1$, $\int P(\mathbf{k})d^3\mathbf{k} = 1$. Thus, $S(\omega)d\omega$ is the probability of having frequency component ω in our field, or the fraction of the total power contained in the vicinity of frequency ω. Similarly, $P(k)d^3k$ is the probability of having component k in the field, or the fraction of the total power contained around wavevector k. Up to a normalization factor, *S and P are the temporal and spatial power spectra* associated with the fields. Thus, the two "effective sinusoids" can be expressed as *ensemble averages*, using $S(\omega)$ and $P(k)$ as weighting functions,

$$\left\langle e^{-i\omega t} \right\rangle_{\omega} = \int S(\omega)e^{-i\omega t}d\omega$$

$$= \Gamma(t)$$

$$(3.1a)$$

$$\left\langle e^{ik \cdot r} \right\rangle_{k} = \int P(\mathbf{k})e^{ik \cdot r}d^3\mathbf{k}$$

$$= W(\mathbf{r})$$

$$(3.1b)$$

Equations (3.1a) and (3.1b) establish that the frequency-average *temporal sinusoid* for a broadband field equals its temporal autocorrelation, denoted by Γ. Similarly, the average *spatial sinusoid* for an inhomogeneous field equals its spatial autocorrelation, denoted by W.

Besides the basic science interest in describing the statistical properties of optical fields, coherence theory can make predictions of experimental relevance. The general problem can be formulated as follows (Fig. 3.1): given the optical field distribution $U(\mathbf{r}, t)$ that varies randomly in space and time, over what *spatiotemporal domain* does the field preserve significant correlations? Experimentally, this question translates into: combining the field $U(\mathbf{r}, t)$ with a replica of itself shifted in both time and space, $U(\mathbf{r}+\boldsymbol{\rho}, t+\tau)$, *on average*, how large can $\boldsymbol{\rho}$ and τ be and still observe "significant" interference?

Intuitively, we expect that monochromatic fields exhibit (infinitely) broad temporal correlations, while plane waves are expected to manifest infinite spatial correlations. This is so because regardless of how much we shift a monochromatic field in time or a plane wave in space, they remain perfectly correlated with their unshifted replicas. On the other hand, it is difficult to picture temporal correlations

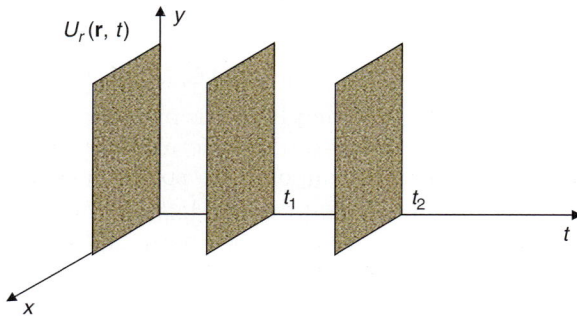

FIGURE 3.1   Spatiotemporal distribution of a real optical field.

decaying over timescales that are shorter than an optical period and spatial correlations that decay over spatial scales smaller than the optical wavelength. In the following we provide a quantitative description of the spatiotemporal correlations.

The statistical behavior of optical fields can be mathematically captured quite generally via a *spatiotemporal correlation function*

$$\Lambda(\mathbf{r}_1,\mathbf{r}_2;t_1,t_2) = \langle U(\mathbf{r}_1,t_1) \cdot U^*(\mathbf{r}_2,t_2) \rangle_{\mathbf{r},t} \tag{3.2}$$

The average $< >$ is performed both *temporally and spatially*, as indicated by the subscripts $\mathbf{r}$ and t. Because common detector arrays capture the spatial intensity distributions in 2D only, we will restrict the discussion to $\mathbf{r} = (x, y)$, without losing generality. These averages are defined in the usual sense as

$$\langle U(\mathbf{r}_1,t_1) \cdot U^*(\mathbf{r}_2,t_2) \rangle_t = \lim_{T \to \infty} \frac{1}{T^2} \int_{-T/2}^{T/2} \int_{-T/2}^{T/2} U(\mathbf{r}_1,t_1) \cdot U^*(\mathbf{r}_2,t_2) dt_1\, dt_2$$

$$\langle U(\mathbf{r}_1,t_1) \cdot U^*(\mathbf{r}_2,t_2) \rangle_{\mathbf{r}} = \lim_{A \to \infty} \frac{1}{A^2} \int_A \int_A U(\mathbf{r}_1,t_1) \cdot U^*(\mathbf{r}_2,t_2) d^2\mathbf{r}_1 d^2\mathbf{r}_2 \tag{3.3}$$

Often, in practice we deal with fields that are both *stationary* (in time) and *statistically homogeneous* (in space). If stationary, the statistical properties of the field (e.g., the average, higher order moments) do not depend on the origin of time. Similarly, for statistically homogeneous fields, their properties do not depend on the origin of space. Wide-sense stationarity is less restrictive and defines a random process with only its first and second moments independent of the choice of origin. For the discussion here, the fields are assumed to be stationary at least in the wide sense. Under these circumstances, the dimensionality of the spatiotemporal correlation function Λ decreases by half, i.e.,

$$\Lambda(t_1,t_2) = \Lambda'(t_2 - t_1)$$

$$\Lambda(\mathbf{r}_1,\mathbf{r}_2) = \Lambda'(\mathbf{r}_2 - \mathbf{r}_1) \tag{3.4}$$

The spatiotemporal correlation function becomes Λ

$$\Lambda(\boldsymbol{\rho},\tau) = \langle U(\mathbf{r},t) \cdot U^*(\mathbf{r}+\boldsymbol{\rho},t+\tau) \rangle_{\mathbf{r},t} \tag{3.5}$$

Note that $\Lambda(\mathbf{0},0) = \langle U(\mathbf{r},t) \cdot U^*(\mathbf{r},t) \rangle_{\mathbf{r},t}$ represents the *spatially averaged irradiance* of the field, which is, of course, a real quantity. However, in general $\Lambda(\boldsymbol{\rho},\tau)$ is complex. Let us define a normalized version of Λ, referred to as the *spatiotemporal complex degree of correlation*

$$\alpha(\boldsymbol{\rho},t) = \frac{\Lambda(\boldsymbol{\rho},\tau)}{\Lambda(\mathbf{0},0)} \tag{3.6}$$

It can be shown that for stationary fields $|\Lambda|$ attains its maximum at $\mathbf{r} = 0$, $t = 0$, thus

$$0 < |\alpha(\mathbf{\rho}, \tau)| < 1 \qquad (3.7)$$

Further, we can define an area $A_C \propto \rho_C^2$ and length $l_c = c\tau_C$, over which $|\alpha(\mathbf{\rho}_C, \tau_C)|$ maintains a significant value, say $|\alpha| > \frac{1}{2}$, which defines a *coherence volume*

$$V_C = A_C \cdot l_C \qquad (3.8)$$

This *coherence volume* determines the maximum domain size over which the fields can be considered correlated. In general an extended source, such as an incandescent filament, may have spectral properties that vary from point to point. Thus, it is convenient to discuss spatial correlations at each frequency, ω, as described in Sec. 3.2.

3.2   Spatial Correlations of Monochromatic Light

3.2.1   Cross-Spectral Density

By taking the Fourier transform of Eq. (3.2) with respect to time, we obtain what we refer to as the spatially averaged *cross-spectral density*. The cross-spectral density is described in more detail in Ref. 4.

$$\begin{aligned} W(\mathbf{\rho}, \omega) &= \int \Lambda(\mathbf{\rho}, \tau) \cdot e^{i\omega\tau} \, d\tau \\ &= \left\langle U(\mathbf{r}, \omega) \cdot U^*(\mathbf{r} + \mathbf{\rho}, \omega) \right\rangle_{\mathbf{r}} \end{aligned} \qquad (3.9)$$

The cross-spectral density function was used previously by Wolf to describe the *second-order statistics* of optical fields, i.e., the Fourier transform of the temporal cross-correlation at two distinct points, $W_{12}(\mathbf{r_1}, \mathbf{r_2}, \omega) = \int \Gamma_{12}(\mathbf{r_1}, \mathbf{r_2}, \tau) \cdot e^{i\omega\tau} \, d\tau$.[2,4] This function describes the similarity in the field fluctuations of two points, like, for example, in the two-slit Young interferometer. Note that two points are always fully correlated if the light is monochromatic, because the field at the two points can differ, at the most, by a constant phase shift. By contrast, across an entire plane, the phase distribution is a random variable. Therefore, in order to capture the spatial correlations in a *ensemble-averaged* sense, which is most relevant to imaging, we will use the spatially averaged version of $W(\mathbf{\rho}, \omega)$, as defined in Eq. 3.9.

One interferometric configuration that allows direct measurement of W is illustrated in Fig. 3.2 via an imaging Mach-Zehnder interferometer. Here the monochromatic field $U(\mathbf{r}, \omega)$ is split in two replicas that are further reimaged at the CCD plane via two 4f lens systems, which induce a relative *spatial shift*, $\mathbf{\rho}$. The question of practical

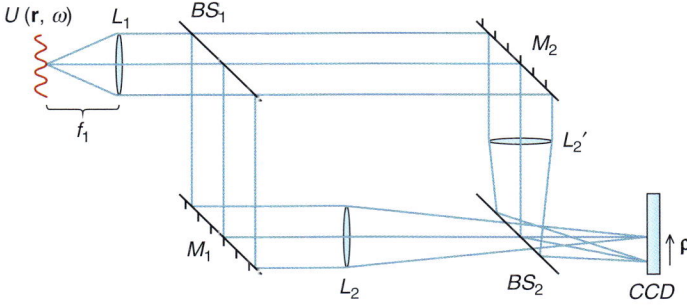

FIGURE 3.2 Mach-Zender interferometry with spatially extended fields.

interest here is: To what extent do we observe fringes, or, more quantitatively, what is the spatially averaged fringe contrast as we vary ρ? For each value of ρ, the CCD records a *spatially resolved* intensity distribution, or an *interferogram*. We can thus compute the *spatial average* of this quantity as

$$I(\rho,\omega) = \frac{1}{A}\int |U(\mathbf{r},\omega)+U(\mathbf{r}+\rho,\omega)|^2 d^2\mathbf{r}$$

$$= \frac{1}{A}\int 2I_1(\mathbf{r},\omega)d^2\mathbf{r} + 2\,\mathrm{Re}\frac{1}{A}\int [U(\mathbf{r},\omega)\cdot U \otimes (\mathbf{r}+\rho,\omega)]d^2\mathbf{r} \qquad (3.10)$$

$$= 2\langle I_1(\mathbf{r},\omega)\rangle_\mathbf{r} + 2\,\mathrm{Re}[W(\rho,\omega)]$$

where we assumed that the interferometer splits the light equally on the two arms. Once the average intensity of each beam, $\langle I_1(\mathbf{r},\omega)\rangle$, is measured separately (e.g., by blocking one beam and measuring the other), the real part of $W(\rho,\omega)$, as defined in Eq. 3.9, can be measured experimentally. Clearly, multiple CCD exposures are necessary corresponding to each ρ. The *complex degree of spatial correlation* at frequency, ω, is defined as

$$\beta(\rho.\omega) = \frac{W(\rho,\omega)}{|W(\mathbf{0},\omega)|} \qquad (3.11)$$

Note that $W(\mathbf{0},\omega)$ is nothing more than the spatially-average optical spectrum of the field,

$$W(\mathbf{0},\omega) = \langle U(\mathbf{r},\omega)\cdot U^*(\mathbf{r},\omega)\rangle_\mathbf{r}$$
$$= \langle S(\mathbf{r},\omega)\rangle_\mathbf{r} \qquad (3.12)$$

Again, it can be shown that $|\beta| \in [0,1]$, where the extremum values of $|\beta|=0$ and $|\beta|=1$ correspond to complete lack of spatial correlation and full correlation, respectively. The area over which $|\beta|$ maintains a *significant value* defines the correlation area at frequency ω, e.g.,

$$A_C = D, \qquad \text{for which } \left|\beta(\rho,\omega)\right|\Bigg|_{(\rho_x \rho_y) \subset D} < \frac{1}{2} \qquad (3.13)$$

Often, we refer to the *coherence area* of a certain field, without referring to a particular optical frequency. In this case, what is understood is the frequency-averaged correlation area, $A = \langle A_c(\omega) \rangle_\omega$. In practice we deal many times with fields that are characterized by a mean frequency, $\omega_0 = \int_0^\infty \omega P(\omega) d\omega / \int_0^\infty P(\omega) d\omega$. In this case, the spatial coherence is fully described by the behavior at this particular frequency. For example, a broad band field is *fully spatially coherent* over a certain domain if $|\beta(\rho,\omega_0)| = 1$, for any ρ in the domain.[5,6]

3.2.2   Spatial Power Spectrum

Since $W(\rho,\omega)$ is a spatial correlation function, it can be expressed via a Fourier transform in terms of a *spatial power spectrum*, $P(\mathbf{k},\omega)$ (see App. B for the correlation theorem),

$$
\begin{aligned}
P(\mathbf{k},\omega) &= \iint W(\rho,\omega) \cdot e^{-i\mathbf{k}\rho} d^2\rho \\
W(\rho,\omega) &= \iint P(\mathbf{k},\omega) \cdot e^{i\mathbf{k}\rho} d^2\mathbf{k}
\end{aligned}
\qquad (3.14)
$$

If the two fields are monochromatic, and of different frequencies, W oscillates sinusoidally in time,

$$
\begin{aligned}
W(\rho,\omega_1,\omega_2) &= \langle U_1(\mathbf{r},\omega_1) \cdot U_2(\mathbf{r}+\rho,\omega_2) \rangle_{\mathbf{r}} \\
&= W_\rho(\rho) \exp[i(\omega_2 - \omega_1)t]
\end{aligned}
\qquad (3.15)
$$

where W_ρ is the spatial component of W.

The physical meaning of $\exp[i(\omega_2 - \omega_1)t]$ is that, performing the spatial correlation measurements in Fig. 3.2 with each field of the interferometer now having two different optical frequencies, ω_1, ω_2 (maybe passing through two different filters), the result oscillates in time and vanishes upon temporal averaging. Thus, maximum contrast fringes in a Mach-Zehnder interferometer like in Fig. 3.2 is obtained by having the same spectral content on both arms of the interferometer.

According to Eq. 3.14, the spatial correlation function $W(\rho,\omega)$ can also be experimentally determined from measurements of the spatial power spectrum, as shown in Fig. 3.3. As discussed in Secs. 2.3 to 2.4, both the far field propagation in free space and propagation through

(a) S(x, y), a/2, a/2, k₁, P(kₓ, k_y), k₀, f, f, CCD

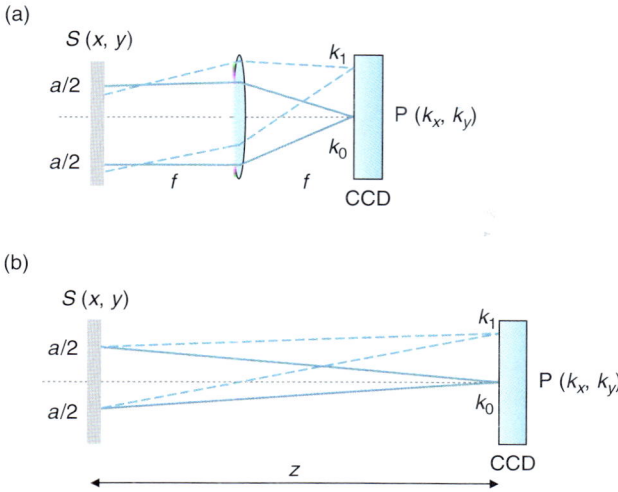

Figure 3.3 Measuring the spatial power spectrum of the field from source S via a (a) lens and (b) Fraunhofer propagation in free space.

a lens can generate the Fourier transform of the source field, as illustrated in Fig. 3.3,

$$\tilde{U}(\mathbf{k},\omega) = \int_A U(\mathbf{r},\omega)\cdot e^{-i\mathbf{k}\mathbf{r}} d^2\mathbf{r} \tag{3.16}$$

The CCD is sensitive to power and, thus, detects the spatial power spectrum, $P(\mathbf{k},\omega) = |\tilde{U}(\mathbf{k},\omega)|^2$.

In Eq. 3.16, the frequency component $\mathbf{k} = (k_x, k_y)$ depends either on the focal distance, for the lens transformation (Fig. 3.3a), or on the propagation distance z, for the Fraunhofer propagation (Fig.3. 3b),

$$\mathbf{k} = \frac{2\pi}{\lambda f}(x',y')$$

$$\mathbf{k} = \frac{2\pi}{\lambda z}(x',y') \tag{3.17}$$

Note that since $x << f$ or $x << z$, the ratios x/f and x/z describe the diffraction angle; therefore sometimes $P(\mathbf{k},\omega)$ is called *angular power spectrum*.

An important question that arises equally often both in astronomy and microscopy is: how does the spatial correlation of the field change upon propagation? For extended sources that are far away from the detection plane, as in Fig. 3.3b, the size of the source may have a significant effect on the Fourier transform in Eq. 3.16. This effect becomes obvious if we replace the source field U with its spatially truncated version, $\underline{U}$, to indicate the finite size of the source

$$\underline{U}(\mathbf{r},\omega) = U(\mathbf{r},\omega) \cdot \Pi\left(\frac{\mathbf{r}}{a}\right) \tag{3.18}$$

where function Π is the typical *2D rectangular function*, here denoting a square of side a. Thus, the far field becomes

$$\underline{\tilde{U}}(\mathbf{k},\omega) = a^2 \tilde{U}(\mathbf{k},\omega) \circledcirc_{k_x k_y} \text{sinc}(a\mathbf{k}) \tag{3.19}$$

where $\circledcirc$ denotes convolution and $sinc(x)$ is the common $\sin(x)/x$ function (see App. B). Thus, the field across the detection plane (x', y'), $\underline{\tilde{U}}(\mathbf{k},\omega)$, is *smooth* over scales given by the width of the *sinc* function. This *smoothness* indicates that the field is *spatially correlated* over this spatial scale. Along x', this correlation distance, x'_c, is obtained by writing explicitly the spatial frequency argument of the *sinc* function. Thus, the highest spatial frequency is $2\pi/a$,

$$\frac{2\pi}{a} = k_x = \frac{2\pi}{\lambda z} \cdot x'_c \tag{3.20}$$

We note that a^2/z^2 describes the solid angle, Ω, subtended by the source and conclude that the correlation area of the field generated by the source in the far zone is of the order of

$$A_C = x_c^2 = \frac{\lambda^2}{\Delta\Omega} \tag{3.21}$$

This simple relationship allowed Michelson to measure interferometrically the angle subtended by stars. For example, the Sun subtends an angle $\theta \approx 10$ mrad, i.e., $\Omega \approx 10^{-4}$ srad. Thus, for the green radiation that is the mean of the visible spectrum, $\lambda = 550$ nm, the coherence area at the surface of the Earth is of the order of $A_C^{\text{Sun}} = 50 \cdot 50$ μm². Measuring this area over which the sun light shows correlations (or generates fringes) provides information about its angular size. For angularly smaller sources, far field spatial coherence is correspondingly higher. *This result is a consequence of the Van Cittert-Zernike theorem, which establishes the relationship between the intensity distribution across a source and the spatial correlations in the far-zone.* Essentially, this is the result of free-space propagation acting as a *spatial low-pass filter*.[7]

It is perhaps not surprising that, aware of this powerful theorem, Zernike himself employed the spatial filtering concept to develop *phase contrast microscopy in the 1930's*.[8,9] It had been known since

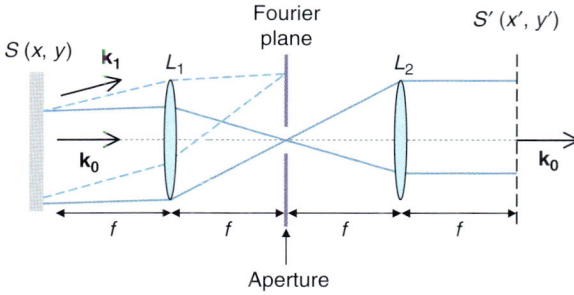

FIGURE 3.4 Spatial filtering via a 4f system.

Abbe, in 1873, that an image can be described as an interference phenomenon.[10] Image formation is the result of simultaneous interference processes that take place at each point in the image. In a quest to turn transparent specimens visible, Zernike employed spatial filtering (in a way that is similar to Fig. 3.4) and extended the coherence area of the illuminating field to exceed the field of view of the microscope. Naturally, in microscopy, this type of illumination is called *coherent*. As a result, an *average field* over the entire image could properly be defined, because now the phase relationship among different points was stable. In this case, the simultaneous interferences at all points that generate the image have a common phase reference, which is the phase associated with the mean field. Thus, controlling the phase delay of the mean field adjusts the contrast of the entire image, like in typical interferometry experiments. This is the underlying idea in phase contrast microscopy, which represents a major breakthrough in microscopy and an important precursor to QPI; it will be discussed in more detail in Chap. 5.

We should mention in passing another major method for intrinsic contrast microscopy (not discussed further in this book), Differential Interference Contrast (DIC, or Nomarski) [12,13], which renders maps of phase gradients across a transparent sample, and is categorized under "incoherent" methods. DIC does use incoherent illumination (no spatial filtering), yet the high-contrast images are generated by interfering an image field with a replica of itself that is slightly shifted transversally. This seems paradoxical. However, the key here is to realize that the numerical aperture of the microscope is finite, i.e., the imaging system itself performs *spatial filtering*. According to Eq. 3.21, the spatial coherence is of the order of

$$A_C = \left(\frac{\lambda}{NA}\right)^2$$

where NA is the numerical aperture of the microscope. This equation simply states that the image field is fully correlated across a region of the order of the diffraction spot. Therefore, in DIC shifting the two replicas of the image field by less than a diffraction spot generates high-contrast fringes. In other words, over the diffraction spot, the image field *is* spatially coherent.

3.2.3   Spatial Filtering

From the properties of Fourier transforms, we infer that higher spatial coherence at frequency ω, i.e., a broader $W(\rho,\omega)$, can be obtained by narrower $P(\mathbf{k},\omega)$. When dealing with extended sources, it is common practice in the laboratory to perform low-pass filtering on $P(\mathbf{k},\omega)$, such that the coherence area extends over the desired field of view (as already mentioned in the context of phase contrast microscopy). This procedure, commonly encountered in QPI experiments, is called, not surprisingly, *spatial filtering*, and is illustrated in Fig. 3.4.

The extended source S emits light at a multitude of frequencies ω, and spatial frequencies $\mathbf{k}$. At a given frequency ω, lens L_1 performs the spatial Fourier transform. If an aperture is placed at this Fourier plane to block the high-spatial frequencies, the field reconstructed by lens L_2 at plane S' (conjugate to S) approximates a plane wave of wave vector $\mathbf{k}_0$. With this procedure, from an extended source, we obtain a highly spatially coherence field. Of course, this procedure is *lossy*, as the energy carried by the high-spatial frequency is lost. Asymptotically, closing down the aperture generates a field that approaches a plane wave at plane S'. Conversely, it should be pointed out that all sources exhibit spatial coherence at least at the scale of the wavelength. This is easily understood by noting that a δ- correlated source, $W(\rho) = \delta(\rho)$ requires $\tilde{W}(\mathbf{k})$ infinitely broad, i.e., $\tilde{W}(\mathbf{k}) = 1$. Clearly, this is impossible, because a planar source can only emit in a 2π srad solid angle. Thus the minimum coherence area for an arbitrary source is of the order of (Eq. 3.21)

$$A_C^{\min} \simeq \frac{\lambda^2}{2\pi} \tag{3.22}$$

Physically, spatial coherence of a field over a given plane describes how close the field is to a plane wave. Alternatively, spatial coherence describes how well can the field be focused to a point (this point, of course, corresponds to a delta-function in the frequency domain). Spatial coherence plays an important role in microscopy and is used to differentiate between two classes of methods: (spatially) coherent vs. incoherent. Of course, quantitative phase imaging requires that the illumination field is spatially coherent, such that a phase shift can be properly defined over the entire field of view of interest.

3.3   Temporal Correlations of Plane Waves

3.3.1   Temporal Autocorrelation Function

Let us have the discussion symmetric to that in Sec. 3.2, where we now investigate the temporal correlations of fields at a particular spatial frequency $\mathbf{k}$ (or, equivalently, a certain direction of propagation). Taking the *spatial Fourier* transform of Λ in Eq. 3.2, we obtain the *temporal correlation function*

$$\Gamma(\mathbf{k},\tau) = \iint \Lambda(\tilde{\rho},\tau) \cdot e^{-i\mathbf{k}\rho} d^2\mathbf{r}$$

$$= \langle U(\mathbf{k},t) \cdot U^*(\mathbf{k},t+\tau) \rangle_t \qquad (3.23)$$

The autocorrelation function Γ is relevant in interferometric experiments of the type illustrated in Fig. 3.5. In a Michelson interferometer, a plane wave from the source is split in two by the beam splitter and subsequently recombined via reflections on mirrors M_1 and M_2. The intensity at the detector has the form (we assume 50/50 beam splitter)

$$I(\mathbf{k},\tau) = \left\langle \left| U(\mathbf{k},t) + U^*(\mathbf{k},t+\tau) \right|^2 \right\rangle_t$$

$$= 2I_1(\mathbf{k}) - 2\,\mathrm{Re}\langle U(\mathbf{k},t) \cdot U^*(\mathbf{k},t+\tau) \rangle \qquad (3.24)$$

where I_1 is the intensity on each arm.

Thus the real part of $\Gamma(\mathbf{k},\tau)$ is obtained by varying the time delay between the two fields. This delay can be controlled by translating one of the mirrors. The *complex degree of temporal correlation* at spatial frequency $\mathbf{k}$ is defined as

$$\gamma(\mathbf{k},\tau) = \frac{\Gamma(\mathbf{k},\tau)}{\left|\Gamma(\mathbf{k},0)\right|} \qquad (3.25)$$

Note that $\Gamma(\mathbf{k},0)$ represents the intensity of the field at wavevector $\mathbf{k}$, i.e.,

$$\Gamma(\mathbf{k},0) = \langle U(\mathbf{k},t) \cdot U \otimes (\mathbf{k},t) \rangle_t$$

$$= I(\mathbf{k}) \qquad (3.26)$$

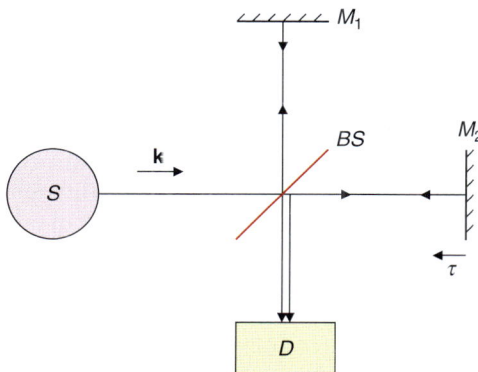

FIGURE 3.5 Michelson interferometry.

The complex degree of temporal correlation has the similar property with its spatial counterpart β, i.e.,

$$0 < |\gamma(\mathbf{k}, \tau)| < 1 \tag{3.27}$$

Similarly, the *coherence time* is defined as the maximum time delay between the fields for which $|\gamma|$ maintains a significant value, say ½.

It is straightforward to show that if we cross-correlate temporally two plane waves of different wave vectors (directions of propagation), the result vanishes unless $\mathbf{k}_1 = \mathbf{k}_2$,

$$\begin{aligned} \Gamma(\mathbf{k}_1, \mathbf{k}_2, \tau) &= \left\langle U_1(\mathbf{k}_1, t) \cdot U_2^*(\mathbf{k}_2, t + \tau) \right\rangle_t \\ &= \Gamma_\tau(\tau) \exp[(\mathbf{k}_2 - \mathbf{k}_1)\mathbf{r}] \end{aligned} \tag{3.28}$$

where Γ_τ is the temporal component of Γ.

Thus, at each moment t, the two plane waves generate fringes parallel to $\mathbf{k}_2 - \mathbf{k}_1$. Symmetric to time domain case, when the spatial correlation at two different frequencies was discussed, if the detector (e.g., a CCD) averages the signal over spatial scales larger than the fringe period, the temporal correlation information is lost. As τ changes, the fringes "run" across the plane such that the contrast averages to 0. For this reason, when measuring the temporal autocorrelation function, the two beams in a typical Michelson interferometer are carefully aligned to be parallel.

3.3.2 Optical Power Spectrum

The temporal correlation, Γ, is the Fourier transform of the power spectrum,

$$\Gamma(\mathbf{k}, \tau) = \int_{-\infty}^{\infty} S(\mathbf{k}, \omega) \cdot e^{-i\omega\tau} d\omega$$

$$\tag{3.29}$$

$$S(\mathbf{k}, \omega) = \int_{-\infty}^{\infty} \Gamma(\mathbf{k}, \tau) \cdot e^{i\omega\tau} d\tau$$

Thus, Γ can be determined via spectroscopic measurements, as exemplified in Fig. 3.6. By using a grating (a prism, or any other dispersive element), we can "disperse" different colors at different angles, such that a rotating detector can measure $S(\omega)$ directly. In order to estimate the coherence time for a broad band field, let us assume a Gaussian spectrum centered at frequency ω_0, and having the r.m.s. width $\Delta\omega$,

$$S(\omega) = S_0 \cdot e^{-\left(\frac{\omega - \omega_0}{\sqrt{2}\Delta\omega}\right)^2} \tag{3.30}$$

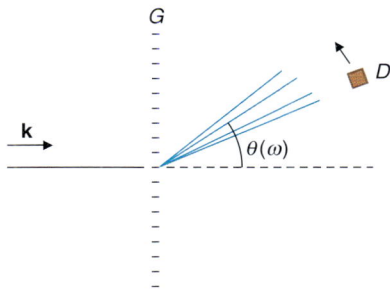

FIGURE 3.6 Spectroscopic measurement using a grating: G grating, D detector, diffraction angle. The dashed line indicates the undiffracted order (zeroth order).

where S_0 is a constant. The autocorrelation function is also a Gaussian, modulated by a sinusoidal function, as a result of the Fourier shift theorem (see App. B)

$$\Gamma(\tau) = \Gamma_0 \cdot e^{-\left(\frac{\Delta\omega\tau}{\sqrt{2}}\right)^2} \cdot e^{i\omega_0\tau} \tag{3.31}$$

From Eq. 3.31, we see that if we define the width of Γ as the coherence time, we obtain

$$\tau_C \propto \frac{1}{\Delta\omega} \tag{3.32}$$

and the coherence length

$$l_C = c\tau_C$$
$$\propto \frac{\lambda^2}{\Delta\lambda} \tag{3.33}$$

The coherence length depends on the spectral bandwidth, in the same way the coherence area dependends on the solid angle (Eq. 3.21). This is not surprising as both types of correlations depend on their respective frequency bandwidth.

3.3.3 Spectral Filtering

The coherence length values can vary broadly, from kilometers for a narrow band laser, to microns for LEDs and white light. Figure 3.7 shows qualitatively the relationship between $\Gamma(\tau)$ and $S(\omega)$. Of course, using narrow band filters has the effect of increasing the coherence length of the field. The short coherence length of a broad band source is the starting point in *low-coherence interferometry* and *optical coherence tomography* [14], as discussed in Chap. 7.

(a)

(b)

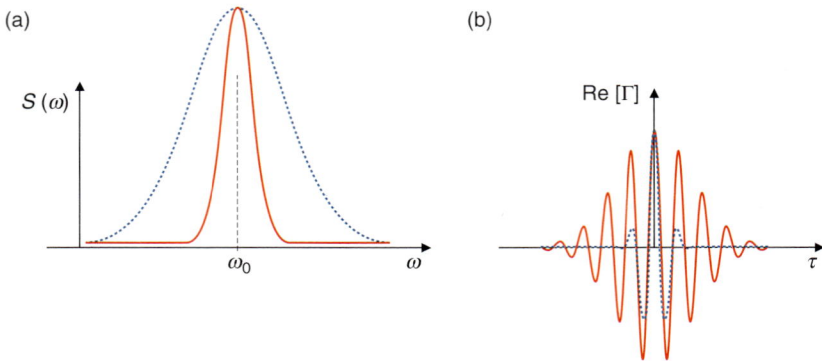

FIGURE 3.7   (a) Broad (dash line) and narrow (solid line) power spectrum; (b) temporal autocorrelation functions associated with the power spectra in (a).

Note that phase can only be defined via correlation functions, in other words there is no absolute origin for measuring a phase shift in time or space. In order to be able to define a *quantitative phase image*, the phase shift itself across the image must be well defined. That is, the image field must be spatially coherent over the field of view. This does not mean that the illumination has to be monochromatic, as long as, on average, each frequency component ω has a correlation area larger than the field of view. Phase contrast microscopy [8,9], which pre-dates lasers by almost three decades, is a notorious white light method where the phase shift is meaningful across the entire field of view.[15] As discussed later in Chap. 12, QPI with white light illumination in fact provides certain advantages over laser illumination.

References

1. J. W. Goodman. *Statistical optics*. (Wiley, New York, 2000).
2. L. Mandel and E. Wolf. *Optical coherence and quantum optics*. (Cambridge University Press, Cambridge and New York, 1995).
3. G. Popescu, Y. K. Park, R. R. Dasari, K. Badizadegan and M. S. Feld. "Coherence properties of red blood cell membrane motions," *Phys. Rev. E.*, 76, 031902 (2007).
4. E. Wolf. "New theory of partial coherence in the space-frequency domain .1. spectra and cross spectra of steady-state sources," *Journal of the Optical Society of America*, 72, 343–351 (1982).
5. E. Wolf. "Solution of the phase problem in the theory of structure determination of crystals from x-ray diffraction experiments," *Phys. Rev. Lett.*, 103, 075501 (2009).
6. L. Mandel and E. Wolf. "Complete coherence in the space-frequency domain," *Optics Communications*, 36, 247–249 (1981).
7. J. W. Goodman. *Introduction to Fourier optics*. (McGraw-Hill, New York, 1996).
8. F. Zernike. "Phase contrast, a new method for the microscopic observation of transparent objects, Part 1," *Physica*, 9, 686–698 (1942).
9. F. Zernike. "Phase contrast, a new method for the microscopic observation of transparent objects," Part 2, *Physica*, 9, 974–986 (1942).

10. E. Abbe. "Beiträge zur Theorie des Mikroskops und der mikroskopischen Wahrnehmung," *Arch. Mikrosk. Anat.*, 9, 431 (1873).
11. E. Wolf. *Introduction to the theory of coherence and polarization of light* (Cambridge University Press, Cambridge, 2007).
12. F. H. Smith. "Microscopic interferometry," *Research* (London) 8, 385 (1955).
13. M. Pluta. *Advanced light microscopy.* (PWN, Warsaw; Elsevier, Amsterdam and New York, 1988).
14. D. Huang, E. A. Swanson, C. P. Lin, J. S. Schuman, W. G. Stinson, W. Chang, M. R. Hee, T. Flotte, K. Gregory, C. A. Puliafito, and J. G. Fujimoto. "Optical coherence tomography," *Science*, 254, 1178–1181 (1991).
15. M. Born and E. Wolf. *Principles of optics: Electromagnetic theory of propagation, interference and diffraction of light.* (Cambridge University Press, Cambridge and New York, 7th edition, 1999).

CHAPTER 4

Image Characteristics

I maging is the process of *mapping* a certain *physical property* of an object and displaying it in a *visual form*.[1] Examples of such physical properties include absorption (e.g., light imaging, x-rays), emission (e.g. fluorescence imaging), scattering (e.g. phase contrast imaging) reflectivity (e.g., ultrasound, common photography), proton density (e.g., MRI), concentration of radionuclides (e.g., nuclear imaging). In microscopy, intrinsic specimen *absorption, emission,* and *scattering* are the quantities of interest. By using exogenous contrast agents, such as stains or fluorescent tags, one can image concentration distributions of certain species (e.g., concentration of a fluorescent dye that binds to the DNA of a cell nucleus). Here we discuss the very basic properties of microscope images, irrespective of the method involved in acquiring the images, and present simple image enhancement operations allowed by numerical image processing.

4.1 Imaging as Linear Operation

In many situations of practical interest, an imaging system, e.g. a microscope, can be approximated by a *linear system*.[2-6] Let us consider the physical property of the sample under investigation is described by the function $S(\mathbf{r})$, $\mathbf{r} = (x, y, z)$. The imaging system outputs an image $I(\mathbf{r})$, which is related to $S(\mathbf{r})$ through a *convolution* operation,

$$I(\mathbf{r}) = S(\mathbf{r}) \otimes h(\mathbf{r}) \tag{4.1}$$

In Eq. 4.1, h defines the *impulse response or Green's function* of the system. In microscopy, h is commonly called *point spread function (PSF)*. Note that here we intend to keep the discussion general, i.e. we are not specific as to what quantity I defines. Clearly, h can be experimentally retrieved by imaging a very small object, i.e., approaching a delta-function in 3D. Thus, replacing the sample distribution $S(\mathbf{r})$ with $\delta(\mathbf{r})$, we obtain

$$I_\delta(\mathbf{r}) = \delta(\mathbf{r}) \otimes h(\mathbf{r}) = h(\mathbf{r}) \tag{4.2}$$

The PSF of the system, $h(\mathbf{r})$, defines how an infinitely small object is blurred through the imaging process. The PSF is, therefore, a measure of the *resolving power* of the system, as detailed next. Note that $I(r)$ can be an intensity distribution or a complex field distribution, depending upon whether the imaging uses (spatially) *incoherent* or *coherent* light, respectively, as discussed in the previous chapter. This distinction is very important because in coherent imaging, the system is linear in complex fields, while in incoherent imaging, the linearity holds in intensities.

4.2   Resolution

The resolution of an imaging system is defined by the minimum distance between two points that are considered "resolved".[4,7-9] What is considered resolved is subject to convention. The half-width half maximum of h in one direction is one possible measure of resolution along that direction. When discussing the image formation, we may encounter other conventions for resolution.

The image can be represented in the spatial frequency domain via a 3D Fourier transform (see App. B)

$$\tilde{I}(\mathbf{k}) = \int_V I(\mathbf{r}) \cdot e^{-i\mathbf{k}\cdot\mathbf{r}} d^3\mathbf{r} \tag{4.3}$$

where $\mathbf{k} = (k_x, k_y, k_z)$ is the (angular) spatial frequency, in units of rad/m. Note that writing I in separable form, $I(\mathbf{r}) = I_x(x) \cdot I_y(y) \cdot I_z(z)$, the 3D Fourier transform in Eq. 4.3 breaks into a product of three 1D integrals,

$$\tilde{I}(\mathbf{k}) = \left(\int I_x(x) \cdot e^{-ik_x \cdot x} dx \right) \cdot \left(\int I_y(y) \cdot e^{-ik_y \cdot y} dy \right) \cdot \left(\int I_z(z) \cdot e^{-ik_z \cdot z} dz \right)$$
$$= \tilde{I}(k_x) \cdot \tilde{I}(k_y) \cdot \tilde{I}(k_z) \tag{4.4}$$

In order to find a relationship between the object and the image in the frequency domain, we Fourier transform Eq. 4.1, and use the convolution theorem (see App. B) and obtain,

$$\tilde{I}(\mathbf{k}) = \tilde{S}(\mathbf{k}) \cdot \tilde{h}(\mathbf{k}) \tag{4.5}$$

with $\tilde{S}$ and $\tilde{h}$ the Fourier transforms of S and h, respectively. The function $\tilde{h}(\mathbf{k})$ is referred to as the *transfer function (TF)*. From the Fourier relationship between PSF and TF, it is clear that *high resolution* (i.e., narrow PSF) demands *broad frequency support* of the system (broad TF), as illustrated in Fig. 4.1. Note that the ideal system for which $h(\mathbf{r}) = \delta(\mathbf{r})$ requires infinite frequency support, which is clearly *unachievable* in practice,

$$\delta(\mathbf{r}) \leftrightarrow 1 \tag{4.6}$$

(a)

(b)

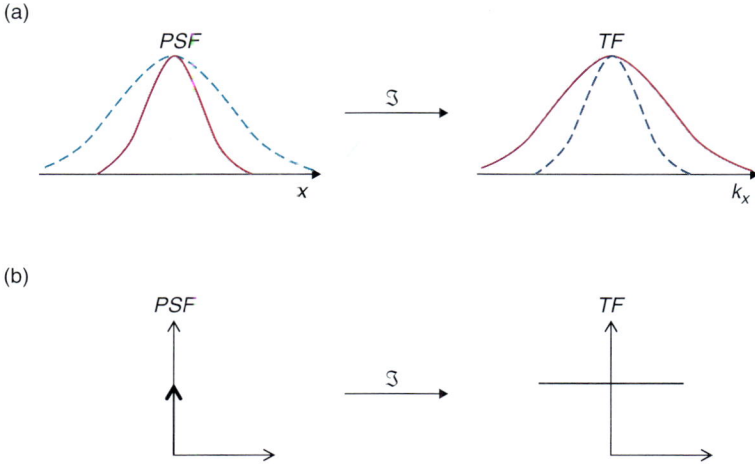

FIGURE 4.1   (*a*) Relationship between the modulus of PFS and TF. (*b*) Infinitely narrow PSF requires infinitely broad TF.

Generally, the physics of the image formation defines the instrument's PSF and TF. We will derive these expressions for coherent and diffraction-limited microscopy later, in Chap. 5. For now, we continue the general discussion regarding image characteristics, irrespective of the type of microscope that generated the image, or of whether the image is an intensity or phase distribution.

4.3   Signal-to-Noise Ratio (SNR)

Any measured image is affected by experimental noise. There are various causes of noise, particular to the light source, propagation medium, and detector, which contribute to the measured signal,

$$\hat{x} = x + \xi \tag{4.7}$$

In Eq. 4.7, $\hat{x}$ is the measured signal, x is the true (noise-free) signal, and ξ the noise. The *variance*, σ^2, of the signal describes how much variation around the average there is over N measurements,

$$\sigma^2 = \frac{1}{N} \sum_{i=1}^{N} \left(\hat{x}_i - \overline{\hat{x}} \right)^2$$

$$\overline{\hat{x}} = \frac{1}{N} \sum_i \hat{x}_i \tag{4.8}$$

Note that the N measurements can be taken over N pixels within the same image, which characterizes the noise *spatially*, or over N values of the same point in a time-series, which defines the noise *temporally*. The *standard deviation* σ is also commonly used and has the benefit of carrying the same units as the signal itself. The SNR is defined in terms of the standard deviation as

$$\text{SNR} = \frac{|x|}{\sigma} \tag{4.9}$$

where we note that the signal is expressed as a modulus, such that SNR > 0. Let us now discuss the effects of averaging on SNR. It can be easily shown that for N *uncorrelated* measurements, the *variance of the sum is the sum of variances*,

$$\sigma_N^2 = N\sigma^2 \tag{4.10}$$

Thus, the standard deviation of the averaged signal is

$$\sigma_N = \frac{\sigma}{\sqrt{N}} \tag{4.11}$$

and the SNR becomes

$$\text{SNR}_N = \sqrt{N} \cdot \text{SNR} \tag{4.12}$$

Equation 4.12 establishes that taking N measurements and averaging the results increases the signal-to-noise ratio by $\sqrt{N}$. Note that this benefit of averaging only exists when there is no correlation between the noise present in different measurements. By contrast, for correlated (coherent) noise, Eq. 4.10 is not valid. To prove this, let us consider the variance associated with the sum of two correlated noise signals a_1 and a_2 (assume $\bar{a}_1 = \bar{a}_2 = 0$ for simplicity),

$$\overline{(a_1 + a_2)^2} = \overline{a_1^2} + \overline{a_2^2} + \overline{2a_1 a_2} \tag{4.13}$$
$$= \sigma_1^2 + \sigma_2^2 + \sigma_{12}^2$$

It is clear that the total variance is the sum of individual variances, $\sigma_1^2 + \sigma_2^2$, only when the cross term vanishes, i.e., $\sigma_{12} = 0$. If, for example, a_1 and a_2 fluctuate randomly in time, and signal a_2 is measured after a delay time τ from the measurement of a_1, the cross term has the familiar form of a cross-correlation,

$$\sigma_{12}(\tau) = \overline{a_1(t) \cdot a_2(t + \tau)} = \frac{1}{\tau} \int a_1(t) a_2(t + \tau) dt \tag{4.14}$$

For stationary random processes, such as most fields encountered in the laboratory, this correlation function decays to zero at long times. Equation 4.14 indicates that if the noise is characterized by some correlation time, τ_c, then averaging different measurements increases the signal to noise only if the interval between measurements τ is larger than τ_c. In other words, the noise becomes uncorrelated at temporal scales larger than τ_c. This is analog to the discussion of temporal coherence in Chap. 3.

4.4   Contrast and Contrast-to-Noise Ratio

The *contrast* of an image quantifies the ability to differentiate between different regions of interest within the image (Fig. 4.2),

$$C_{AB} = |S_A - S_B| \tag{4.15}$$

where $S_{A,B}$ stands for the signal associated with regions A and B. The modulus in Eq. 4.15 ensures that the contrast has always positive values.

Unlike resolution, which is established solely by the microscope itself, contrast is a property of the instrument-sample combination. For example, the same microscope, characterized by a certain resolution, renders superior contrast for stained than unstained tissue slices.

In practical situations, the simple definition of contrast in Eq. 4.15 is insufficient because it ignores the effects of noise. It is easy to

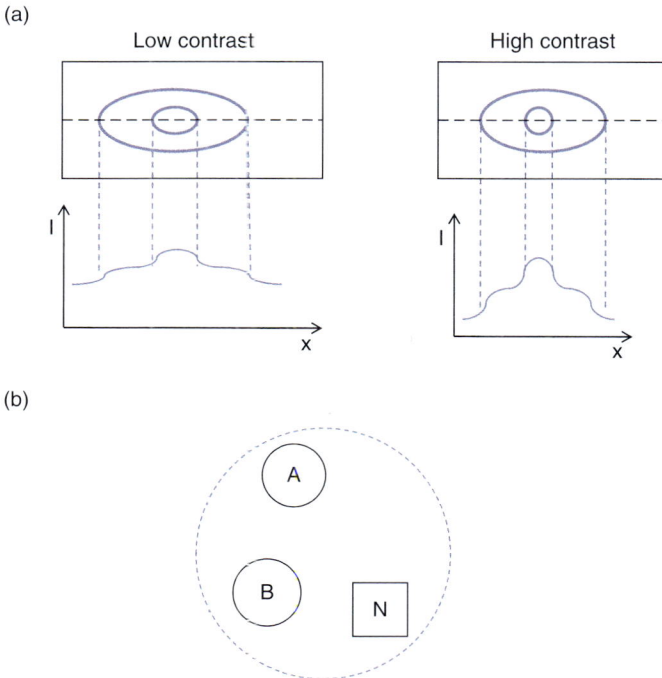

FIGURE 4.2   (*a*) Illustration of a low- and high-contrast image. (*b*) Region of interest A, B, and noise region N.

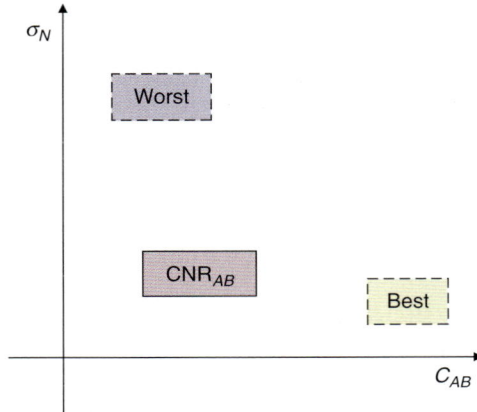

Figure 4.3 Contrast-to-noise ratio vs. contrast C_{AB} and noise standard deviation σ_N.

imagine circumstances where the noise itself is of very high contrast across the field of view, which by the sole definition of Eq. 4.15, generates high values of C_{AB}. Thus, a better quantity to use for real, noisy images is the *contrast-to-noise ratio* (CNR),

$$\mathrm{CNR}_{AB} = \frac{C_{AB}}{\sigma_N}$$

$$= \frac{|S_A - S_B|}{\sigma_N}$$

(4.16)

In Eq. 4.16, σ_N is the standard deviation associated with the noise in the image. Of course, the best-case scenario from a measurement point of view happens at low noise and high contrast (Fig. 4.3).

4.5 Image Filtering

Filtering is a generic term that relates to a multiplication operation taking place in the frequency domain of the image,

$$\tilde{I}(\mathbf{k}) = \tilde{I}_0(\mathbf{k}) \cdot \tilde{H}(\mathbf{k})$$

(4.17)

where $\tilde{I}_0$ is the Fourier transform of the image and H is the *filter* function. The *filtered image* $I(\mathbf{r})$ is obtained by Fourier transforming back $\tilde{I}(\mathbf{k})$ in the space domain,

$$I(\mathbf{r}) = I_0(\mathbf{r}) \otimes H(\mathbf{r})$$

(4.18)

Equation 4.18 establishes that the filtered image I is the convolution between the *unfiltered image* I_0 and the Fourier transform of the filter function H. Note that, in fact, any measured image is a filtered version of the object (the transfer function, TF, of the instrument is the filter), as stated in Eq. 4.5.

Typically, filtering is used to enhance (or diminish) certain features in the image. Depending on the frequency range that they allow to pass, we distinguish three types of filters: *low-pass, band-pass,* and *high-pass.*

4.5.1 Low-Pass Filtering

Let us consider the situation where the noise in the measured image exhibits high-frequency fluctuations (Fig. 4.4a). This noise contribution is effectively diminished if a filter that blocks the high frequencies (passes the low frequencies) is used (Fig. 4.4b). The resulting filtered image has better SNR (Fig. 4.4c). However, in the process, resolution is diminished.

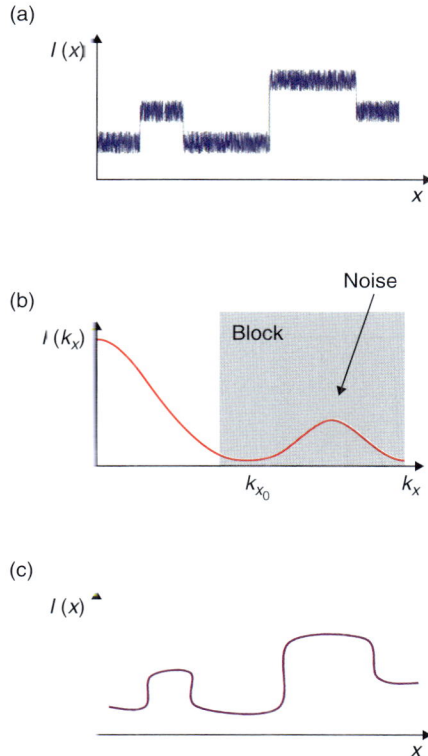

FIGURE 4.4
Low-pass filtering:
(a) noisy signal,
(b) filtering high
frequencies, and
(c) low-passed
signal.

4.5.2   Band-Pass Filter

In this case, the filter function H passes a certain range (band) of frequencies from the image. This type of filtering is useful when structures of particular range of sizes are to be enhanced in the image.

4.5.3   High-Pass Filter

Using a filter that selectively passes the high frequencies reveals finer details within the image. One particular application of high-pass filtering is *edge detection*. This application is particularly useful in selecting (*segmenting*) an object of interest (e.g., a cell) from the image. Thus the gradient of the image along one dimension can significantly enhance the edges in an image (Fig. 4.5).

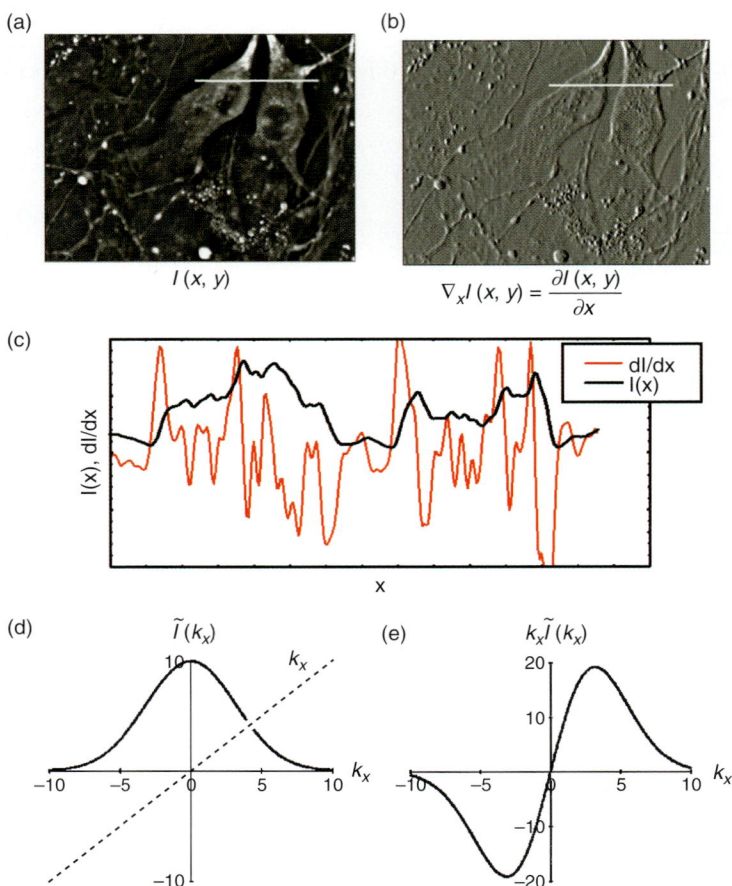

FIGURE 4.5   Edge enhancement: (*a*) original image; (*b*) gradient along one direction; (*c*) profiles along the lines shown in (*a*) and (*b*); (*d*) frequency domain profile associated with (*a*); (*e*) frequency domain profile associated with (*b*), i.e., $k_x I(k_x)$.

Thus, the new, edge-enhanced image is

$$I_x(\mathbf{r}) = \nabla_x[I(\mathbf{r})] = \frac{\partial I(\mathbf{r})}{\partial x} \tag{4.19}$$

Taking the Fourier transform of Eq. 4.19, we find

$$\tilde{I}_x(\mathbf{k}) = ik_x \cdot \tilde{I}(\mathbf{k}) \tag{4.20}$$

In Eq. 4.20, i indicates a $\pi/2$ shift that occurs upon differentiation (i.e., sines become cosines). Most importantly, it can be seen that the new frequency content of the gradient image $\tilde{I}_x$ is enhanced at high frequencies due to the multiplication by k_x (Fig. 4.5). Clearly, the gradient image suffers from "shadow" artifacts due to the change in sign of the first-order derivative across an edge (Fig. 4.5b). This artifact is commonly known in DIC microscopy, where the gradient is computed "optically".[7]

Taking the Laplacian of the image removes these anisotropic artifacts for the simplest case of isotropic samples, related to the change in sign of the first-order derivative. Thus,

$$\nabla^2 I(\mathbf{r}) = \frac{\partial^2 I}{\partial x^2} + \frac{\partial^2 I}{\partial y^2} \tag{4.21}$$

$$\nabla^2 I(\mathbf{r}) \rightarrow -k^2 \cdot I(\mathbf{k})$$

where $k^2 = k_x^2 + k_y^2$.

Figure 4.6a shows the results of taking the Laplacian of the same image. Even finer details are now visible in the image, as the high-pass filter is stronger due to the k^2 multiplication of $I(\mathbf{k})$. Further, the shadow artifacts are less disturbing because they do not change sign across an edge.

(a)

(b)

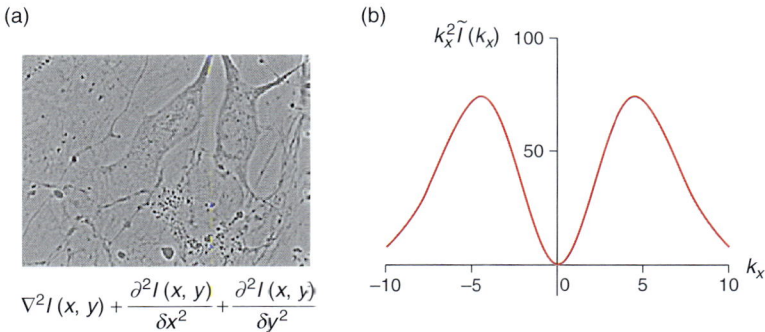

$$\nabla^2 I(x, y) + \frac{\partial^2 I(x, y)}{\delta x^2} + \frac{\partial^2 I(x, y)}{\delta y^2}$$

FIGURE 4.6 Laplacian (a) and frequency domain profile (b) associated with the image in Fig. 4.5a.

Of course, there are many, much more sophisticated filters and algorithms for image enhancement and restoration (see, for example,[10, 11]) that are beyond the scope of this discussion.

References

1. A. R. Webb. *Introduction to biomedical imaging.* (Wiley, Hoboken, NJ, 2003).
2. R. N. Bracewell. *The Fourier transform and its applications.* (McGraw Hill, Boston, MA, 2000).
3. J. D. Gaskill. *Linear systems, Fourier transforms, and optics.* (Wiley, NY, 1978).
4. J. W. Goodman. *Introduction to Fourier optics.* (McGraw-Hill, New York, 1996).
5. A. Papoulis. *The Fourier integral and its applications.* (McGraw-Hill, New York, 1962).
6. A. Papoulis. *Systems and transforms with applications in optics.* (McGraw-Hill, New York, 1968).
7. M. Pluta. *Advanced light microscopy.* (PWN; Elsevier: Distribution for the USA and Canada, Elsevier Science Publishing Co., Warszawa Amsterdam; New York, 1988).
8. D. B. Murphy. *Fundamentals of light microscopy and electronic imaging.* (Wiley-Liss, New York,, 2001).
9. M. Born and E. Wolf. *Principles of optics: Electromagnetic theory of propagation, interference and diffraction of light.* (Cambridge University Press, Cambridge; New York, 1999).
10. J. C. Russ. *The image processing handbook* (CRC Press, Boca Raton, FL, 1998).
11. A. C. Kak and M. Slaney. *Principles of computerized tomographic imaging.* (Society for Industrial and Applied Mathematics, Philadelphia, PA, 2001).

CHAPTER 5

Light Microscopy

Here we use the framework of Fourier optics established in Chap. 2 to describe *coherent image formation*, i.e., imaging obtained by illuminating the specimen with spatially coherent light. Specifically, we describe resolution, contrast, and phase-sensitive methods to enhance contrast.

5.1 Abbe's Theory of Imaging

We found in Sec. 2.4 that a convergent lens has the remarkable property of producing at its *back* focal plane the Fourier transform of the field distribution at its *front* focal plane.[1,2] Thus, one way to describe an imaging system (e.g., a microscope) is in terms of a system of two lenses that perform two successive Fourier transforms.

Figure 5.1 shows the geometrical optics image formation through this microscope. A system in which the back focal plane of the first lens (objective, Ob) overlaps with the front focal plane of the second (tube lens, TL) is called *telecentric*.

The image field U_3 is the Fourier transform of U_2, which is the Fourier transform of U_1. It can be easily shown that applying two forward Fourier transforms recovers the original function, up to a sign change in the coordinates (rotation of the image by 180°).

$$F\{F[f(x,y)]\} = f(-x,-y) \tag{5.1}$$

This explains why the image A'B' is inverted with respect to the object AB. However, note that because the two lenses have different focal distances, the image field is also scaled by a factor M, called *transverse magnification*. We can calculate this magnification by evaluating the field U_3 as a function of U_1,

$$U_3(k_{x2}, k_{y2}) = \int_{-\infty}^{\infty}\int_{-\infty}^{\infty} U_2(x_2, y_2) \cdot e^{-i(k_{x2} \cdot x_2 + k_{y2} \cdot y_2)} dx_2 dy_2 \tag{5.2a}$$

$$U_2(k_{x1}, k_{y1}) = \int_{-\infty}^{\infty}\int_{-\infty}^{\infty} U_1(x_1, y_1) \cdot e^{-i(k_{x1} \cdot x_1 + k_{y1} \cdot y_1)} dx_1 dy_1 \tag{5.2b}$$

65

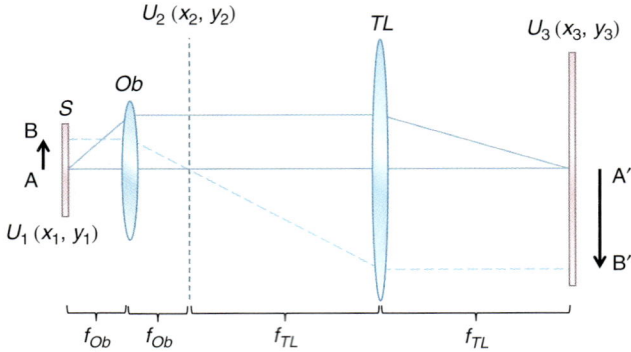

FIGURE 5.1 Coherent image formation in a microscope: Ob, objective lens, TL, tube lens.

In Eq. (5.2), the spatial frequencies are defined as

$$k_{x1} = \frac{2\pi x_2}{\lambda f_{Ob}} ; \; k_{y2} = \frac{2\pi y_2}{\lambda f_{Ob}}$$

$$k_{x2} = \frac{2\pi x_3}{\lambda f_{TL}} ; \; k_{y2} = \frac{2\pi y_3}{\lambda f_{TL}} \tag{5.3}$$

Plugging Eq. (5.2b) in Eq. (5.2a), one finds the final expression that relates the image and the object

$$U_3(x_3, y_3) = U_1\left(-\frac{x_1}{M}, -\frac{y_1}{M}\right) \tag{5.4}$$

where the magnification is given by

$$M = \frac{f_{TL}}{f_{Ob}} \tag{5.5}$$

Potentially, the ratio f_{TL}/f_{Ob} can be made arbitrarily large (for instance, by cascading many imaging systems). However, this does not mean that the microscope is able to resolve arbitrarily small objects. We have already encountered the limited resolution in extracting the structure of inhomogeneous objects via scattering experiments (Sec. 2.5). Clearly, the microscope obeys the same limits. Thus, unlike magnification, resolution is fundamentally limited by the laws of physics, as described in the next section.

Figure 5.2 provides a physical explanation for the image formation, originally formulated by Abbe in 1873. This theory is quite

(a)

(b)

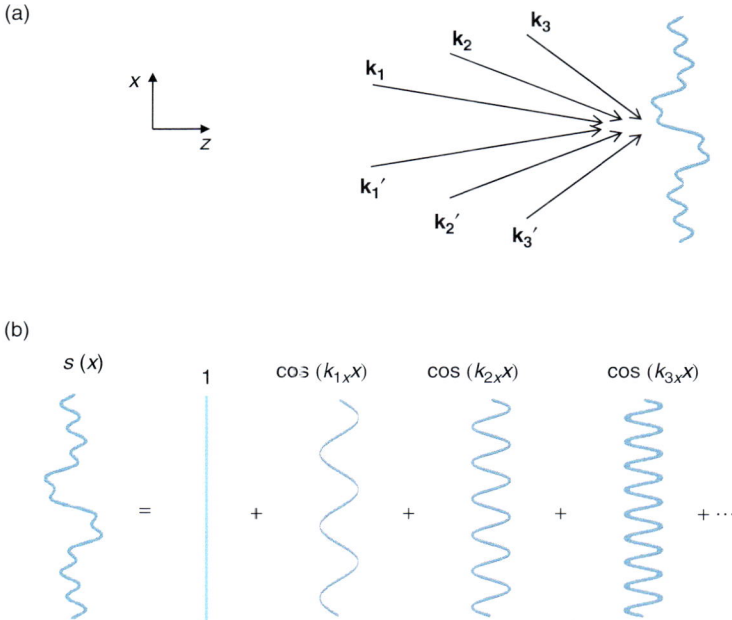

FIGURE 5.2 (a) Abbe's concept of imaging as an interference phenomenon: the pairs of wave vectors $\mathbf{k}_{1,2,3}$-$\mathbf{k'}_{1,2,3}$ generate standing waves of different frequencies along the x-axis; (b) frequency decomposition of the resulting field.

well summarized by Abbe's own words: "The microscope image is the interference effect of a diffraction phenomenon".[3] Thus, a given image field is formed by the interference between plane waves propagating along different directions (Fig. 5.2a). The resulting field can therefore be decomposed into sinusoids of various frequencies and phase shifts (Fig. 5.2b).

Of course, the same picture applies at the sample plane, where each spatial frequency generates pairs of plane waves (diffraction orders) propagating symmetrically with respect to the optical axis (Fig. 5.3). As this frequency increases, the respective diffraction angle reaches the point where it exceeds the maximum allowed by the objective. This framework allowed Abbe to derive his famous formula for the resolution limit, which we will now discuss.

Let us express quantitatively the resolution of the microscope. Figure 5.4 illustrates how the apertures present in the microscope objective limit the maximum angle associated with the light scattered by the specimen. The effect of the objective is that of a *low-pass filter*, with the cut-off frequency in 1D given by

$$k_M = \frac{2\pi}{\lambda_{Ob}} \cdot x_M = \frac{2\pi}{\lambda} \cdot \theta_M \tag{5.6}$$

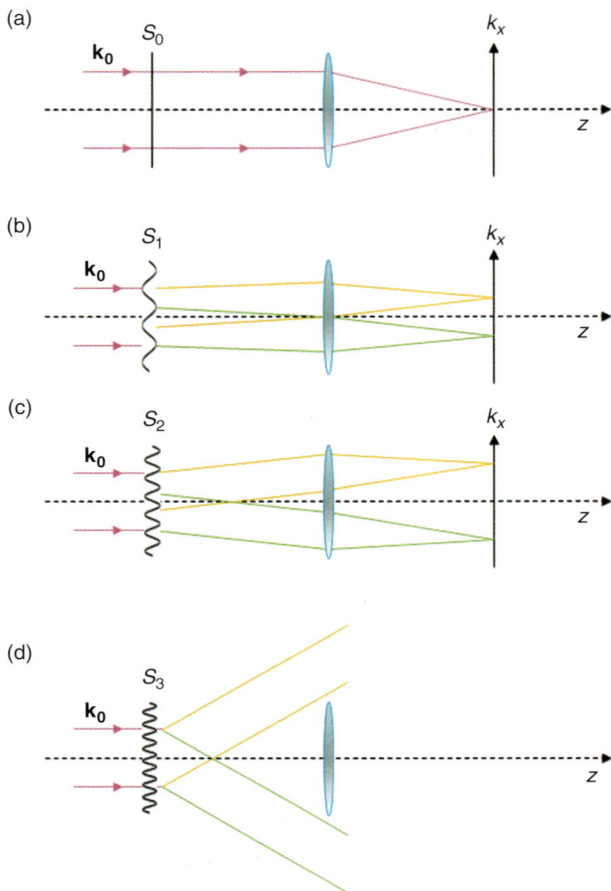

FIGURE 5.3   Low-pass filtering effect by the microscope objective.

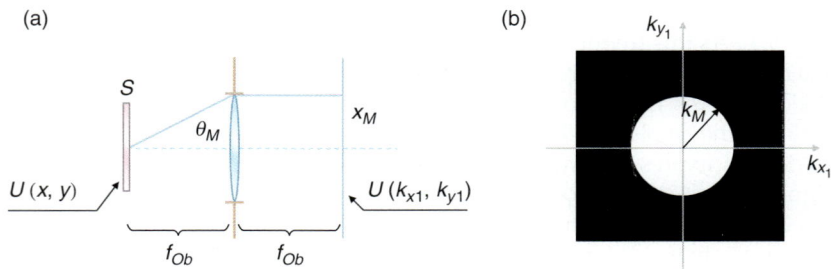

FIGURE 5.4   Frequency cut-off in a light microscope: (*a*) maximum half-angle subtended by the entrance pupil from the specimen; (*b*) entrance pupil.

where θ_M is the maximum half-angle subtended by the *entrance pupil* from the specimen. *Qualitatively*, it is clear that this low-pass filter has the effect of smoothing-down the details in the sample field, i.e., limiting the spatial resolution of the instrument. *Quantitatively*, we need to find the relationship between the ideal (infinite resolution) sample field $U_1(x_1, y_1)$ and the smooth (image) field, $U_3(x_3, y_3)$. Note that this calculation can be done in two equivalent ways: either (1) express the sample field as the Fourier transform of the field at the *entrance pupil*, or (2) express the image field as the Fourier transform of the *exit pupil* (fields at the entrance and exit pupil are merely scaled versions of each other, i.e., they are the image of one another). Following the second path, the image field can be expressed in terms of the Fourier transform of $U_2(x_2, y_2)$. Thus, rewriting Eq. (5.2a), we obtain

$$U_3(x_3, y_3) = \int_{-\infty}^{\infty}\int_{-\infty}^{\infty} U_2(k_{x2}, k_{y2}) \cdot e^{i(k_{x2} \cdot x_3 + k_{y2} \cdot y_3)} dk_{x2} dk_{y2} \qquad (5.7)$$

In Eq. (5.7), $U_2(k_{x2}, k_{y2})$ is the frequency domain field that is *truncated* by the exit pupil function. Mathematically, the pupil function, $P(k_{x2}, k_{y2})$, is the transfer function of the microscope,

$$U_2(k_{x2}, k_{y2}) = U_2(k_{x2}, k_{y2}) \cdot P(k_{x2}, k_{y2}) \qquad (5.8)$$

where U_2 is the unrestricted (of infinite support) Fourier transform of the sample field U_1. Combining Eqs. (5.7) and (5.8), we obtain the expression for the image field, U_3, as the Fourier transform of a *product* between the ideal field, U_2, and the pupil function, P. Thus, the image field U_3 can be written as the convolution between U_1 and the Fourier transform of P.

$$U_3(x_3, y_3) = \int_{-\infty}^{\infty}\int_{-\infty}^{\infty} U_3(x_3', y_3') \cdot g(x_3 - x_3', y_3 - y_3') dx_3' dy_3'$$
$$= U_1(x_3/M, y_3/M) \otimes g(x_3, y_3) \qquad (5.9)$$

In Eq. (5.9), function g is the Green's function or PSF of the instrument, and is defined as

$$g(x_3, y_3) = \int_{-\infty}^{\infty}\int_{-\infty}^{\infty} P(k_{x2}, k_{y2}) \cdot e^{i(k_{x2} \cdot x_3 + k_{y2} \cdot y_3)} dk_{x2} dk_{y2} \qquad (5.10)$$

Clearly, if in Eq. (5.9) the input (sample) field U_1 is a point, expressed by a δ-function, the imaging system blurs it to a spot that in the imaging plane is of the form $\delta(x_1, y_1) \otimes g(x_1, y_1) = g(x_1, y_1)$. Thus, g is also called sometimes the *impulse response* of the coherent imaging instrument. In other words, a point in the sample plane is "smeared" by the instrument into a spot whose size is given by the width of g.

Interestingly, from Eq. (5.10), we see that the impulse response g is merely the field diffracted by the exit pupil (see Fig. 5.5a); hence the phrase "diffraction limited resolution". In order to obtain an actual expression for the impulse response g, we need to know the pupil function P. Most commonly, the pupil function is a disk, defined as

$$P(k_{x2}, k_{y2}) = \begin{vmatrix} 1, & \text{if } k_{x2}^2 + k_{y2}^2 \le k_M^2 \\ 0, & \text{otherwise} \end{vmatrix}$$

(5.11)

$$= \Pi(k/2k_M)$$

(a) Entrance pupil   Exit pupil

(b)

$$\left(\frac{J_1(x)}{x} \right)^2$$

−10   −5   0   5   10

$3.83 = 1.22\pi$

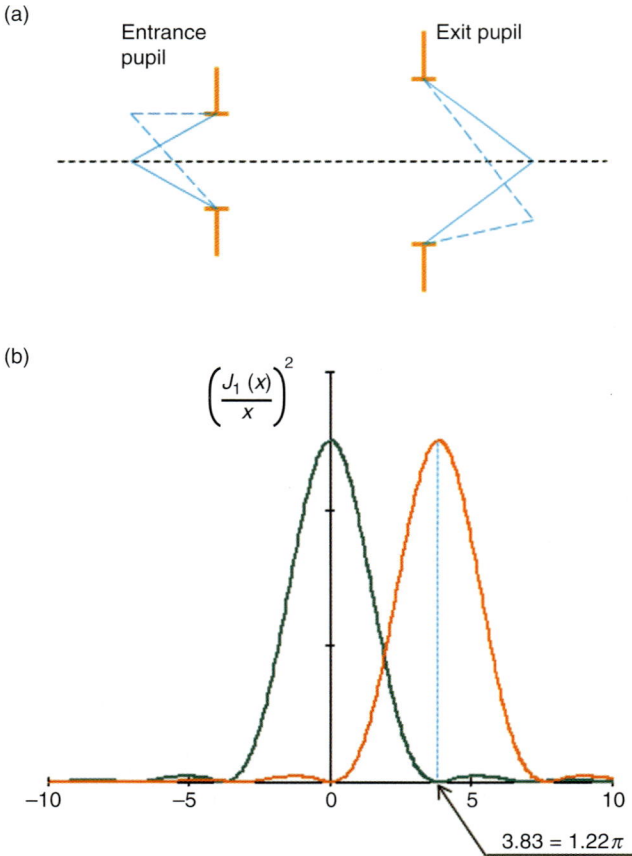

FIGURE 5.5   (a) Entrance and exit pupils of an imaging system. (b) Two points considered resolved by the Rayleigh criterion.

where, like before (see Sec. 2.5., also App. B) we used function Π to denote a "rectangular" function in polar coordinates, with $k = \sqrt{k_{x2}^2 + k_{y2}^2}$. In this case, it is easy to show that the Fourier transform of P yields g of the form (see App. B)

$$g(\rho) = \frac{J_1(k_M\rho)}{k_M\rho}$$

$$\rho^2 = x^2 + y^2 \qquad (5.12)$$

$$k_M = 2\pi\frac{r_M}{\lambda f_{Ob}}$$

where J_1 is the Bessel function of first kind and order and r_M is the radius of the exit pupil. The intensity profile, $|g(\rho)|^2$, is depicted in Fig. (5.5b). Gaskill[1] and others refer to the function $2J_1(x)/x$ as "Sombrero" function, because of its 2D surface plot resembling a Mexican hat.

The Rayleigh criterion for resolution postulates that two points are considered *resolved* if the maxima associated with their diffraction patterns are separated by at least the first root of the function (normalized coordinate x_0 in Fig. 5.5). In other words, the maximum of one function overlaps with the root of the second function. This root occurs at $x_0 = 1.22\pi$. Thus the resolution, ρ_0, is obtained by solving

$$2\pi\frac{r_M}{\lambda f_{Ob}} \cdot \rho_0 = 1.22\pi \qquad (5.13)$$

Note that r_M/f_{Ob} represents the maximum half-angle subtended by the entrance pupil (Fig. 5.4); this quantity is referred to as the *numerical aperture* of the objective, NA. Thus, finally we obtain the well-known result for resolution defined by the Rayleigh criterion,

$$\rho_0 = 0.61\frac{\lambda}{NA} \qquad (5.14)$$

Therefore, resolution can be increased either by using higher NA values or shorter wavelengths. Without the use of immersion liquids, the numerical aperture is limited to $NA = 1$; thus, at best, the microscope can resolve features of the order of $\lambda/2$.

Recently, very exciting research has brought the concept of limited resolution into question. In particular, it has been shown that if instead of the *linear* interaction between light and the specimen presented here, a *nonlinear* mechanism is employed, the resolving power of microscopes can be extended, virtually indefinitely.[4] However, this discussion is beyond the scope of this book.

5.2 Imaging of Phase Objects

As shown in Chap. 4, *resolution* is a property of the instrument itself, while *contrast* depends on both the instrument and sample. Here, we analyze a special class of samples, which do not absorb or scatter light significantly. Thus, they only affect the phase of the illuminating field and not its amplitude; these are generally known as *phase objects*.

Consider a plane wave, $e^{ik_0 z}$, incident on a specimen, characterized by a *complex transmission function* of the form $A_s e^{i\phi_s(x,y)}$. An ideal imaging system generates at the image plane an identical (i.e., phase and amplitude) replica of the sample field, up to a scaling factor defined by the magnification. Clearly, since A_s is not a function of spatial coordinate (x,y), the image field amplitude, A_i, is also a constant. Because the detector is only responsive to intensities, the measurement at the image plane yields no information about the phase,

$$I_i = \left| A_i \cdot e^{i\phi(x,y)} \right|^2 = A_i^2 \tag{5.15}$$

Equation 5.15 states that imaging a phase object produces an intensity image that is constant across the plane, i.e., the image has *zero contrast*. For this reason, imaging transparent specimens such as live cells is very challenging. Developing clever methods for generating contrast of phase objects has been driving the microscopy field since the beginning, four centuries ago. The *numerical* filtering shown in Sec. 4.5 will not help, because essentially the intensity has no structure at all. However, there are *optical* techniques that can be employed to enhance contrast, as described below.

Consider the intensity profile along one direction for a transparent sample (Fig. 5.6a). As discussed in Sec. 4.4., the low contrast is expressed by the small deviation from the mean, of only few percentage points, of the intensity fluctuations, $\left[I(x) - \langle I(x) \rangle \right] / \langle I(x) \rangle$. Ideally, for highest contrast, these normalized fluctuations approach unity. An equivalent manifestation of the low contrast is the narrow histogram of the pixel values, which indicates that the intensity at all points is very similar. To achieve high contrast, this distribution must be broadened.

One straightforward way to increase contrast optically is to simply remove the low-frequency content of the image, i.e., DC component, before the light is detected. For coherent illumination, this *high-pass* operation can be easily accomplished by placing an obstruction on-axis at the Fourier plane of the objective (see Fig. 5.7). Note that, in the absence of the specimen, the incident plane wave is focused on axis and, thus, entirely blocked (no light reaches the detector). Not

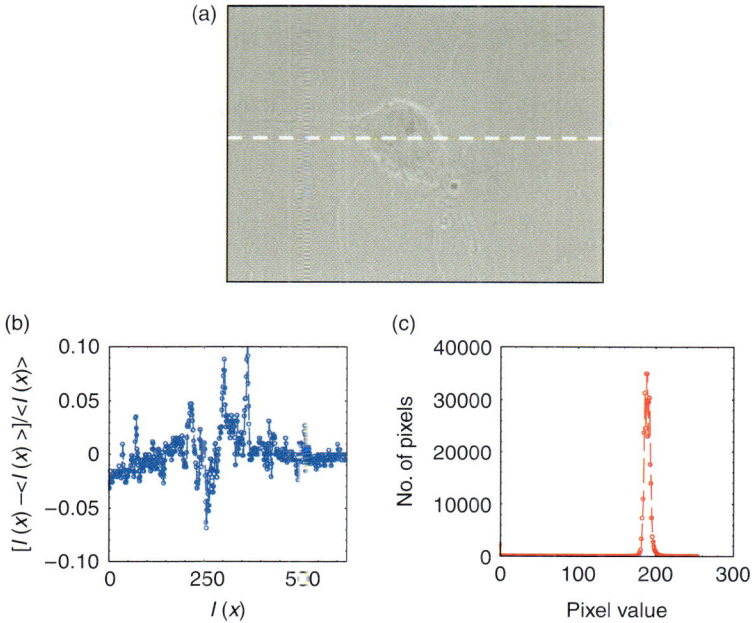

FIGURE 5.6 (a) Low-contrast image of a neuron. (b) Intensity profile along the line shown in (a). (c) Histogram of intensity distribution in (a).

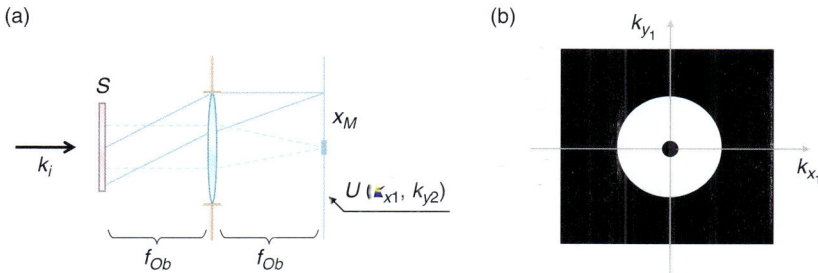

FIGURE 5.7 Dark field microscopy: (a) the unscattered component is blocked; (b) the entrance pupil showing the low-frequency obstruction.

surprisingly, this type of "zero-background" imaging is called *dark field microscopy*. This is one of the earliest modalities of generating contrast. Originally, this idea was implemented with the illumination beam propagating at an angle higher than allowed by the NA of the objective. Again, this oblique illumination is such that, without a sample, all the light is blocked by the entrance pupil of the system (hence, also a dark field approach).

5.3   Zernike's Phase Contrast Microscopy

Phase contrast microscopy (PCM) represents a major breakthrough in the field of light microscopy. Developed in the 1930s by the Dutch physicist Frits Zernike, for which he received the Nobel Prize in physics in 1953, PCM strikes with its simplicity and yet powerful capability. Much of what is known today in cell biology can be traced back to this method as it allows *label-free, noninvasive investigation* of live cells.

The principle of PCM exploits the early theory of image formation attributed to Abbe.[3] As already described above, the image field is the result of the *superposition* of fields originating at the specimen. For coherent illumination, this image field, U, can be conveniently decomposed into its spatial average, U_0, and fluctuating component, $U_1(x,y)$,

$$U(x,y) = U_0 + [U(x,y) - U_0]$$
$$= U_0 + U_1(x,y)$$

(5.16)

In Eq. (5.16), the average field, $U_0 = \langle U(x,y) \rangle_{x,y}$, can be expressed as

$$U_0 = \frac{1}{A} \int\int U(x,y)dxdy$$

(5.17)

where A is the area of the image.

Note that this average field can be properly defined only when the coherence area of the field is larger than the field of view (see discussion in Sec. 3.2.). The summation of complex fields over areas larger than the coherence area is meaningless, because the phase is not properly defined.

Taking the Fourier transform of Eq. (5.16) provides another interpretation of the fields U_0 and U_1,

$$\tilde{U}(k_x, k_y) = \delta(0,0) + \tilde{U}_1(k_x, k_y)$$

(5.18)

Clearly, the average field U_0 is the unscattered field, which is focused on axis by the objective, while U_1 corresponds to the scattered, high-frequency, component. It follows that the decomposition in Eq. (5.16) describes the image field as the interference between the scattered and unscattered components. In the spatial domain, the resulting image intensity is that of an interferogram,

$$I(x,y) = |U(x,y)|^2$$
$$= |U_0|^2 + |U_1(x,y)|^2 + 2|U_0| \cdot |U_1(x,y)| \cdot \cos[\Delta\phi(x,y)]$$

(5.19)

where $\Delta\phi$ is the phase difference between the scattered and unscattered field.

For the optically thin specimens of interest here, the phase $\Delta\phi$ exhibits small variations, for which the corresponding intensity change is insignificant (e.g., Fig. 5.6a).

By contrast, the intensity is very sensitive to $\Delta\phi$ changes around $\Delta\phi = \pi/2$, or, equivalently, if we replace the cosine term with a sine. This can also be understood by realizing that the Taylor expansion around zero gives a quadratic function of the phase in the case of the cosine, $\cos(x) \simeq 1 - x^2/2$, which is negligible for small x values, and a linear dependence for sine, $\sin(x) \simeq x$.

Zernike understood that by simply shifting the phase of the unscattered light by $\pi/2$, the image intensity will suddenly exhibit great contrast. Before we show how this was achieved optically, let us investigate another variable in generating contrast, namely the ratio between the amplitudes of the two interfering beams.

Like in common interferometry, we can define the *contrast* of this interference pattern as

$$\gamma(x,y) = \frac{I_{max} - I_{min}}{I_{max} + I_{min}}$$

$$= \frac{2|U_0| \cdot |U_1(x,y)|}{|U_0|^2 + |U_1(x,y)|^2} \tag{5.20}$$

$$= \frac{2\beta(x,y)}{1 + \beta(x,y)^2}$$

where β is the ratio between the amplitudes of the two fields, $\beta = |U_1| / |U_0|$.

The contrast γ is a quantitative measure for how intensity across the image varies as a function of $\Delta\phi$ when it covers the entire trigonometric circle, i.e., $-1 < \cos(\Delta c) < 1$. Figure 5.8 shows the behavior of γ vs. β. It can be seen that the maximum contrast is achieved when $\beta = 1$, i.e., when the two amplitudes are equal, which is a well known result in interferometry. Unfortunately, for transparent samples, $|U_0| \gg |U_1|$, i.e., the unscattered light is much stronger than the scattered light, which is another reason for low contrast ($\beta \ll 1$). Therefore, in addition to the $\pi/2$ phase shift, attenuating $|U_0|^2$ (the unscattered light) is beneficial for improving the contrast. Let us now discuss Zernike's optical implementation of these ideas.

Placing a small metal film that covers the DC component in the Fourier plane of the objective can both *attenuate* and *shift* the phase of the unscattered field (see Fig. 5.9). Let us consider that the transmission function of this *phase contrast filter* is $a \cdot e^{i\alpha}$, where a describes its

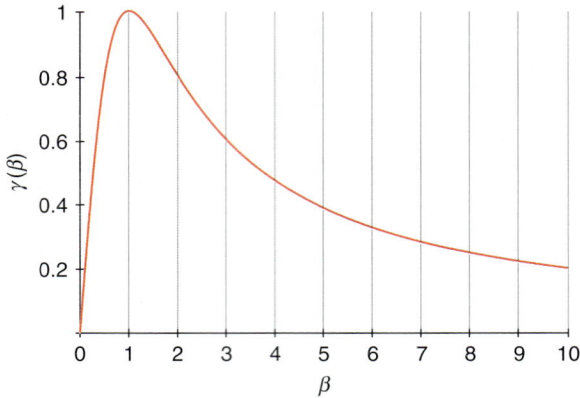

Figure 5.8   Contrast in the intensity image vs. the ratio between the scattered and unscattered field amplitude, $\beta = |U_1|/|U_0|$ [see Eq. (5.20)].

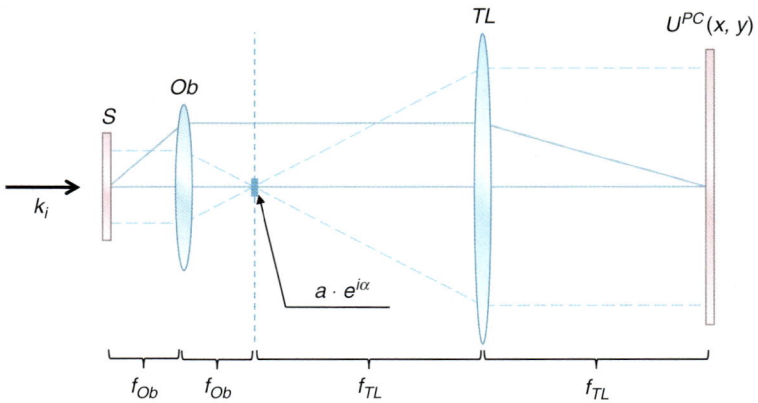

Figure 5.9   Phase contrast microscope.

attenuation and α the phase shift.[5] In the following, we find the values for a and α that produce optimal contrast.

For simplicity, we consider that the specimen under investigation is a phase object and that the image field is of unit amplitude, i.e.,

$$U(x,y) = e^{i\phi(x,y)} = 1 + U_1(x,y) \tag{5.21}$$

where $U_0 = 1$ is the unscattered field. With the phase contrast filter, the new field becomes

$$U^{PC}(x,y) = a \cdot e^{i\alpha} + U_1(x,y)$$

$$= a \cdot e^{i\alpha} + e^{i\phi(x,y)} - 1 \tag{5.22}$$

The intensity of the *phase contrast image* has the form

$$I^{PC}(x,y) = a^2 + 1 + 1 + 2[a\cos(\alpha + \phi) - a\cos\alpha - \cos\phi]$$

$$= a^2 + 2[1 - a\cos\alpha - \cos\phi + a\cos(\alpha + \phi)] \tag{5.23}$$

In Eq. (5.23), we can neglect $1 - \cos\phi$ for small ϕ. Further, if we choose $\alpha = \pm \frac{\pi}{2}$ (thus *positive* and *negative* phase contrast), we readily obtain

$$I^{PC}(x,y) = a^2 \pm 2a\sin\phi(x,y)$$

$$\simeq a^2 \pm 2a\phi(x,y) \tag{5.24}$$

Clearly, the $\pm \frac{\pi}{2}$ shift of the unscattered light produces an intensity that is now linear in ϕ, as intended. The contrast is further improved for an attenuation of the order $a = 2\phi$.

Since the attenuation factor is fixed and ϕ is sample dependent, it is impossible to satisfy the condition of maximum contrast for all points simultaneously. Nevertheless, a practical value of the order of $a = 2 - 3$ is typically used, corresponding to phase shifts of the order of 1–1.5 rad, which is common for live cells.

The powerful capability of PCM is illustrated in Fig. 5.10. Phase contrast microscopy is significantly more effective in enhancing contrast than the dark field method. Instead of removing the unscattered light completely, in phase contrast some DC field is maintained, but is placed in quadrature $\left(\frac{\pi}{2} \text{ out of phase}\right)$ with the scattered component. Perhaps an entertaining analogy can be drawn between phase contrast and martial arts: While carefully blocking the opponent's strikes makes a good approach for defense, managing to turn the opponent's energy against him/herself is perhaps even better (and energy efficient). In microscopy, the unscattered (direct or DC component) light is, of course, the enemy.

Figure 5.10b reveals the well-known glowing edges associated with phase contrast, generally known as "halo" effects. This

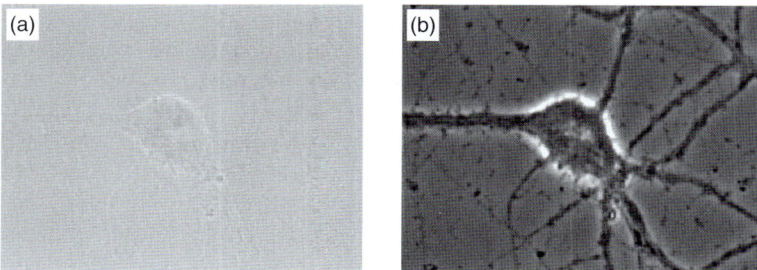

FIGURE 5.10 Bright field (a) and phase contrast; (b) image of an unstained neuron.

interesting phenomenon happens whenever light from such edges scatter strongly such that portions of this high-frequency field end up reaching the phase contrast filter and receive the treatment of the DC field. At the image plane, we now have two types of high-frequency fields, phase shifted by $\pi/2$ and not, which generate high-contrast fringes (bright or dark halos, depending on the sign of the phase shift filter). Essentially, the existence of the halo indicates the failure of our assumption for a perfect 2D Fourier transform at the back focal plane of the objective. This is a manifestation of the third dimension of the object. Thus, the quantitative derivation of the halo intensity distribution requires taking into account a 3D scattering model of the object, which is beyond the scope of our discussion here. Nevertheless, recent advances in QPI have impacted the halo artifacts as well. Thus, we will discuss in Sec. 12.2 a new method (spatial light interference microscopy) that transforms a common phase contrast into a QPI instrument and, in the process, reduces the halo effect.

References

1. J. D. Gaskill. *Linear systems, Fourier transforms, and optics* (Wiley, New York, 1978).
2. J. W. Goodman. *Introduction to Fourier optics.* (McGraw-Hill, New York, 1996).
3. E. Abbe. "Beiträge zur Theorie des Mikroskops und der mikroskopischen Wahrnehmung." *Arch. Mikrosk. Anat.*, 9, 431 (1873).
4. S. W. Hell. "Far-field optical nanoscopy." *Science,* 316, 1153–1158 (2007).
5. M. Born and E. Wolf. *Principles of optics: Electromagnetic theory of propagation, interference and diffraction of light* (Cambridge University Press, Cambridge; New York, 1999).

Further Reading

1. M. Born and E. Wolf. *Principles of optics: Electromagnetic theory of propagation, interference and diffraction of light.* (Cambridge University Press, Cambridge; New York, 1999).
2. J. W. Goodman. *Introduction to Fourier optics.* (McGraw-Hill, New York, 1996).
3. S. Inoué. *Collected works of Shinya Inoué : Microscopes, living cells, and dynamic molecules.* (World Scientific, Hackensack, NJ, 2008).
4. M. Pluta. *Advanced light microscopy.* (PWN; Elsevier: Distribution for the USA and Canada, Polish Scientific Publishers, Warszawa, 1988).
5. T. Wilson and C. Sheppard. *Theory and practice of scanning optical microscopy.* (Academic Press, London, 1984).
6. J. Mertz. *Introduction to optical microscopy.* (Roberts; Greenwood Village, Colorado, 2010).

CHAPTER 6

Holography

6.1 Gabor's (In-Line) Holography

In 1948, Dennis Gabor introduced "A new microscopic principle",[1] which he termed *holography* (from Greek *holos,* meaning "whole" or "entire," and *grafe,* "writing"). The name was chosen to indicate that the method records the entire field information (i.e., amplitude *and* phase) not just the usual intensity. Initially Gabor proposed this technique to "read" optically electron micrographs that suffered from severe spherical aberrations.[1] Nevertheless, the proof of principle demonstration was performed entirely in the optical domain; in fact, holography has remained since largely connected with optical fields. In 1971. Gabor was awarded the Nobel Prize in Physics "for his invention and development of the holographic method."

Holography is a two-step process: (1) *writing* the hologram, which involves recording on film the amplitude and phase information, and (2) *reading* the hologram, by which the hologram is illuminated with a reference field similar to that in step 1. Gabor's original setup for writing the hologram is described in Fig. 6. 1.

A point source of monochromatic light is collimated by a lens and the resulting beam illuminates the semitransparent object. Note that Gabor's experiments predate the invention of lasers by more than 12 years. Thus, the light source used in these initial experiments was a mercury lamp, with appropriate spatial (angular) and temporal (color) filtering to increase the spatial and temporal coherence, respectively. The film records the Fresnel diffraction pattern of the field emerging from the object. As in phase contrast microscopy (Sec. 5.4), the light passing through a semitransparent object consists of the scattered (U_1) and unscattered field (U_0). At a distance z behind the object, the detector (photographic film during Gabor's time) records

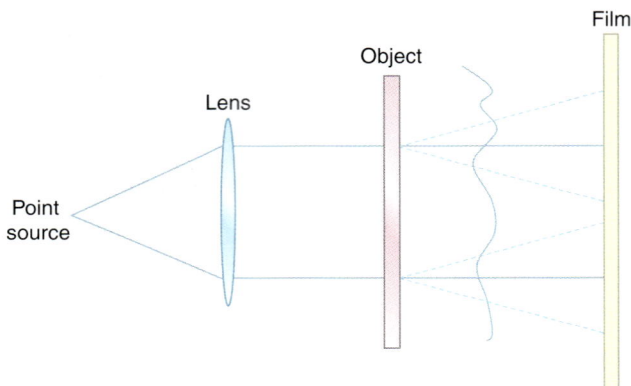

FIGURE 6.1   In-line optical setup for writing Fresnel holograms.

an intensity distribution generated by the interference of these two fields,

$$I(x,y) = \left| U_0 + U_1(x,y) \right|^2$$

$$= \left| U_0 \right|^2 + \left| U_1(x,y) \right|^2 + U_0 \cdot U_1^*(x,y) + U_0^* \cdot U_1(x,y) \tag{6.1}$$

Assuming a *linear response* to intensity associated with the photographic film, its transmission function has the form

$$t(x,y) = a + bI(x,y) \tag{6.2}$$

where a and b are constants. Thus, the hologram is now *written* and all the necessary information about the object is in the transmission function t.

Reading the hologram essentially means illuminating it as if it is a new object (Fig. 6.2). The field scattered from the hologram is the product between the illuminating plane wave (assumed to be U_0) and the transmission function,

$$U(x,y) = U_0 \cdot t(x,y)$$

$$= U_0 \left(a + b \left| U_0 \right|^2 \right) + b U_0 \cdot \left| U_1(x,y) \right|^2 \tag{6.3}$$

$$+ b \left| U_0 \right|^2 \cdot U_1(x,y) + b \left| U_0 \right|^2 \cdot U_1^*(x,y)$$

In Eq. (6.3), the first term is spatially constant, and the second term, $b U_0 \left| U_1(x,y) \right|^2$ is negligible compared to the last two terms because, for a transparent object, the scattered field is much weaker than the unscattered field, $\left| U_0 \right| \gg \left| U_1(x,y) \right|$. Thus, the last two terms of Eq. (6.3) are the relevant ones. Remarkably, these terms contain the complex field U_1 and its U_1^*. Therefore, an observer positioned behind the hologram will see at position z behind the transparency an image

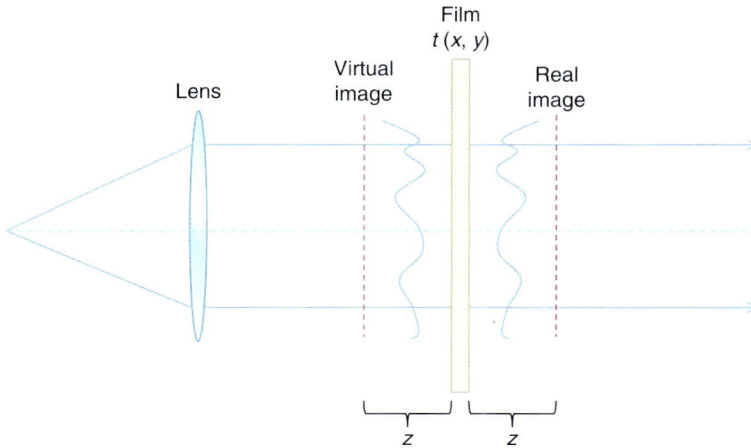

FIGURE **6.2** Reading an in-line hologram.

that resembles the original object (field U_1). Note that U_1 is the high-frequency part of the original field U and not the field itself.

Field U_1^* indicates "backward" propagation, such that a second (*virtual*) image is formed at a distance z in front of the film. If the observer focuses on the plane of the first (real) image, she/he will see an overlap between the in-focus image and the out-of-focus ("twin") image due to propagation over a distance $2z$. This overlap significantly degrades the signal to noise of the reconstruction and represents the main drawback of in-line holography. This is the reason why Gabor apparently abandoned holography by the mid-1950s.[2]

In summary, in-line holography can be summarized as the process of recording the Fresnel diffraction pattern of the object onto a photosensitive film. The visualization is the reverse process by which the hologram is illuminated with a plane wave and the resulting field observed at the same Fresnel distance away. The existence of the twin images, in essence, is due to the hologram being a real signal, the Fourier transform of which must be an even function, i.e., symmetric with respect to the film position. In the following section, we discuss the method that circumvented the obstacle posed by the twin image formation and turned holography into a main stream technique.

6.2 Leith and Upatnieks's (Off-Axis) Holography

The advancement of holography, from Gabor's initial work to the more practical implementation using the off-axis method is well captured by Adolf W. Lohmann[3]

"*To a large extent the success of holography is associated with the invention of the off-axis reference hologram by Emmett Leith and Juris Upatnieks.*[4,5]

The evolution from Gabor's inline hologram[1] to off-axis holography, however, is marked by important intermediate steps, for instance, single-sideband holography."[6,7] Lohman's own work on holography predates Leith's, but his 1956 paper has remained less known perhaps because it was published in German.[6] In his 1962 paper, Leith acknowledges Lohman's contributions:[4] *"A discussion of various similar techniques for eliminating the twin image is given by Lohmann, Optica Acta (Paris) **3**, 97 (1956). These are likewise developed by use of a communication theory approach."*

Leith and Upatnieks's pioneering paper on off-axis holography was titled "Reconstructed Wavefronts and Communication Theory,"[4] suggesting upfront the transition from describing holography as a visualization method to a way of transmitting information. In full analogy to the methods of radio communication, off-axis holography essentially adds spatial modulation (i.e., *carrier frequency*) to the optical field of interest. Interestingly, Gabor himself, like many electrical engineers at the time, was familiar with concepts of theory of communication and, in fact, published on the subject before his 1948 holography paper.[8]

The principle of writing an off-axis hologram is described in Fig. 6.3. The object is illuminated by a monochromatic plane wave, U_0, and

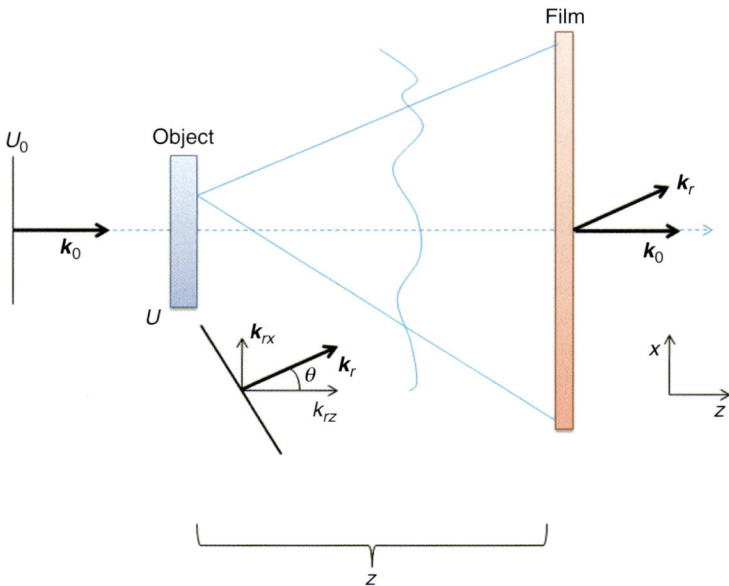

FIGURE 6.3 Off-axis setup for writing a hologram; k_0 and k_r are the wavevectors of the incident and reference fields.

the transmitted field reaches the photographic film at a distance z. The field distribution across the film, i.e., the Fresnel diffraction pattern, $U_F(x,y)$, is a convolution between the transmission function of the object, U, and the Fresnel diffraction kernel (see Sec. 2.2. and Eqs. 2.20 and 2.21),

$$U_F(x,y) = U(x,y) \circledv e^{\frac{ik_0(x^2+y^2)}{2z}} \tag{6.4}$$

In Eq. (6.4), we ignored the irrelevant prefactors that do not depend on x and y.

In contrast to in-line holography, here the reference field, U_r, is delivered at an angle θ (hence "off-axis") with respect to the object beam. The total field at the film plane is

$$U_t(x,y) = U_F(x,y) + |U_r| \cdot e^{ik_r \cdot \mathbf{r}}$$
$$= U_F(x,y) + |U_r| \cdot e^{i(k_{rx} \cdot x + k_{rz} \cdot z)} \tag{6.5}$$

where $k_{rx} = k_0 \cdot \sin\theta$ and $k_{rz} = k_0 \cdot \cos\theta$.

Note that the z-component of the reference wavevector produces a constant phase shift, $k_{rz} \cdot z$, which can be ignored because it is constant across the x-y plane. Thus the resulting transmission function associated with the hologram is proportional to the intensity, i.e.,

$$t(x,y) = |U_F(x,y)|^2 - |U_r|^2$$
$$+ U_F(x,y) \cdot |U_r| \cdot e^{-ik_{rx} \cdot x} + U_F^*(x,y) \cdot |U_r| \cdot e^{ik_{rx} \cdot x} \tag{6.6}$$

Reading the hologram is described in Fig. 6.4.

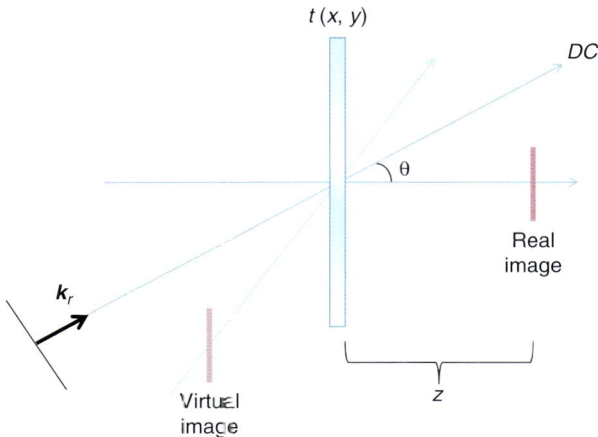

FIGURE 6.4   Reading the off-axis hologram.

Illuminating the hologram with a reference plane wave, U_r, the field at the plane of the film becomes

$$U_h(x,y) = |U_r| \cdot e^{i\mathbf{k}_r \cdot \mathbf{r}} \cdot t(x,y)$$

$$= |U_F(x,y)|^2 \cdot |U_r| \cdot e^{i\mathbf{k}_r \cdot \mathbf{r}} + |U_r|^3 \cdot e^{i\mathbf{k}_r \cdot \mathbf{r}} \quad\quad (6.7)$$

$$+ U_F(x,y) \cdot |U_r|^2 + U_F^*(x,y) \cdot |U_r|^2 \cdot e^{i2k_{rx} \cdot x}$$

Equation 6.7 establishes that, along the optical axis, the observer has access to the complex field $U_F(x,y)$ (third term in Eq. [6.7]), which, at a distance z from the film, reconstructs the identical replica of the object field. In other words, in reading the hologram, the free space performs the inverse operation of that in Eq. (6.4), that is, a *deconvolution*. This is the *real image*. The last term in Eq. (6.7) is modulated at a frequency $2k_{rx}$. Observing along this direction gives access to U_F^*, which reconstructs the object field behind the film, due to the complex conjugation. This is the *virtual image*.

As anticipated, the main accomplishment of this configuration is that the two images are now observed along different directions, without obstructing each other. Note that the first two terms in Eq. (6.7), the DC component, propagates along the direction of $\mathbf{k}_r$, which is also convenient. With the proper off-axis angle for writing/reading, the real image can be obtained *unobstructed*. In practice, the modulation frequency, k_{rx}, has to be carefully chosen to ensure the desired resolution in the final reconstruction, i.e., it must satisfy the Nyquist theorem applied to this problem. We will revisit this aspect later, when discussing off-axis methods for quantitative phase imaging (Sec. 8.3. and Chap. 9).

6.3   Nonlinear (Real Time) Holography or Phase Conjugation

In the 1970s, researchers realized that by exploiting the nonlinear response of materials, the writing and reading steps of holography can be combined into one.[9–11] This process has been termed *phase conjugation* and proposed as a way to correct imperfections (aberrations) in imaging systems. Here we briefly review its principle.

Yariv showed that nonlinear *four-wave mixing* can be interpreted as *real-time holography*.[11] The idea relies on the third-order nonlinearity response of the material used as writing/reading medium. Let us consider two strong (pump) fields, U_1 and U_2, that are the time reverse of each other (i.e., counter-propagating) and incident on a $\chi^{(3)}$ material, as shown in Fig. 6.5. If an *object*

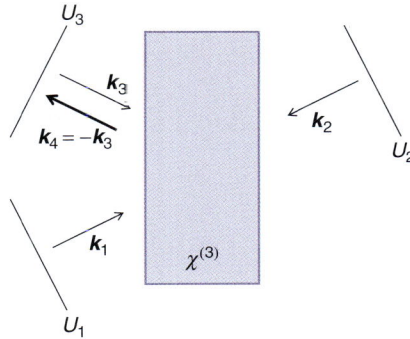

FIGURE 6.5 The four-wave mixing process: emerging field U_4 is phase conjugated to U_3.

field, U_3, is applied simultaneously, the *nonlinear* induced polarization can be written as

$$P^{(NL)}(-\omega_4 = \omega_1 + \omega_2 - \omega_3) = \frac{1}{2}\chi^{(3)}U_1U_2U_3^* e^{i[\omega_3 t - (k_1 - k_1 + k_3)r]}$$

$$= \frac{1}{2}\chi^{(3)}|U_1|^2 \cdot U_3^* \cdot e^{-i\omega_4 t + ik_4 \cdot r} \quad (6.8)$$

Clearly, the field emerging from the material, U_4, is the time-reversed version of U_3, i.e. $\omega_4 = -\omega_3$ and $k_4 = -k_3$, as indicated by the complex conjugation (U_3^*).

The efficiency of generating this phase conjugated field U_4 can be calculated from the *nonlinear wave equation*, which yields two coupled differential equations for the fields U_3 and U_4.[11] In principle, the time-reversed field U_4 can exceed in power the incident object field U_3, by converting some of the pump power. Phase conjugation or real-time holography has been used for various applications during the past decade.[12] Recently, this concept has received renewed attention in the context of tissue scattering removal.[13]

6.4 Digital Holography

Advancement in the theory of information and computing opened the door for a new era in the field of holography. Soon after the off-axis solution to the twin image problem was proposed, it was realized that either *writing* or *reading* the hologram could be performed *digitally* rather than optically. In the following sections, we discuss the basic principles of both these approaches.

6.4.1 Digital Hologram Writing

Vander Lugt showed in 1964 that *matched filters* can be recorded optically and used for applications such as character recognition.[14] The

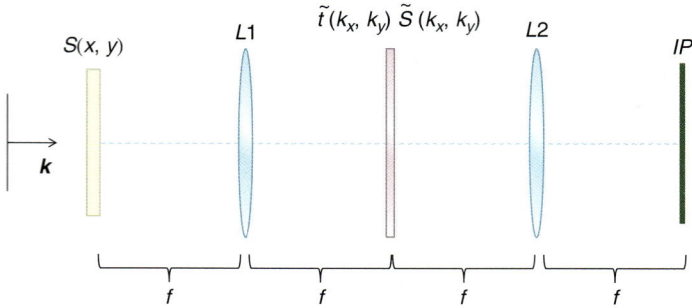

FIGURE 6.6 4f optical system with digitally printed hologram, $t(k_x,k_y)$, in the Fourier plane; $S(x,y)$ unknown signal, L1, L2 lenses, IP image plane.

idea is to calculate the cross-correlation between the known signal of interest and an unknown signal which, as a result, determines (i.e., recognizes) the presence of the first in the second.

In 1966, Brown and Lohmann of IBM showed for the first time that such a matched filter can be *written* digitally using a computer controlled plotter.[15] The goal was to calculate the autocorrelation operation in the Fourier domain, where it becomes a product (see App. B for the correlation theorem). Optically, the experimental setup overlaps (multiplies) the Fourier transform of the unknown signal with that of the signal of interest, as shown in Fig. 6.6. The unknown signal, S, is illuminated by a plane wave and Fourier transformed by lens L_1, where the resulting field, $\tilde{S}$, overlaps with the Fourier transform of the signal of interest, $\tilde{t}$.

Remarkably, Brown and Lohmann plotted the *complex* (i.e., phase and amplitude) Fourier transform of the desired signal using a binary mask, i.e., a 2D array of small dots, as used in the printing industry. The transmission function can be approximated by

$$t(k_x,k_y) = t_0 + \tilde{t}_1(k_x,k_y) \cdot e^{ik_x \cdot x_0} \tag{6.9}$$

In Eq. (6.9), t_0 is the DC component and $\tilde{t}_1$ is the exact Fourier transform of the signal of interest. The modulation, $e^{ik_x \cdot x_0}$, carries the spirit of off-axis holography, as it removes the twin image overlap. The field at the Fourier plane, $\tilde{U}(k_x,k_y)$, is the product between the incoming field and the mask, which acts as a diffracting object,

$$\tilde{U}(k_x,k_y) = \tilde{S}(k_x,k_y) \cdot [t_0 + \tilde{t}_1(k_x,k_y) \cdot e^{ik_x \cdot x_0}] \tag{6.10}$$

At the image plane (IP), the resulting field contains the cross-correlation of the fields. This can be easily seen by, taking the Fourier transform of Eq. (6.10), we obtain

$$U(x,y) = t_0 S(x,y) + S(x,y) \otimes t_1(x - x_0, y) \tag{6.11}$$

Note that due to the *shift theorem* (see App. B), find the modulation by $e^{ik_x \cdot x_0}$ shifts the correlation term away from the DC; this DC term is, up to a constant, the original signal S (the first term on the right-hand side in Eq. 6.11). Figure 6.7 illustrates this process, assuming that the signal of interest is the letter A, and the unknown signal is made of 9 letters placed in a 3×3 matrix. A digital mask containing

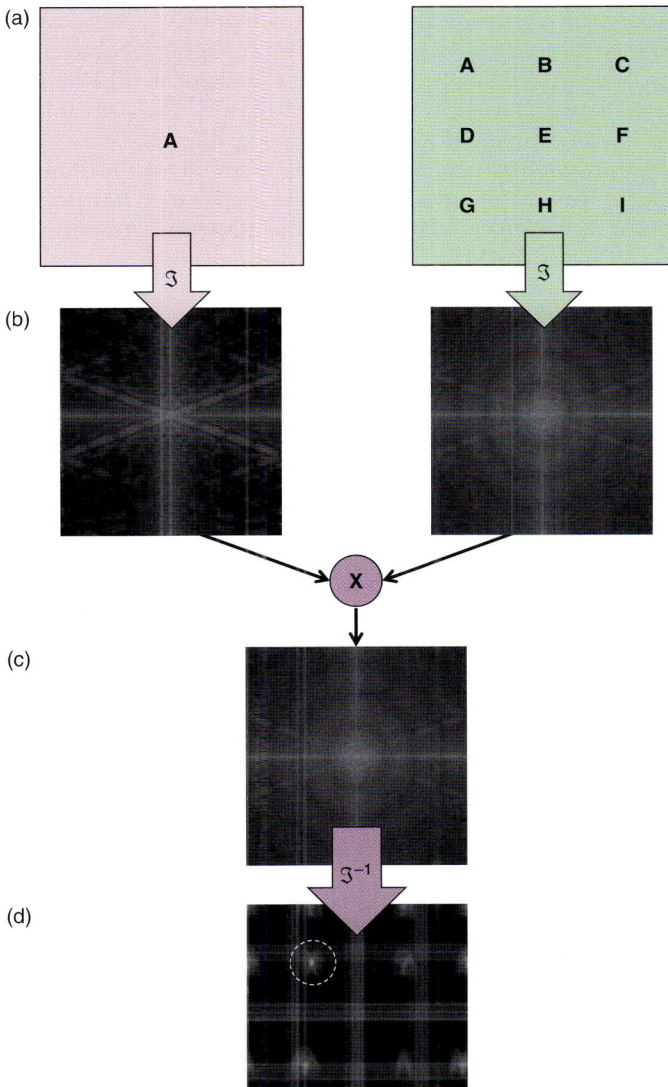

FIGURE 6.7 Character recognition using a digitally imprinted and Lugt's matched filter.

the Fourier transform of the letter A is multiplied by the Fourier transform of the unknown signal. The Fourier transform of this product, i.e., the *cross-correlation*, is shown in the bottom panel. Here the highest signal corresponds to the autocorrelation function of signal A with itself. Hence, character A is recognized within the unknown signal.

6.4.2 Digital Hologram Reading

In 1967, J. W. Goodman and R. W. Lawrence reported "Digital image formation from electronically detected holograms."[16] A *vidicon* (camera tube) was used to record an off-axis hologram. Numerical processing was based on the fast Fourier transform (FFT) algorithm proposed 2 years earlier by Cooley and Tukey.[17] The principle of this pioneering measurement is described in Fig. 6.8.

The transparency containing the signal of interest is illuminating by a plane wave. The emerging field, U_0, is Fourier transformed by the lens at its back focal plane, where the 2D array is positioned. The off-axis reference field U_r is incident on the detector at an angle θ. The total field at the detector is

$$U(x', y') = \tilde{U}_0(k_x, k_y) + |U_r| \cdot e^{ik_r \cdot r'}$$

$$k_x = 2\pi x'/\lambda f; \quad k_y = 2\pi y'/\lambda f \tag{6.12}$$

where $\tilde{U}_0$ denotes the Fourier transform of U_0.

The intensity that is recorded has the form

$$I(x', y') = |\tilde{U}(k_x, k_y)|^2$$

$$= |\tilde{U}_0(k_x, k_y)|^2 + |U_r|^2 + \tag{6.13}$$

$$+ \tilde{U}_0(k_x, k_y) \cdot |U_r| \cdot e^{-ik_{rx} \cdot x'} + \tilde{U}_0^*(k_x, k_y) \cdot |U_r| \cdot e^{ik_{rx} \cdot x'}$$

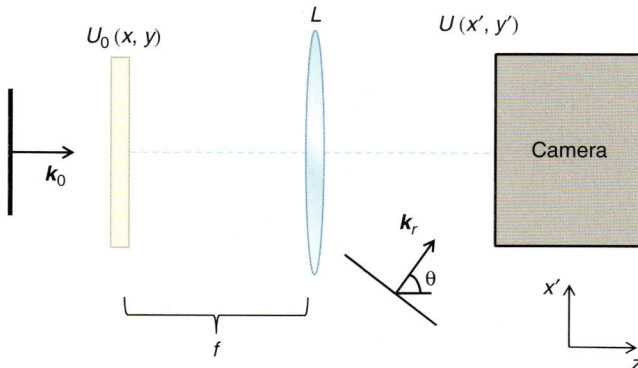

FIGURE 6.8   Digital recording of a Fourier hologram.

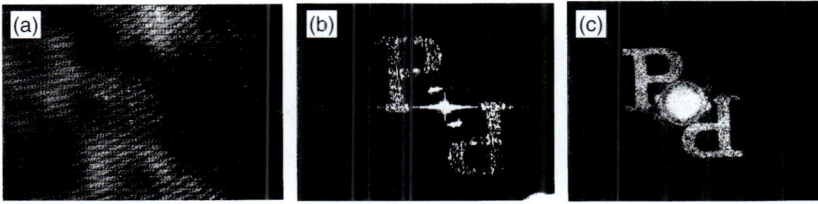

Figure 6.9   (a) Hologram stored in computer memory; (b) display of the image reconstructed by digital computation; (c) image obtained optically from a photographic hologram, shown for comparison. (Reprinted with permission from J. W. Goodman and R. W. Lawrence, *Appl. Phys. Lett.* 11, 77 (1967). Copyright 1967, American Institute of Physics.)

This intensity distribution, which was recorded *digitally*, is now Fourier transformed *numerically*. Note that, due to the modulation $e^{ik_{rx} \cdot x'}$, the last two terms in Eq. (6.13) generate Fourier transforms that are shifted symmetrically with respect to the origin. The first two terms in Eq. (6.12) are not modulated and, thus, represent the DC component of the signal. Figure 6.9 illustrates this procedure using Goodman and Lawrence's original results.[16]

Of course this numerical reconstruction is analogous to the optical one described in Sec. 6.4.1, where we discussed Vander Lugt's matched filter. Thus, if instead of the plane wave reference we had an arbitrary field interfering with the object field, the numerical reconstruction would yield the correlation between the Fourier transform of the two fields. However, since the reference field is a plane wave, the Fourier transform of which is a δ-function, this correlation operation returns the original signal.

In sum, in this chapter we presented a succinct review of the holography concept development, to be later used as background for the more recent developments in QPI. Since its invention, holography has become a vast field with many applications (see for example several books on the subject, Refs. 18–20). In particular, holographic microscopy has been recognized early on as a potentially powerful tool for biology (see, for example, a review in Vol. 2, Chap. 11 of Ref. 21). In the following chapters, we will focus mainly on the recent developments made possible by technological progress in digital recording devices. The applications of interest here are in biology and medicine.

References

1. D. Gabor. "A new microscopic principle." *Nature*, 161, 777 (1948).
2. J. Hecht. "Holography and the laser." *Optics and Photonics News*, July/August (2010).
3. M. Testorf and A. W. Lohmann. "Holography in phase space," *Applied Optics*, 47, A70–A77 (2008).

4. E. Leith and J. Upatnieks, "Reconstructed wavefronts and communication theory." *JOSA*, 52, 1123-1128 (1962).
5. E. N. Leith and J. Upatnieks, Wavefront reconstruction with continuous-tone objects, *JOSA*, 53, 1377 (1963).
6. A. Lohmann. "Optische Einseitenbandübertragung angewandt auf das Gabor-Mikroskop." *Opt. Acta*, 3, 97 (1956).
7. O. Bryngdah and A. Lohmann. "Single-sideband holography. *J. Opt. Soc. Am.*, 58, 620-& (1968).
8. D. Gabor. "Theory of Communicaton." *J. Inst. Electr. Eng.*, 93, 329 (1946).
9. B. Y. Zeldovich, V. I. Popovichev, V. V. Ragulsky, and F. S. Faizullov. *JETP Lett.*, 15, 109 (1972).
10. R. W. Hellwarth. "Generation of time-reversed wavefront by nonlinear refraction." *J. Opt. Soc. Am.*, 67, 1 (1977).
11. A. Yariv and D. M. Pepper. "Amplified reflection, phase conjugation, and oscillation in degenerate four-wave mixing." *Opt. Lett.*, 1, 17 (1977).
12. B. I. A. Zel*dovich, N. F. Pilipe*t*ski*i, and V. V. Shkunov. *Principles of phase conjugation.* (Springer-Verlag, Berlin; New York, 1985).
13. Z. Yaqoob, D. Psaltis, M. S. Feld, and C. Yang. "Optical phase conjugation for turbidity suppression in biological samples." *Nature Photonics*, 2, 110–115 (2008).
14. A. Vanderlugt. "Signal-detection by complex spatial-filtering." *IEEE Transactions on Information Theory.* 10, 139-& (1964).
15. B. R. Brown and A. W. Lohmann. "Complex spatial filtering with binary masks." *Appl. Opt.* 5, 967 (1966).
16. J. W. Goodman and R. W. Lawrence. "Digital image formation from electronically detected holograms." *Appl. Phys. Lett.* 11, 77 (1967).
17. J. W. Cooley and J. W. Tukey, "An algorithm for the machine calculation of complex fourier series." *Mathematics of Computation*, 19, 297 (1965).
18. P. Hariharan. *Basics of holography.* (Cambridge University Press, Cambridge, UK; New York, 2002).
19. A. Y. Pasmurov, J. S. Zinoviev, and Institution of Electrical Engineers. *Radar imaging and holography* (Institution of Electrical Engineers, London, 2005).
20. J. R. Vacca. *Holograms & holography: Design, techniques, & commercial applications.* (Charles River Media, Hingham, MA, 2001).
21. M. Pluta. *Advanced light microscopy.* (PWN; Elsevier: Distribution for the USA and Canada, Elsevier Science Publishing Co., Warszawa, Amsterdam; New York, 1988).

CHAPTER 7

Point-Scanning QPI Methods

In this chapter, we review the principle of quantitative phase methods based on point measurements, in which the image is reconstructed via scanning. These types of measurements can be regarded as an extension to optical coherence tomography (OCT), which is an intensity-based interferometric technique. OCT is a label-free (i.e., *intrinsic contrast*) method that enables 3D imaging of tissues. Developed in the Fujimoto Laboratory at Massachusetts Institute of Technology,[1] the principle of OCT relies on *low-coherence interferometry (LCI)*, i.e., interferometry with broadband light. Below we review the basics of LCI and OCT, as they apply to the more recent developments in QPI.

7.1 Low-Coherence Interferometry (LCI)

Let us consider a typical Michelson interferometer, in which a broadband source is used for illumination (Fig. 7.1a). The light is split by the beam splitter (BS) and directed toward two mirrors, M1 and M2, which reflect the field back. Mirror M2 has an adjustable position, such that the phase delay between the two fields can be tuned. Finally, the two beams are recombined at the detector, via the BS (note that power is lost at each pass through the BS).

Examples of low-coherence sources include light-emitting diodes (LEDs) superluminescent diodes (SLDs), Ti:Saph lasers, and even white light lamps. Of course, the term "low-coherence" is quite vague, especially since even the most stabilized lasers are of finite coherence length. Generally, by low coherence we understand a field that has a coherence length of the order of tens of microns (wavelengths) or less. We found previously (Sec. 3.3) that the *coherence length* of a field with central wavelength λ_0 and bandwidth $\Delta\lambda$ is of the order of $\tau_c \simeq \lambda_0^2/\Delta\lambda$. A qualitative comparison between the optical spectrum of a broadband source and that of a laser is shown in Fig. 7.1b.

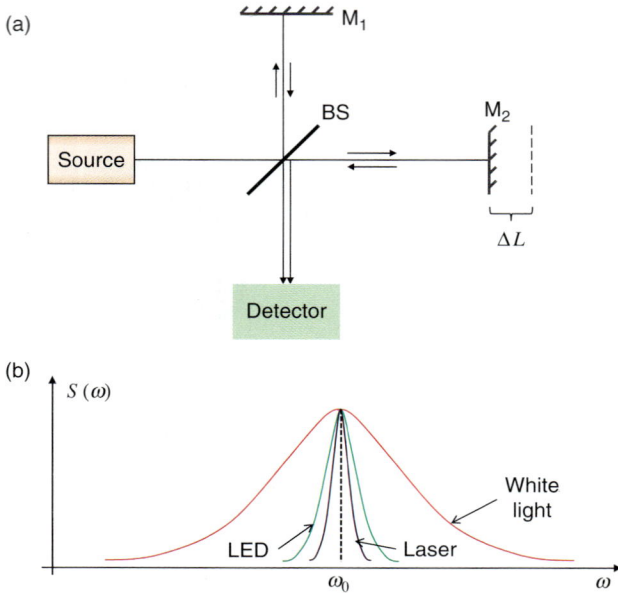

FIGURE 7.1   (a) LCI using a Michelson interferometer. Mirror M2 adjusts the path-length delay between the two arms. (b) Illustration of optical spectrum for various sources.

Let us now calculate the LCI signal in this Michelson interferometer. The total *instantaneous* field at the detector is the sum of the two fields,

$$U(t) = U_1(t) + U_2(t + \tau) \tag{7.1}$$

where τ is the delay introduced by the mobile mirror (M_2), $\tau = 2\Delta L/c$ (the factor of 2 stands for the double pass through the intereferometer arm). The intensity at the detector is the *modulus-squared average* of the field,

$$I(\tau) = \left\langle |U(t)|^2 \right\rangle$$

$$= I_1 + I_2 + \left\langle U_1(t) \cdot U_2^*(t + \tau) \right\rangle + \left\langle U_1^*(t) \cdot U_2(t + \tau) \right\rangle \tag{7.2}$$

$$= I_1 + I_2 + 2\,\mathrm{Re}[\Gamma_{12}(\tau)]$$

where the angular brackets denote temporal averaging.

In Eq. (7.2) we recognize the temporal cross-correlation function Γ_{12}, defined as (see Sec. 3.3)

$$\Gamma_{12}(\tau) = \left\langle U_1(t) \cdot U_2^*(t + \tau) \right\rangle \tag{7.3}$$

From Eq. (7.3), we see that the cross-correlation function can be measured experimentally by simply scanning the position of mirror M_2. Using the (generalized) Wiener-Kintchin theorem, Γ_{12} can be expressed as a Fourier transform

$$\Gamma_{12}(\tau) = \int W_{12}(\omega) \cdot e^{-i\omega\tau} d\omega \qquad (7.4)$$

In Eq. (7.4), W_{12} is the *cross-spectral density*,[2] defined as

$$W_{12}(\omega) = U_1(\omega) \cdot U_2^*(\omega) \qquad (7.5)$$

Note that W_{12} at each frequency ω is obtained via measurements that are *time-averaged* over time scales of the order of $1/\omega$. Thus, W_{12} is inherently an average quantity.

For simplicity, we assume that the fields on the two arms are identical, i.e., the beam splitter is 50/50. We will study later the case of dispersion on one arm and the effects of the specimen itself. For now, if $U_1(\omega) = U_2(\omega)$, $W_{12}(\omega)$ reduces to the spectrum of light, S, and Γ_{12} reduces to the autocorrelation function, Γ,

$$\Gamma(\tau) = \int S(\omega) \cdot e^{-i\omega\tau} d\omega \qquad (7.6)$$

The intensity measured by the detector has the form shown in Fig. 7.2a. Subtracting the signal at large τ (the DC component), which equals $2I_1$ for perfectly balanced interferometers, the real part of Γ is obtained as shown in Fig. 7.2b. Therefore, as discussed in App. A, we can calculate the complex analytic signal associated with this measured signal. Using the Hilbert transformation, the imaginary part of Γ can be obtained as,

$$\text{Im}[\Gamma(\tau)] = -\frac{1}{\pi} P \int \frac{\text{Re}[\Gamma(\tau')]}{\tau - \tau'} d\tau' \qquad (7.7)$$

where P indicates a principal value integral. Thus, the *complex analytic signal* associated with the measured signal, $\text{Re}[\Gamma(\tau)]$, is the autocorrelation function Γ, characterized by an amplitude and phase, as illustrated in (Fig. 7.2c).

For a spectrum centered at ω_0, $S(\omega) = S'(\omega - \omega_0)$, it can be easily shown (via the shift theorem, see App. B) that the autocorrelation function has the form

$$\begin{aligned}
\Gamma(\tau) &= \int S'(\omega - \omega_0) \cdot e^{-i\omega\tau} d\omega \\
&= e^{i\omega_0\tau} \int S'(\omega) \cdot e^{-i\omega\tau} d\omega \\
&= |\Gamma(\tau)| \cdot e^{i\omega_0\tau}
\end{aligned} \qquad (7.8)$$

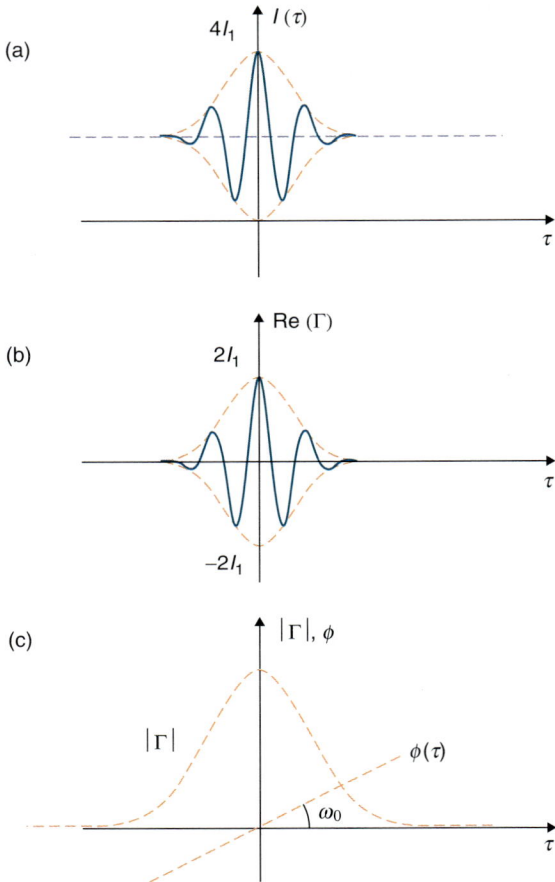

(a)

$4I_1$ $I(\tau)$

τ

(b)

$2I_1$ Re (Γ)

$-2I_1$

τ

(c)

$|\Gamma|, \phi$

$|\Gamma|$

$\phi(\tau)$

ω_0

τ

FIGURE 7.2   Low coherence interferometry with a perfectly balanced interferometer.

Equation (7.8) establishes that the envelope of Γ equals the Fourier transform of the shifted spectrum, S'. If we assumed a symmetric spectrum, the envelope is a real function. Further, the phase (modulation) of Γ is linear with τ, $\phi(\tau) = \omega_0 \tau$, where the slope is given by the mean frequency, ω_0.

This type of LCI measurement forms the basis for time-domain optical coherence tomography (Sec. 7.3.). The name "time-domain" indicates that the measurement is performed in time, via scanning the position of one of the mirrors. Alternative, frequency domain measurements will be also discussed in Sec. 7.4.

7.2   Dispersion Effects

In practice, the fields on the two arms of the interferometer are rarely identical. While the amplitude of the two fields can be matched via attenuators, making the optical pathlength identical is more difficult. Here, we will study the effect of dispersion due to the two beams passing through different lengths of dispersive media, such as glass. This is always the case when using a thick beam splitter in the Michelson interferometer (Fig. 7.3). Typically, the beam splitter is made of a piece of glass half-silvered on one side.

It can be seen in Fig. 7.3 that field U_1 passes through the glass three times, while field U_2 only once. Therefore, the phase difference between U_1 and U_2 has the frequency dependence

$$\phi(\omega) = 2k(\omega)L = 2n(\omega)k_0L \tag{7.9}$$

where L is the thickness of the beam splitter, k_0 the vacuum wavenumber, and n the frequency-dependent refractive index. This *spectral phase* can be expanded in Taylor series around the central frequency,

$$\phi(\omega) = \phi(\omega_0) + 2L\frac{dk(\omega)}{d\omega}\bigg|_{\omega=\omega_0}(\omega-\omega_0) + L\cdot\frac{d^2k(\omega)}{d\omega^2}\bigg|_{\omega=\omega_0}(\omega-\omega_0)^2 + \cdots$$

$$\tag{7.10}$$

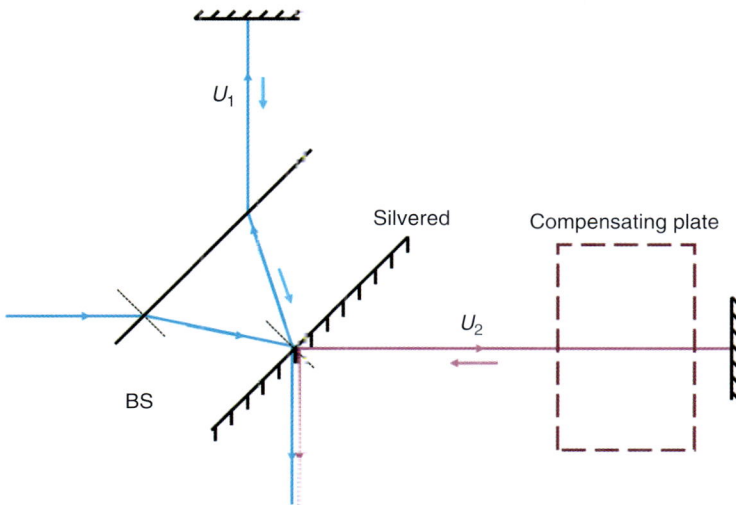

FIGURE 7.3   Transmission of the two beams through a thick beam splitter.

In Eq. (7.10), the individual terms of the Taylor expansion correspond to the following quantities:

- phase shift of mean frequency,

$$\phi(\omega_0) = \frac{2\omega_0}{c} \cdot L \tag{7.11}$$

- group velocity,

$$v_g = \frac{d\omega}{dk}\bigg|_{\omega=\omega_0} \tag{7.12}$$

- group velocity dispersion (GVD)

$$\beta_2 = \frac{d^2k}{d\omega^2}\bigg|_{\omega=\omega_0} = \frac{\partial}{\partial\omega}\left(\frac{1}{v_g}\right) \tag{7.13}$$

GVD has units of $s^2\!/\!m = s\!/\!Hz \cdot m$ and defines how a light pulse spreads in the material due to dispersion effects (delay per unit frequency bandwidth, per unit length of propagation). Note that GVD is sometimes defined as a derivative with respect to wavelength.

We can now express the cross-spectral density as

$$W_{12}(\omega) = U_1(\omega) \cdot U_1^*(\omega) \cdot e^{i\phi(\omega)}$$

$$= S(\omega) \cdot e^{i\phi(\omega)} \tag{7.14}$$

$$= S'(\omega - \omega_0) \cdot e^{i\phi(\omega_0)} \cdot e^{i\frac{2L}{v_g}(\omega-\omega_0)} \cdot e^{iL\beta_2(\omega-\omega_0)^2}$$

In Eq. (7.14), we assume that the two fields are of equal amplitude, and differ only through the phase shift due to the unbalanced dispersion, like, for example, due to the thick beam splitter shown in Fig. 7.3. The temporal cross-correlation function is obtained by taking the Fourier transform of Eq. (7.14). Let us consider first the case of negligible GVD, i.e., $\beta_2 = 0$,

$$\Gamma_{12}(\tau) = e^{i\phi(\omega_0)} \int S'(\omega - \omega_0) \cdot e^{i\frac{2L}{v_g}(\omega-\omega_0)} \cdot e^{-i\omega\tau} \, d\omega$$

$$\tag{7.15}$$

$$= e^{i\omega_0\tau} \cdot e^{i\phi(\omega_0)} \int S'(\omega - \omega_0) \cdot e^{-i(\omega-\omega_0)\left(\tau-\frac{2L}{v_g}\right)} \, d(\omega - \omega_0)$$

Note that we can denote $\omega - \omega_0$ as a new variable, to make it evident that the integral in Eq. (7.15) amounts to the shifted autocorrelation envelope,

$$\Gamma_{12}(\tau) = \left| \Gamma\left(\tau - \frac{2L}{v_g}\right) \right| \cdot e^{i[\omega_0 \tau + \phi(\omega_0)]} \tag{7.16}$$

Equation (7.16) establishes that in the absence of GVD there is no shape change in either the amplitude or the phase of the original correlation function. The phase shift, $\phi(\omega_0)$, is due to the zeroth order (*phase velocity*) term in the expansion of Eq. (7.10), while the envelope shift, or *group delay*, is caused by the first (*group velocity*) term (Fig. 7.4).

Note that the envelope shift, $\tau_0 = 2L/v_g$, can be conveniently compensated by adjusting the position of mobile mirror of the interferometer.

FIGURE 7.4

(*a*) Autocorrelation function for a perfectly balanced detector. (*b*) Group delay (first order) effects. (*c*) GVD (second order) effects.

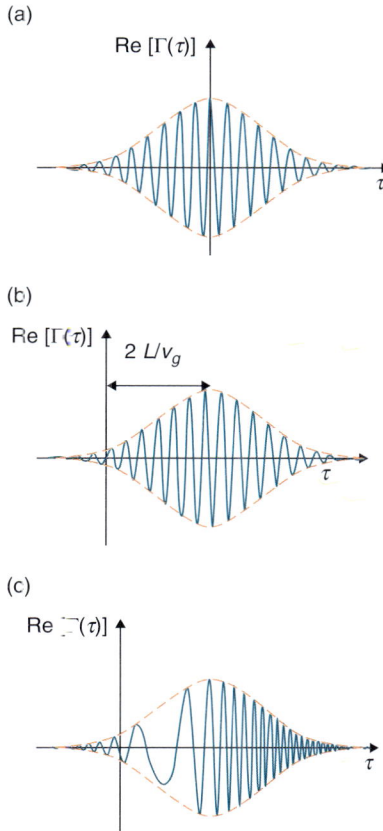

(a)

Re $[\Gamma(\tau)]$

(b)

Re $[\Gamma(\tau)]$

$2\,L/v_g$

(c)

Re $[\Gamma(\tau)]$

We conclude that if the beam splitter material has no GVD, the interferometer operates as if it is perfectly balanced.

Let us now investigate separately the effect of the GVD itself. The cross-correlation has the form

$$\Gamma_{12}(\tau) = \int S'(\omega - \omega_0) \cdot e^{iL\beta_2(\omega - \omega_0)^2} \cdot e^{-i\omega\tau} d\omega \qquad (7.17)$$

Note that the Fourier transform in Eq. (7.17) yields a convolution between the Fourier transform of S' and that of $e^{iL\beta_2\omega^2}$, i.e.,

$$\Gamma_{12}(\tau) = \Gamma(\tau) \circledcirc e^{-i\frac{\tau^2}{4\beta_2 L}} / \sqrt{2\beta_2 L} \qquad (7.18)$$

Equation (7.18) establishes that the cross-correlation function Γ_{12} is broader than Γ due to the convolution operation. Roughly, convolving two functions gives a function that has the width equal to the sum of the two widths (for Gaussian functions, this relationship is exact). Thus, if Γ has a width of τ_c (i.e., coherence time of the initial field), the resulting Γ_{12} has a width of the order of $\tau_{12} = \tau + \sqrt{2\beta_2 L}$. Somewhat misleadingly, it is said in this case that the *coherence time* (or length) "increased," or that dispersion changes the coherence of light. In fact, the autocorrelation function for each field of the interferometer is unchanged. It is only their cross-correlation that is sensitive to unbalanced dispersion. Perhaps a more accurate description is to say that, in the presence of dispersion, the cross-correlation time is larger than the autocorrelation (coherence) time.

This dispersion effect ultimately degrades the axial resolution of OCT images, as we will see in the next section. In practice, great effort is devoted towards compensating for any unbalanced dispersion in the interferometer. Since Michelson's time, this effect was well known; compensating for a thick beam splitter was accomplished by adding an additional piece of the same glass in the interferometer, such that both fields undergo the same number (three) of passes through glass.

7.3   Time-Domain Optical Coherence Tomography

Optical coherence tomography (OCT) is typically implemented in fiber optics, where one of the mirrors in the interferometer is replaced by a 3D specimen (Fig. 7.5). In this geometry, the depth-information is provided via the LCI principle discussed previously and the x-y resolution by a 2D scanning system, typically comprised of galvo-mirrors. The transverse (x-y) resolution is straightforward, as it is given by the numerical aperture of the illumination. In the following, we discuss in more detail the depth resolution and its limitation.

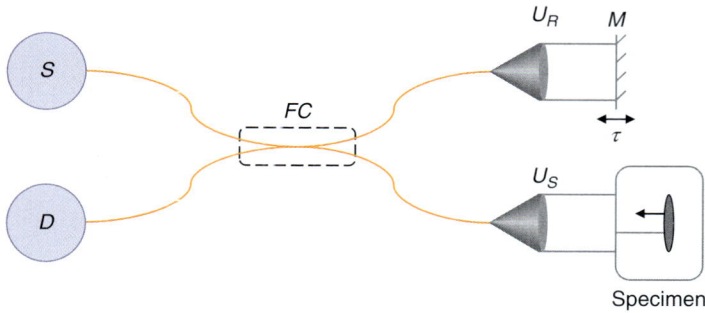

FIGURE 7.5 Fiber optics, time domain OCT.

At each point (x, y), the OCT signal consists of the cross-correlation Γ_{12} between the reference field U_R and specimen field U_S; its Fourier transform is the cross-spectral density

$$W_{12}(\omega) = U_S(\omega) \cdot U_R^*(\omega)$$

$$= S(\omega) \cdot \tilde{h}(\omega) \tag{7.19}$$

where $S(\omega) = U_R(\omega) \cdot U_R^*(\omega)$ is the spectrum of the source and $\tilde{h}(\omega)$ the *spectral modifier,* which is a complex function characterizing the spectral response of the specimen,

$$U_S(\omega) = U_R(\omega) \cdot \tilde{h}(\omega) \tag{7.20}$$

The two fields are initially identical, i.e., the interferometer is *balanced,* but the specimen is modifying the incident field via $\tilde{h}(\omega)$. The resulting cross-correlation obtained by measurement is a convolution operation, as obtained by Fourier transforming Eq. (7.19),

$$\Gamma_{12}(\tau) = \Gamma(\tau) \circledv h(\tau) \tag{7.21}$$

where h is the time response function of the sample, the Fourier transform of $\tilde{h}(\omega)$,

$$h(\tau) = \int \tilde{h}(\omega) \cdot e^{-i\omega\tau} d\omega \tag{7.22}$$

7.3.1 Depth-Resolution in OCT

Note that the LCI configuration depicted in Fig. 7.1, i.e., having a mirror as object, is mathematically described by introducing a δ response function, $h(\tau) = \delta(\tau - \tau_0)$. In this case, the cross-correlation reduces to (from Eq. 7.21)

$$\Gamma_{12}(\tau) = \Gamma(\tau - \tau_0) \tag{7.23}$$

where $\tau_0 = 2z/c$ represents the time delay due to the depth location z of the reflector (Fig. 7.6). In other words, by scanning the reference mirror,

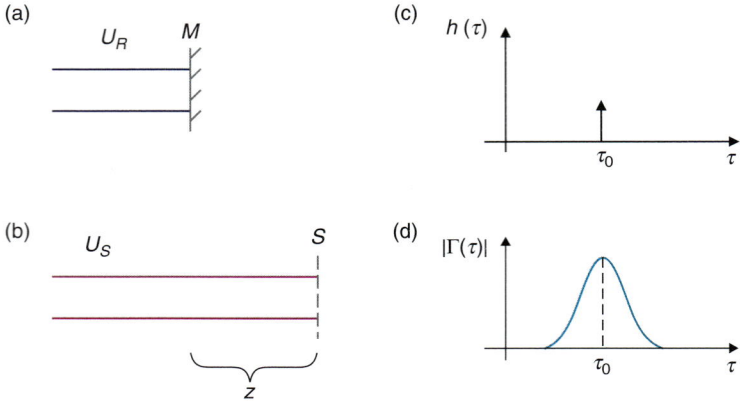

FIGURE 7.6   Response from a reflector at depth z: (*a*) the reference arm; (*b*) the sample arm; (*c*) the ideal response function from the mirror in (*b*); (*d*) the measured response from the mirror in (*b*).

the position of the second mirror is measured experimentally with an accuracy given by the width of the cross-correlation function, $\Gamma_{12}(\tau)$.

OCT images are obtained by retaining the modulus of the cross-correlation, $\Gamma_{12}(\tau) = |\Gamma_{12}(\tau)| \cdot e^{i[\omega_0\tau + \phi(\tau)]}$. The high frequency component (carrier), $e^{i\omega_0\tau}$, is filtered via demodulation (low-pass filtering). Thus, the *impulse response* function of OCT is $|\Gamma(\tau)|$. As illustrated in Fig. 7.6*d*, the width of the envelope of Γ establishes the ultimate resolution in locating the reflector's position. This resolution in time is nothing more than the coherence time of the source, τ_C, provided that the interferometer is balanced (otherwise it is the cross-correlation time). In terms of depth, this resolution limit is the *coherence length*, $l_C = v\tau_C$, with v the speed of light in the medium. This result establishes the well-known need for broadband sources in OCT, as $\tau_C \propto \dfrac{1}{\Delta\omega}$.

It is important to realize that the coherence length is the absolute best resolution that OCT can deliver. As mentioned in the previous section, the performance can be drastically reduced by dispersion effects due to an unbalanced interferometer. Further, note that the specimen itself does "unbalance" the interferometer. Consider a reflective surface buried in a medium characterized by dispersion (e.g., a tumor located at a certain depth in the tissue), as illustrated in Fig. 7.7. Even when the attenuation due to depth propagation is negligible (weakly scattering, low-absorption medium), the spectral phase accumulated is different for the two depths (see Eq. [7.18]),

$$\Gamma_{12}(\tau, L_{1,2}) = \Gamma(\tau) \circledv e^{-i\frac{\tau^2}{4\beta_2 L_{1,2}}} / \sqrt{2\beta_2 L_{1,2}} \tag{7.24}$$

where β_2 is the GVD of the medium (e.g., tissue) above the reflector.

FIGURE **7.7**
Depth-dependent
resolution in OCT.

(a)

U_S

L_1

(b)

U_S

L_2

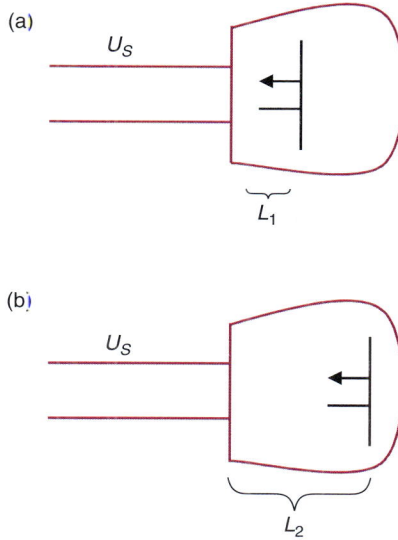

Therefore, the impulse response is broader from the reflector that is placed deeper ($L_2 > L_1$) in the medium. This is to say that in OCT the axial resolution degrades with depth.

7.3.2 Contrast in OCT

Contrast in OCT is given by the difference in reflectivity between different structures, i.e., their refractive index contrast. OCT contrast in a transverse plane (x-y, or *en face*) also depends on depth. Figure 7.8 illustrates the change in SNR with depth. As can be seen, the scattering properties of the surrounding medium are affecting SNR in a depth-dependent manner. Thus, the background backscattering ultimately limits the *contrast-to-noise ratio* in an x-y image.

The signal to noise can be defined at each depth, as

$$SNR(\tau) = \frac{|\Gamma_{12}(\tau)|}{\sigma_N(\tau)} \qquad (7.25)$$

where $\sigma_N(\tau)$ is the standard deviation of the noise around the time delay, τ. This noise component is due to mechanical vibrations, source noise, detection/electronic noise, but, most importantly, due to the scattering from the medium that surrounds the structure of interest.

In an *en face* image, we can define the contrast-to-noise ratio as (Fig. 7.9)

$$CNR(\tau) = \frac{\left\| \Gamma_A(\tau) \right| - \left| \Gamma_B(\tau) \right\|}{\sigma_N(\tau)} \qquad (7.26)$$

where A and B are two structures of interest.

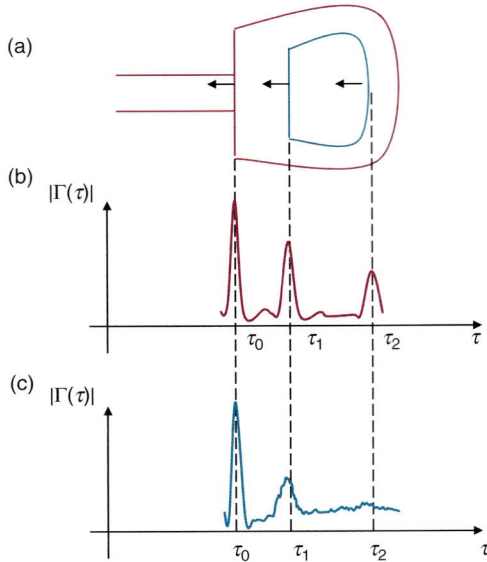

(a)

(b) $|\Gamma(\tau)|$

τ_0 τ_1 τ_2 τ

(c) $|\Gamma(\tau)|$

τ_0 τ_1 τ_2 τ

FIGURE 7.8 Signals from an object surrounded by scattering medium, illustrated in (a). (b) Depth-resolved signals for *weakly scattering* medium. (c) Depth-resolved signals for *strongly scattering* medium.

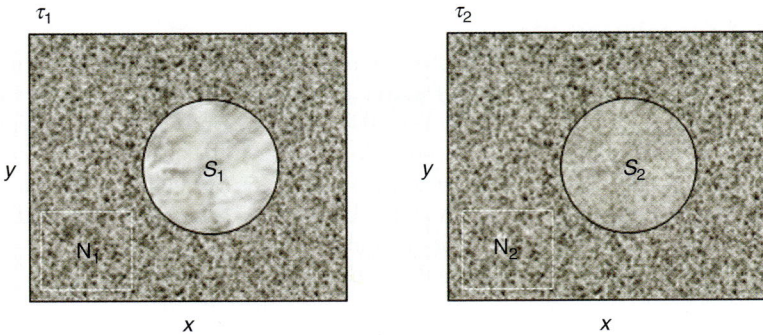

τ_1

y S_1

N_1

x

τ_2

y S_2

N_2

x

FIGURE 7.9 En-face images corresponding to the two time-delays shown in Fig. 7.8.

7.4 Fourier Domain and Swept Source OCT

The need for mechanical scanning in time domain OCT limits the acquisition rate and is a source of noise that ultimately affects the phase stability in the measurements (note that generally OCT is *not* a common path system). Fortunately, there is a faster method for measuring the cross-correlation function, $\Gamma_{12}(\tau)$. Thus, it is entirely equivalent to obtain Γ_{12} by measuring first the cross-spectral density, $W_{12}(\omega)$, and take its Fourier transform numerically, i.e., exploiting the generalized Wiener-Kintchin theorem,

$$\Gamma_{12}(\tau) = \int W_{12}(\omega) \cdot e^{-i\omega\tau} d\omega \qquad (7.27)$$

In 1995, Fercher et al., from University of Vienna, applied this idea to obtain depth scans of a human eye.[3] They also pointed out that $W_{12}(\omega)$ can be measured either by a spectroscopic measurement, where the colors are dispersed onto a detector array, or by *sweeping* the wavelength of the source and using a single (point) detector.

The principle of these frequency domain measurements is illustrated in Fig. 7.10. The total field at the spectrometer in Fourier domain OCT (Fig. 7.10*a*) is

$$U(\omega) = U_R(\omega) + U_S(\omega) \qquad (7.28)$$

where U_R and U_S are the reference and specimen fields, respectively. As before, the frequency response of the object can be described by a complex function, the *spectral modifier, $\tilde{h}(\omega)$*, such that the field returned from the specimen can be written as

$$U_S(\omega) = U_R(\omega)\tilde{h}(\omega)e^{-i\omega s_0/c} \qquad (7.29)$$

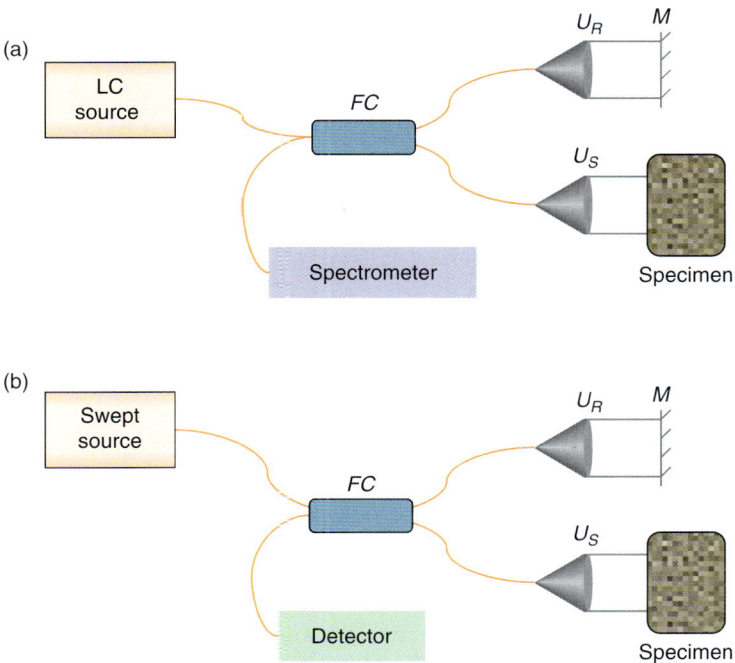

FIGURE 7.10 (*a*) Fourier domain OCT. (*b*) Swept source OCT.

In Eq. (7.29), the phase factor $e^{-i\omega s_0/c}$ indicates that the two arms of the interferometer are mismatched by a pathlength s_0, which is fixed. The intensity vs. frequency detected by the spectrometer results from combining Eqs. (7.28) and (7.29),

$$I(\omega) = |U(\omega)|^2$$

$$= S(\omega) + S(\omega)\left|\tilde{h}(\omega)\right|^2 + 2S(\omega)\left|\tilde{h}(\omega)\right|\cos[\omega s_0/c + \phi(\omega)]$$

(7.30)

In Eq. 7.30, $\phi(\omega)$ is the spectral phase associated with the object, i.e., the argument of the frequency modifier $\tilde{h}(\omega)$. We can rewrite Eq. (7.30) to better emphasize the DC and modulated terms,

$$I(\omega) = a(\omega)\{1 + b(\omega) \cdot \cos[\omega s_0/c + \phi(\omega)]\}$$

(7.31)

where

$$a(\omega) = S(\omega)\left[1 + \left|\tilde{h}(\omega)\right|^2\right]$$

$$b(\omega) = \frac{2\left|\tilde{h}(\omega)\right|}{1 + \left|\tilde{h}(\omega)\right|^2}$$

(7.32)

A typical signal $I(\omega)$ is illustrated in Fig. 7.11. Note that the modulation of the signal in Fig. 7.11 is described by the (real) function $b(\omega)$ (Eq. [7.32b]). This function can be easily shown to satisfy the relationships

$$0 \le b(\omega) \le 1$$

$$b(\omega) = 1 \text{ if } \left|\tilde{h}(\omega)\right| = 1$$

(7.33)

Thus, not surprisingly, the highest contrast of modulation, $b = 1$, is obtained when the intensities of the two arms of the interferometer are equal, $I_R(\omega) = I_S(\omega)$ and the specimen only affects the phase of the incident field, i.e., $\left|\tilde{h}(\omega)\right| = 1$.

In order to obtain the time domain response, $I(\tau)$, the measured signal $I(\omega)$ is Fourier transformed numerically. Since $I(\omega)$ is a real signal, its Fourier transform is even (Fig. 7.11b). Let us consider for simplicity that the specimen has a "flat" frequency response in amplitude, $\left|\tilde{h}(\omega)\right| = $ const, which is to say that the reflectivity of the object is not frequency dependent. In this case, the DC term, i.e., the Fourier transform of $a(\omega)$, is nothing more than the autocorrelation function $\Gamma(\tau)$. Further, let us consider that the object consists of a single reflective surface at depth z_0. In this case the modulated term in Eq. (7.31) has

(a)

(b)

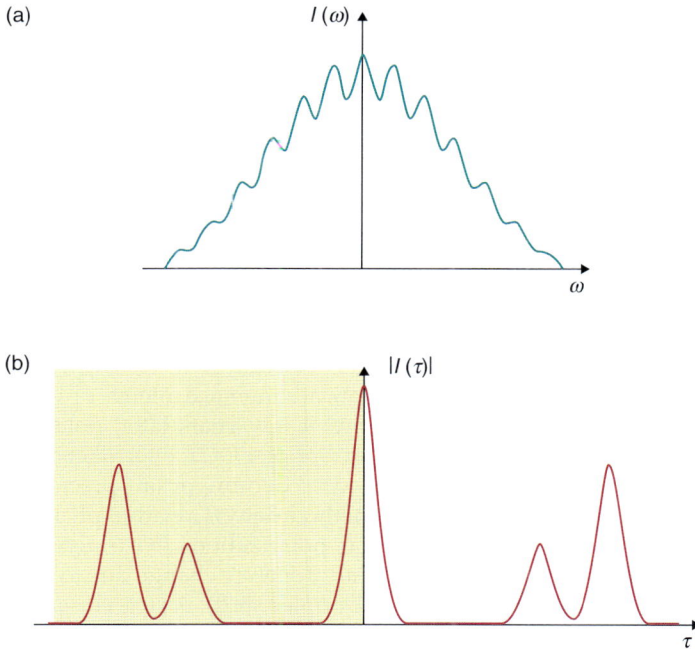

FIGURE 7.11 (a) Typical row signal in Fourier domain OCT. (b) Symmetric depth-resolved signal obtained by Fourier transforming the raw data.

the form $S(\omega) \cdot \cos[\omega s_0/c - 2\omega z_0/c]$. The Fourier transform of this AC component has the form

$$I_{AC}(\tau) \propto \Gamma(\tau) \circledv \{\delta[(s_0 - 2z_0)/c] + \delta[(s_0 + 2z_0)/c]\} \qquad (7.34)$$

Clearly, according to Eq. (7.34), the depth position of the object, z_0, can be experimentally retrieved from either side bands (delta functions), as shown in Fig. 7.11b. Like in time-domain OCT, the accuracy of this depth measurement is limited by the width of $\Gamma(\tau)$, which convolves the entire signal, i.e., the coherence length of the illuminating field. The Fourier domain measurement is fully equivalent with its time domain counterpart.

Note that the pathlength mismatch, s_0, is fixed and sets the upper value of the depth that can be accessed in the specimen, $z_{max} = s_0/4n$, where n is the refractive index of the specimen. This can be easily understood by noting that the modulation of the $I(\omega)$ signal must be at least twice the maximum frequency of interest, i.e., $s_0\omega/c \geq 2s_{max}\omega/c = 4nz_{max}\omega/c$.

The description above holds valid for swept-source OCT as well, the only difference being that the frequency components of $I(\omega)$ are measured in succession rather than simultaneously. Recent developments in laser sources allow sweeping broad spectra at very high speeds, up to *100 kHz*.[4]

In summary, Fourier domain OCT offers a fast alternative to acquiring depth-resolved signals. The detection via spectrometer makes the measurements *single shot*, such that the phase information across the measured signal $I(\omega)$ is stable. This feature is the premise for using such a configuration in achieving point-scanning QPI. In the following sections, we describe the main developments associated with phase-sensitive imaging. Depending on whether or not the phase map associated with an object is retrieved quantitatively (in radians), we divide the discussion in qualitative (Sec. 7.5.) and quantitative methods (Sec. 7.6.).

7.5   Qualitative Phase-Sensitive Methods

In 1994, Izatt et al., at Massachusetts Institute of Technology, showed that, due to its coherence gating, optical coherence microscopy (OCM) is superior to confocal microscopy.[5] A few years later, polarization sensitive OCT by de Boer et al.[6] and Doppler OCT, by Izatt et al.[7] have been reported and can be regarded as two important precursors to OCT-based QPI. Thus, the first method detects relative phase shifts between fields with perpendicular polarizations (*birefringence*) and the second measures the *time derivative of phase* change due to moving objects, i.e., a frequency shift (or Doppler shift). A step closer to QPI has been achieved by using relative phase measurements from two adjacent points in the sample and from two different colors at the same point on the sample. We discuss these approaches below.

7.5.1   Differential Phase-Contrast OCT

Using two polarization channels to encode relative phase information, Hitzenberger and Fercher demonstrated "Differential phase contrast in optical coherence tomography".[8] As the title suggests, this method retrieves phase difference information from different points across the object. The experimental setup is shown in Fig. 7.12.[8]

The broadband light from a superluminescent diode passes through a polarizer and further enters a Michelson interferometer. On the sample arm, the quarter waveplate (QWP) makes the light circularly polarized, such that after passing through the Wallaston prism, two replicas of the beam are created, with mutually perpendicular polarizations. The two beams are formed by the lens at the object plane, with a shift in position, Δx. Propagating back through the lens and Wallaston prism, the polarization of the sample beam becomes elliptical due to the phase shift induced by the sample, $\phi(x+dx)-\phi(x)$. In order to detect this ellipticity and, thus, the phase information of interest, the reference beam is made circularly polarized by double propagation through a QWP at 22.5°. The polarizing beam splitter directs the total (reference and sample) field toward two detectors, which measure the intensity of the mutually orthogonal polarizations, I_1 and I_2. With proper phase bias, the system can operate in a linear regime, in which $\phi(x+dx)-\phi(x)=I_2/I_1$. Clearly, this signal is proportional to the *phase gradient* along one direction,

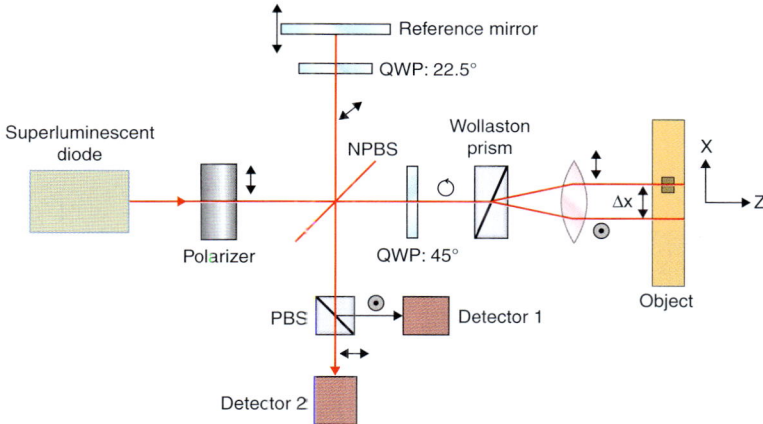

FIGURE 7.12 Schematic of the OCT setup for differential phase contrast tomography: QWP, quarter waveplate, NPBS, nono-polarizing beam splitter, PBS, polarizing beam splitter. [Reproduced with permission from C. K. Hitzenberger and A. F. Fercher, *Opt. Lett.*, 24 (9), 622 (1999). Copyright Optical Society of America 1999.]

$d\phi/dx = \lim_{\delta x \to 0} [\phi(x + \delta x) - \phi(x)]/\delta x$, rather than the phase itself and is similar to what is measured in differential interference contrast microscopy (see Chap. 7, Vol. 2 in Ref. 9).

Figure 7.13 shows a demonstration of this method, where a layered sample has been mapped using differential phase measurement.[8] As the authors point out, if the sample itself exhibits birefringence, the interpretation of the signal becomes cumbersome. Nevertheless, this technique is sensitive to phase shifts and can be employed to image objects that are essentially transparent. This principle was implemented with optical fibers by Dave and Milner.[10] Later this phase-contrast approach was further applied to image single layers of cells by Sticker et al.[11] and Rylander et al.[12]

7.5.2 Interferometric Phase-Dispersion Microscopy

Yang et al. in Michael Feld's group, at Massachusetts Institute of Technology, developed "interferometric phase-dispersion microscopy," whereby the phase difference between the fundamental and second-harmonic light, both interacting with the specimen, is quantified.[13] The experimental setup is described in Fig. 7.14.[13] The measurement involves a Michelson interferometer, like in OCT, and the illumination consists of a composite 800 nm/400 nm beam, obtained from a *Ti: sapphire* laser and its second harmonic. Unlike with common OCT, the sample under investigation is illuminated in transmission, i.e., the light passes through the object twice. The reference mirror is translated at constant speed, which induces a Doppler frequency shift and generates a heterodyne signal for each of the wavelengths. Because the interferometer is not common path, each signal is noisy. However, the noise in the

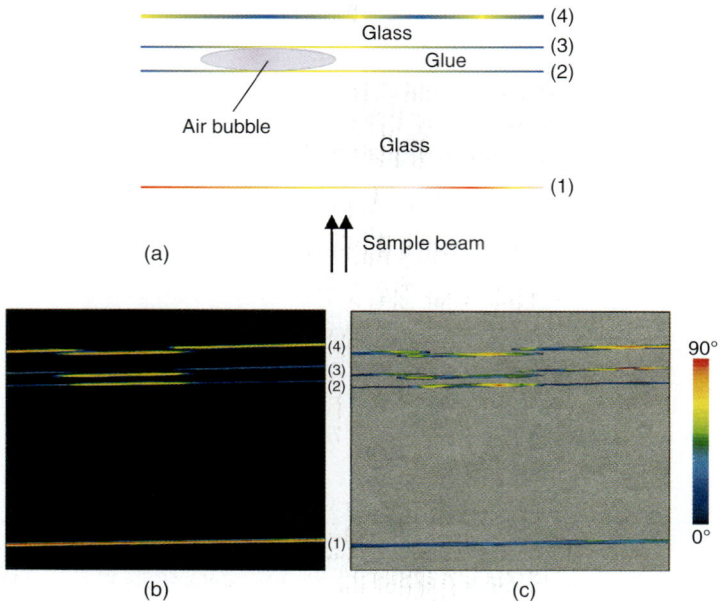

FIGURE 7.13 OCT images of a test object consisting of a microscope carrier glass, a single-component glue, and a cover glass. For easy identification the interfaces are numbered (1) to (4). Illumination is from the bottom of the object. (*a*) Schematic of the test object (not drawn to scale), (*b*) intensity image, (*c*) phase image. The dimensions of the tomograms are 3 mm horizontal and 2.5 mm vertical (optical distance). [Reproduced with permission from C. K. Hitzenberger and A. F. Fercher, *Opt. Lett.,* 24 (9), 622 (1999). Copyright Optical Society of America 1999. (Ref. 8).]

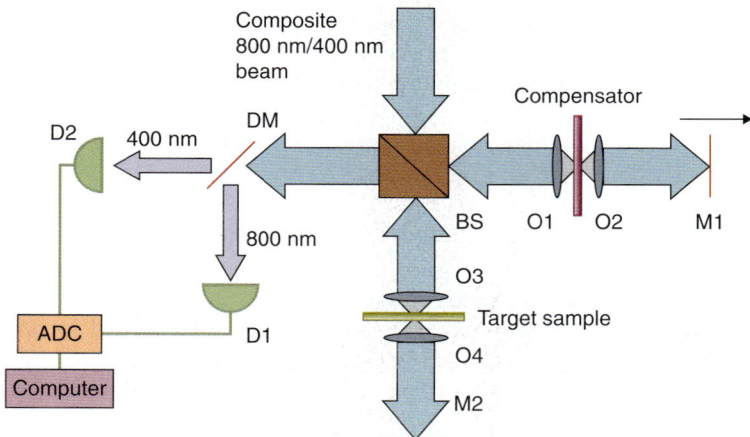

FIGURE 7.14 Experimental setup: M1, M2, mirrors (M1 is the moving reference mirror); BS, beam splitter; O1–O4, microscope objectives; D1, D2, photodetectors; DM, 400–800-nm dichroic mirror; ADC, analog–digital converter. [Reproduced with permission from C. H. Yang, A. Wax, I. Georgakoudi, E. B. Hanlon, K. Badizadegan, R. R. Dasari and M. S. Feld, Interferometric phase-dispersion microscopy, *Opt. Lett.,* 25, 1526–1528 (2000). Copyright Optical Society of America 2000.]

two signals is correlated and can be subtracted to a good extent by taking the difference of the two phase measurements. Thus, a jitter in the system that generates a random displacement, Δx, introduces a phase shift

$$\phi_{1,2} = k_{1,2}\Delta x \tag{7.35}$$

where $k_{1,2} = \dfrac{2\pi}{\lambda_{1,2}}$ is the wave number for wavelength $\lambda_1 = 800$ nm and $\lambda_2 = 400$ nm. Note that $\dfrac{\phi_1}{k_1} = \dfrac{\phi_2}{k_2}$. Therefore, the pathlength difference between the two measurements is noise-free,

$$\Delta s = \left(\Delta n_2 \cdot L + \dfrac{\phi_2}{k_2} \right) - \left(\Delta n_1 \cdot L + \dfrac{\phi_1}{k_1} \right) \tag{7.36}$$

$$= L(\Delta n_2 - \Delta n_1)$$

where $\Delta n_{1,2}$ is the refractive index of the specimen at wavelength $\lambda_{1,2}$, and L is the sample thickness. Therefore, this method provides a stable measurement of the refractive index difference, i.e., the *dispersion*, at two harmonic wavelengths.

Figure 7.15 illustrates the capability of this dispersion measurement to distinguish between a droplet of water and one of DNA.[13]

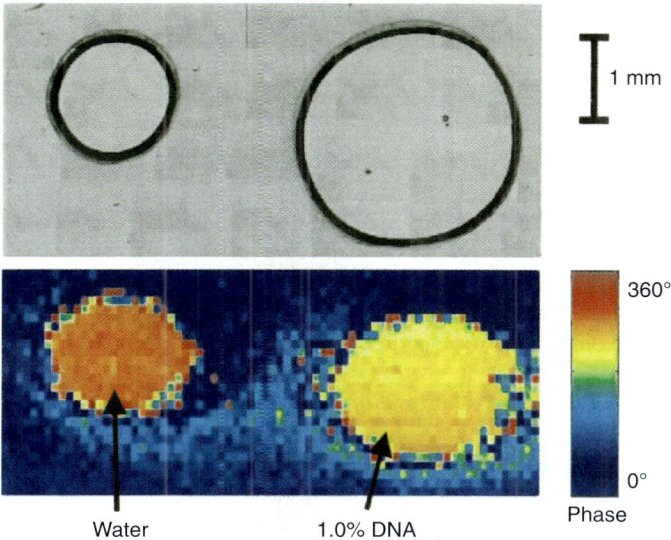

FIGURE 7.15 Comparison of images from (top) PCM and (bottom) PDM of a drop of water and a drop of 1.0% DNA solution sandwiched between two coverslips. The refractive-index dispersion. $\Delta n_{400\,nm} - \Delta n_{800\,nm}$ of the DNA solution is $(1.3 +/- 0.2) \times 10^{-4}$. [Reproduced with permission from C. H. Yang, A. Wax, I. Georgakoudi, E. B. Hanlon, K. Badizadegan, R. R. Dasari and M. S. Feld, Interferometric phase-dispersion microscopy, *Opt. Lett.*, 25, 1526–1528 (2000). Copyright Optical Society of America 2000.]

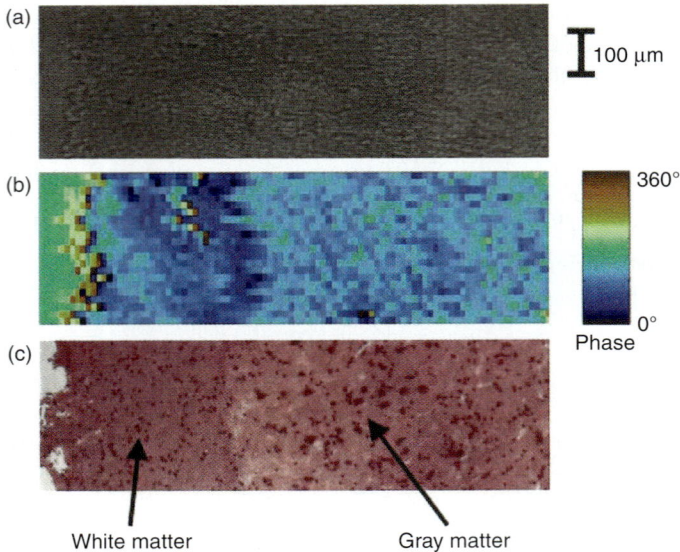

FIGURE 7.16   Images of a white matter–gray matter interface in a 16-mm-thick brain sample. (*a*) PCM image; (*b*) PDM image; (*c*) an adjacent frozen section stained with hemotoxylin and eosin. [Reproduced with permission from C. H. Yang, A. Wax, I. Georgakoudi, E. B. Hanlon, K. Badizadegan, R. R. Dasari and M. S. Feld, Interferometric phase-dispersion microscopy, *Opt. Lett.*, 25, 1526–1528 (2000). Copyright Optical Society of America 2000.]

While the droplets appear indistinguishable under phase contrast microscopy, which is sensitive only to the wavelength-averaged, or group refractive index, this dispersion-resolved measurement is sensitive enough to tell them apart.

Further, this method was applied to imaging brain slices, as shown in Fig. 7.16.[13] While this method uses dispersion as an efficient way to cancel the phase noise and cannot retrieve stable phase information at a single wavelength, it demonstrates that phase dispersion can be used as an intrinsic marker for chemical specificity. Later, Yang et al. exploited this principle further, in combination with OCT depth sectioning, to allow tomographic measurements.[14]

7.6   Quantitative Methods

Here we present methods that advanced the QPI field by providing both *stable* and *quantitative* phase measurements. The essence of these measurements is to create a reference field that does not pass through the sample. This approach contrasts with the techniques presented in Sec. 7.5, in which both interfering beams interacted with the specimen.

7.6.1   Phase-Referenced Interferometry

In 2001, Yang et al., with Massachusetts Institute of Technology, demonstrated "phase-referenced interferometer with subwavelength and subhertz sensitivity applied to the study of cell membrane dynamics."[15] This approach builds on the previous method of stabilizing the measurement by using two harmonically related wavelengths (Sec. 7.5.2). The novelty here is that now one of the beams is reflected by the bottom of the coverslip supporting the specimen and, thus, is not affected by the sample, which allows for *quantitative* phase measurements.

The experimental geometry is shown in Fig. 7.17.[15] A Michelson interferometer is illuminated via a composite low-coherence 775 nm (Ti: Saph laser) and CW 1550 nm (superluminescent diode). The CW light (1550 nm) is focused in such a way as to maximize the reflection at the coverslip-air interface. The low-coherence (LC) field is passing into the sample and carries the phase information of interest. The two (noisy) phase signals have the form

$$\phi_{CW} = k_{CW} \cdot \Delta x$$
$$\phi_{LC} = k_{LC} \cdot \Delta x + \phi_S$$

$$(7.37)$$

where, like in Sec. 7.5.2, Δx is the noise jitter, $k_{LC} = 2k_{CW}$, and ϕ_S is the phase shift due to the specimen and the quantity of interest. Note that

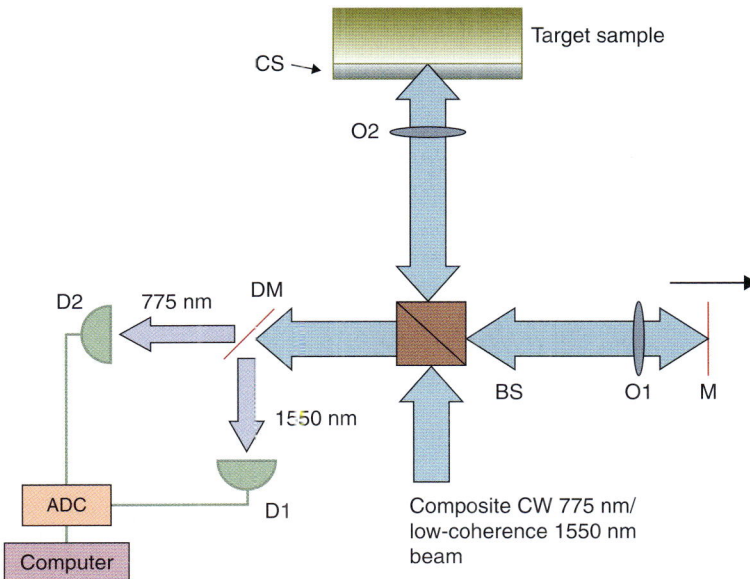

FIGURE **7.17**   Experimental setup: M, reference mirror; BS, beam splitter; O1, O2, microscope objectives; D1, D2, photodetectors; DM, 775–1550-nm dichroic mirror; CS, coverslip; ADC, analog–digital converter. [Reproduced with permission from C. Yang, A. Wax, M. S. Hahn, K. Badizadegan, R. R. Dasari and M. S. Feld, Phase-referenced interferometer with subwavelength and subhertz sensitivity applied to the study of cell membrane dynamics, *Opt. Lett.*, 26, 1271–1273 (2001). Copyright Optical Society of America 2001. (Ref. 15).]

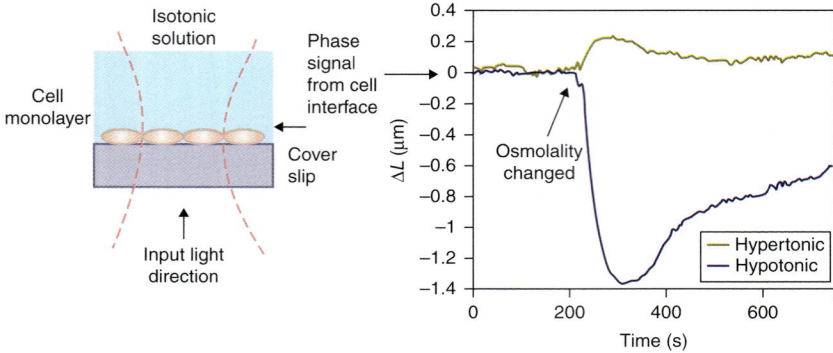

FIGURE 7.18   Phase-referenced interferometry analysis of a live HT29 cell monolayer grown on a coverslip. The graph shows changes in cell thickness when the osmolality is changed from its normal value to 85% (hypotonic) and 115% (hypertonic) at $t = 230$ s. The dashed curves indicate the beam profile. [Reproduced with permission from C. Yang, A. Wax, M. S. Hahn, K. Badizadegan, R. R. Dasari and M. S. Feld, Phase-referenced interferometer with subwavelength and subhertz sensitivity applied to the study of cell membrane dynamics, *Opt. Lett.*, 26, 1271–1273 (2001). Copyright Optical Society of America 2001.]

the main difference with respect to the dispersion measurement is that now the phase information associated with the specimen is present in only one of the beams (LC in this case). Thus, ϕ_S is obtained with high stability following a simple subtraction,

$$\phi_S = \phi_{LC} - 2\phi_{CW} \qquad (7.38)$$

This *phase referencing* provides long-term stability and enables sensitive measurements of slow biological processes. Figure 7.18 shows a demonstration of this capability, where hypotonic and hypertonic responses of live cells are measured.[15] Despite being limited to a single point measurement and slow acquisition rates, this method demonstrated that phase-resolved measurements can enable new biological studies. Further advances in frequency domain OCT methods provided an avenue for much faster measurements, as described below.

7.6.2   Spectral-Domain QPI

In 2005, two independent groups reported on using spectral domain OCT to achieve quantitative phase imaging: "Spectral-domain Phase Microscopy" by Choma et al., with the Izatt group at Duke University,[16] and "Spectral-domain Optical Coherence Phase Microscopy for Quantitative Phase-Contrast Imaging" by Joo et al., with the de Boer group at Wellman Center for Photomedicine.[17]

The main idea is to combine the *single-shot* ability of spectral domain OCT with the *common-path* phase reference interferometry. Choma et al. demonstrated this idea with both a swept-source and Fourier domain interferometer.[16] These two experimental configurations

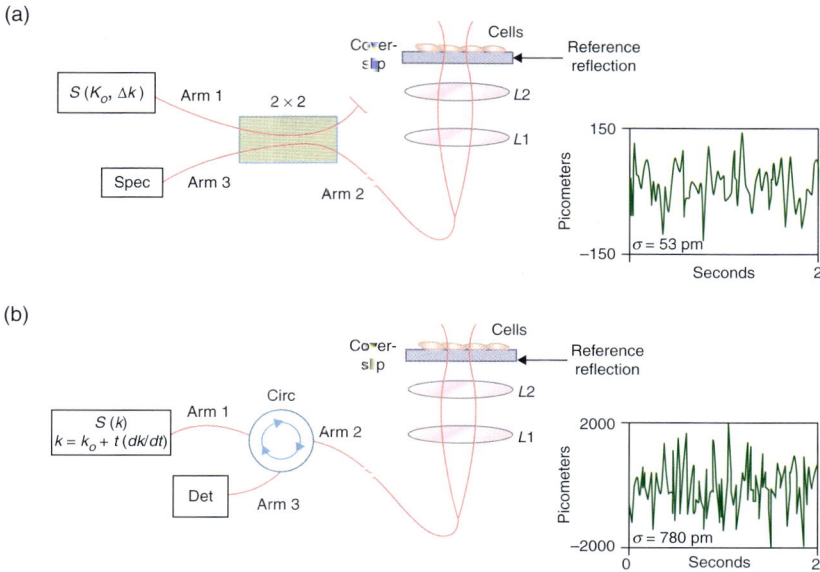

FIGURE 7.19 (a) Fourier-domain (FD) spectral-domain phase microscopy (SDPM) interferometer. The source is a 5 mW with a center wavelength and a 3-dB bandwidth of 830 and 45 nm, respectively. The spectrometer (spec) has a 25-ms readout rate and a 5-ms integration time. (b) Swept-source (SS) SDPM interferometer. The narrow-linewidth source is swept through a 130-nm bandwidth over 5 ms with a center wavelength of 1310 nm and an average power of 3 mW. The insets show the displacement signals recorded from a clean coverslip. [Reproduced with permission from M. A. Choma, A. K. Ellerbee, C. H. Yang, T. L. Creazzo and J. A. Izatt, Spectral-domain phase microscopy, *Opt. Lett.*, 30, 1162–1164 (2005). Copyright Optical Society of America 2005.]

are shown in Fig. 7.19.[16] The reference field is provided by the specular reflection from the bottom of the coverslip supporting the specimen, as discussed in the previous section.

In Sec. 7.4, we found that the spectral-domain signal contains the phase information of interest in a single shot,

$$I(\omega) \propto U_R(\omega) \cdot U_S(\omega) \cdot \cos[\omega s_0/c + \omega \delta s/c] \qquad (7.39)$$

In Eq. (7.39), U_R and U_S are reference and sample fields, s_0 is the fixed mismatch in pathlength between the two fields, and δs is the pathlength shift due to the sample. The remarkable stability of the system allows for detecting nanometer scale changes in a coverslip thickness as it cools by $1.2°C$[16] (see Fig. 7.20).

Using a similar geometry, Joo et al. demonstrated 25-pm pathlength stability and imaged human epithelial chick cells with high transverse resolution[17] (see Fig. 7.21). Subcellular structures are clearly revealed by this phase image, which required an acquisition time of 3.6 seconds.

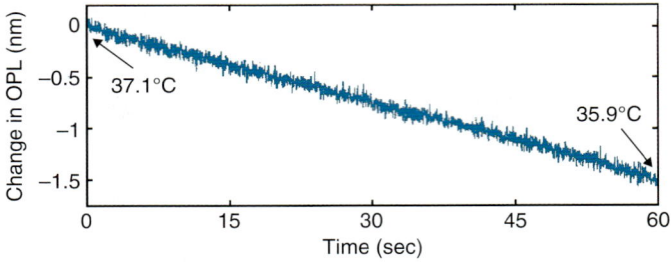

FIGURE 7.20 Change in the optical pathlength (OPL) of a 213-µm borosilicate coverslip as the water bath is cooled by 1.2°C. [Reproduced with permission from M. A. Choma, A. K. Ellerbee, C. H. Yang, T. L. Creazzo and J. A. Izatt, Spectral-domain phase microscopy, *Opt. Lett.,* 30, 1162–1164 (2005). Copyright Optical Society of America 2005.]

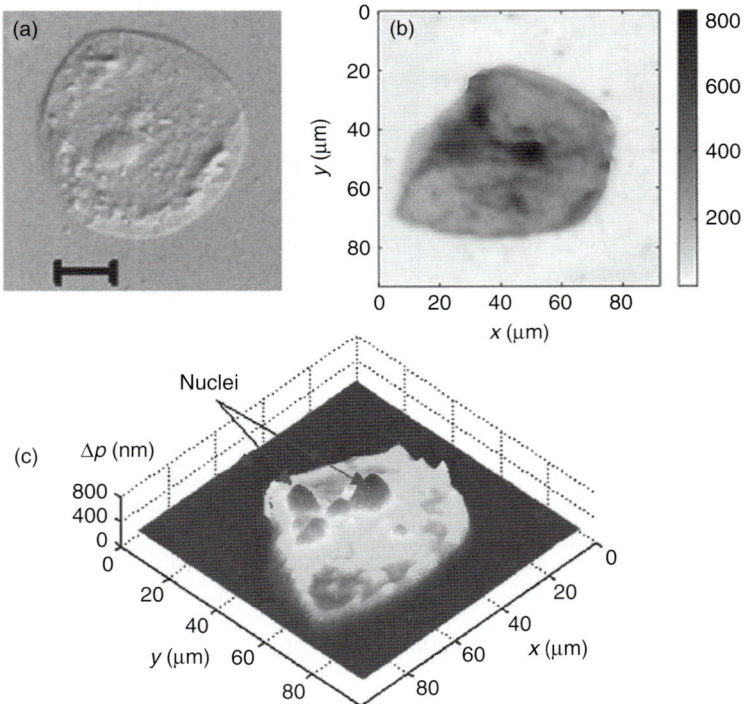

FIGURE 7.21 Images of human epithelial cheek cells. (*a*) Image recorded by a Nomarski microscope (10; N.A., 0.3); the bar represents 20 µm. (*b*) Spectral domain optical coherence microscopy (SD-OCPM) image, along with the gray scale denoting the OPL in nanometers. (*c*) Surface plot of (*b*), showing optically thick structures such as nuclei and subcellular structures in the cell. [Reproduced with permission from C. Joo, T. Akkin, B. Cense, B. H. Park and J. E. de Boer, Spectral-domain optical coherence phase microscopy for quantitative phase-contrast imaging, *Opt. Lett.,* 30, 2131–2133 (2005). Copyright Optical Society of America 2005.]

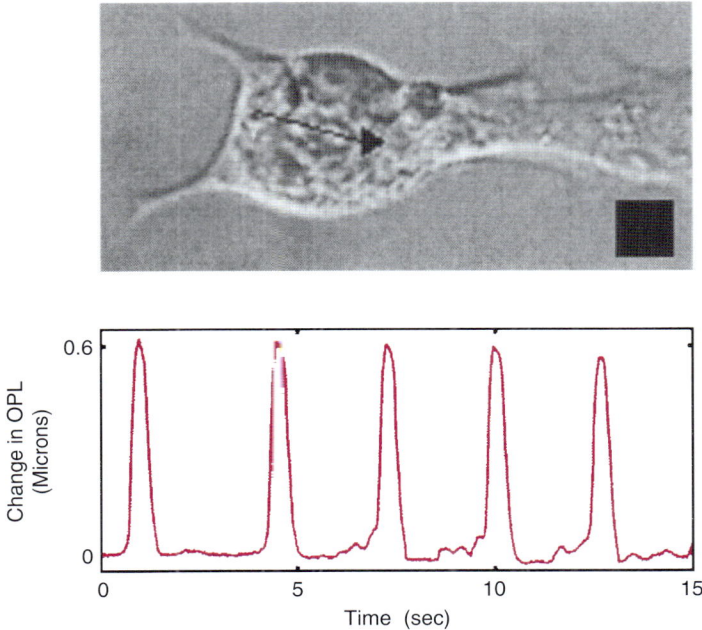

FIGURE 7.22   Top, photomicrograph of an isolated cardiomyocyte from a chick embryo. The tip of the arrow denotes the location of the beam. The box is 10 μm × 10 μm. Bottom, change in thickness of beating cardiomyocyte measured with spectral domain phase microscopy. [Reproduced with permission from M. A. Choma, A. K. Ellerbee, C. H. Yang, T. L. Creazzo and J. A. Izatt, Spectral-domain phase microscopy, *Opt. Lett.*, 30, 1162–1164 (2005). Copyright Optical Society of America 2005.]

This Fourier domain method was further applied by Choma et al. to study the dynamics of beating chick embryo cardiomyocytes.[16] The cell shown in Fig. 7.22 was beating spontaneously at 0.3 Hz, which is well resolved by the phase measurement.

Clearly, *spectral-domain methods*, i.e., *Fourier domain and swept source techniques*, offer important advantages over time-domain when performing phase imaging. While the common path referencing and raster scanning are not trivial to implement without loss of stability, this approach shows significant promise for further biological studies. Later, swept-source phase measurement was implemented in a full-field geometry, such that the raster scanning was eliminated.[18]

7.7   Further Developments

The last decade witnessed very active research in point-scanning phase-sensitive methods and their novel applications to biology.[14,16–84] Of course, it is impossible to describe all the techniques here. Instead, we presented the unifying principle of low-coherence interferometry (Sec. 7.1.), with both time-domain (Sec. 7.2.) and frequency-domain implementations (Sec. 7.3.). The various phase-referencing techniques that allow for QPI share the main ideas described in Secs. 7.5. and 7.6.

More recently, the applications of point-scanning QPI methods have diversified and gone progressively deeper into biology. Many applications deal with quantifying flow velocities via phase-sensitive Doppler OCT.[20,24,25,30,40,45,52,55,72,77,78] Several point-measurement techniques have been applied for investigating the structure and dynamics of live cells.[12,13,15–17,85,86] Polarization-sensitive OCT was used to quantify phase retardation in the retinal nerve fiber.[87] Electrokinetic[88] and thermorefractive[89] properties of tissue and tissue phantoms have been measured by differential phase OCT. Phase-sensitive OCT-type measurements have also been performed for studying static cells[12] and for monitoring electric activity in nerves.[85,86] Spectral-domain phase microscopy was successfully used to measure flows in the cytoplasm of live cells[45] and asses cytoskeleton microrheology.[56]

References

1. D. Huang, E. A. Swanson, C. P. Lin, J. S. Schuman, W. G. Stinson, W. Chang, M. R. Hee, T. Flotte, K. Gregory, C. A. Puliafito, and J. G. Fujimoto. "Optical coherence tomography." *Science,* 254, 1178–1181 (1991).
2. E. Wolf. "New theory of partial coherence in the space-frequency domain .1. spectra and cross spectra of steady-state sources." *Journal of the Optical Society of America,* 72, 343–351 (1982).
3. A. F. Fercher, C. K. Hitzenberger, G. Kamp, and S. Y. Elzaiat. "Measurement of intraocular distances by backscattering spectral interferometry." *Optics Communications,* 117, 43–48 (1995).
4. D. C. Adler, Y. Chen, R. Huber, J. Schmitt, J. Connolly, and J. G. Fujimoto, "Three-dimensional endomicroscopy using optical coherence tomography." *Nature Photonics,* 1, 709–716 (2007).
5. J. A. Izatt, M. R. Hee, G. M. Owen, E. A. Swanson, and J. G. Fujimoto. "Optical coherence microscopy in scattering media." *Optics Letters,* 19, 590–592 (1994).
6. J. F. deBoer, T. E. Milner, M. J. C. vanGemert, and J. S. Nelson, "Two-dimensional birefringence imaging in biological tissue by polarization-sensitive optical coherence tomography." *Optics Letters,* 22, 934–936 (1997).
7. J. A. Izatt, M. D. Kulkami, S. Yazdanfar, J. K. Barton, and A. J. Welch. "In vivo bidirectional color Doppler flow imaging of picoliter blood volumes using optical coherence tomograghy." *Optics Letters,* 22, 1439–1441 (1997).
8. C. K. Hitzenberger and A. F. Fercher. "Differential phase contrast in optical coherence tomography." *Optics Letters,* 24, 622–624 (1999).
9. M. Pluta. *Advanced Light Microscopy.* (PWN; Elsevier: Distribution for the USA and Canada, Elsevier Science Publishing Co., Warszawa Amsterdam; New York, 1988).
10. D. P. Dave and T. E. Milner. "Optical low-coherence reflectometer for differential phase measurement." *Optics Letters,* 25, 227–229 (2000).
11. M. Sticker, M. Pircher, E. Gotzinger, H. Sattmann, A. F. Fercher, and C. K. Hitzenberger. "En face imaging of single cell layers by differential phase-contrast optical coherence microscopy." *Optics Letters,* 27, 1126–1128 (2002).
12. C. G. Rylander, D. P. Dave, T. Akkin, T. E. Milner, K. R. Diller, and A. J. Welch, "Quantitative phase-contrast imaging of cells with phase-sensitive optical coherence microscopy," *Optics Letters,* 29, 1509–1511 (2004).

13. C. H. Yang, A. Wax, I. Georgakoudi, E. B. Hanlon, K. Badizadegan, R. R. Dasari, and M. S. Feld, "Interferometric phase-dispersion microscopy." *Optics Letters,* 25, 1526–1528 (2000).

14. C. H. Yang, A. Wax, R. R. Dasari, and M. S. Feld. "Phase-dispersion optical tomography." *Optics Letters,* 26, 686–688 (2001).

15. C. Yang, A. Wax, M. S. Hahn, K. Badizadegan, R. R. Dasari, and M. S. Feld, "Phase-referenced interferometer with subwavelength and subhertz sensitivity Appl. to the study of cell membrane dynamics." *Optics Letters,* 26, 1271–1273 (2001).

16. M. A. Choma, A. K. Ellerbee, C. H. Yang, T. L. Creazzo, and J. A. Izatt, "Spectral-domain phase microscopy." *Optics Letters,* 30, 1162–1164 (2005).

17. C. Joo, T. Akkin, B. Cense, B. H. Park, and J. F. de Boer. "Spectral-domain optical coherence phase microscopy for quantitative phase-contrast imaging." *Optics Letters,* 30, 2131–2133 (2005)

18. M. V. Sarunic, S. Weinberg, and J. A. Izatt. "Full-field swept-source phase microscopy." *Optics Letters,* 31, 1462–1464 (2006).

19. C. K. Hitzenberger and A. F. Fercher. "Differential phase contrast in optical coherence tomography." *Optics Letters,* 24, 622–624 (1999).

20. Y. H. Zhao, Z. P. Chen, C. Saxer, S. H. Xiang, J. F. de Boer, and J. S. Nelson. "Phase-resolved optical coherence tomography and optical Doppler tomography for imaging blood flow in human skin with fast scanning speed and high velocity sensitivity." *Optics Letters,* 25, 114–116 (2000).

21. C. K. Hitzenberger, E. Gotzinger, M. Sticker, M. Pircher, and A. F. Fercher. "Measurement and imaging of birefringence and optic axis orientation by phase resolved polarization sensitive optical coherence tomography." *Optics Express,* 9, 780–790 (2001).

22. C. K. Hitzenberger, M. Sticker, R. Leitgeb, and A. F. Fercher. "Differential phase measurements in low-coherence interferometry without 2 pi ambiguity." *Optics Letters,* 26, 1864–1866 (2001).

23. M. Sticker, C. K. Hitzenberger, R. Leitgeb, and A. F. Fercher. "Quantitative differential phase measurement and imaging in transparent and turbid media by optical coherence tomography." *Optics Letters,* 26, 518–520 (2001).

24. Z. H. Ding, Y. H. Zhao, H. W. Ren, J. S. Nelson, and Z. P. Chen. "Real-time phase-resolved optical coherence tomography and optical Doppler tomography." *Optics Express,* 10, 236–245 (2002).

25. H. W. Ren, K. M. Brecke, Z. H. Ding, Y. H. Zhao, J. S. Nelson, and Z. P. Chen. "Imaging and quantifying transverse flow velocity with the Doppler bandwidth in a phase-resolved functional optical coherence tomography." *Optics Letters,* 27, 409–411 (2002).

26. H. W. Ren, Z. H. Ding, Y. H. Zhao, J. J. Miao, J. S. Nelson, and Z. P. Chen. "Phase-resolved functional optical coherence tomography: simultaneous imaging of in situ tissue structure, blood flow velocity, standard deviation, birefringence, and Stokes vectors in human skin." *Optics Letters,* 27, 1702–1704 (2002).

27. K. Takada. "Phase error measurement of an arrayed-waveguide grating in the 1.3-mu m wavelength region by optical low coherence interferometry." *IEEE Photonics Technology Letters,* 14, 965–967 (2002).

28. R. A. Leitgeb, C. K. Hitzenberger, A. F. Fercher, and T. Bajraszewski. "Phase-shifting algorithm to achieve high-speed long-depth-range probing by frequency-domain optical coherence tomography." *Optics Letters,* 28, 2201–2203 (2003).

29. M. Sato, J. Iwasaki, T. Ohotaki, Y. Hashimoto, and N. Tanno. "New phase-shifting method using two images." *Optical Review,* 10, 456–461 (2003).

30. T. Q. Xie, Z. G. Wang, and Y. T. Pan. "High-speed optical coherence tomography using fiberoptic acousto-optic phase modulation." *Optics Express,* 11, 3210–3219 (2003).

31. E. Gotzinger, M. Pircher, M. Sticker, A. F. Fercher, and C. K. Hitzenberger. "Measurement and imaging of birefringent properties of the human cornea with phase-resolved, polarization-sensitive optical coherence tomography." *Journal of Biomedical Optics,* 9, 94–102 (2004).

32. G. Lamouche, M. L. Dufour, B. Gauthier, and J. P. Monchalin. "Gouy phase anomaly in optical coherence tomography." *Optics Communications,* 239, 297–301 (2004).

33. M. Pircher, E. Goetzinger, R. Leitgeb, and C. K. Hitzenberger. "Transversal phase resolved polarization sensitive optical coherence tomography." *Phys Med Biol*, 49, 1257–1263 (2004).

34. K. Takada and S. Satoh. "Measurement of slowly varying component in phase error distribution of a large-channel-spacing arrayed-waveguide grating." *Electronics Letters*, 40, 1486–1487 (2004).

35. S. A. Telenkov, D. P. Dave, S. Sethuraman, T. Akkin, and T. E. Milner. "Differential phase optical coherence probe for depth-resolved detection of photothermal response in tissue," *Phys. Med. Biol.*, 49, 111–119 (2004).

36. L. Wang, W. Xu, M. Bachman, G. P. Li, and Z. P. Chen. "Phase-resolved optical Doppler tomography for imaging flow dynamics in microfluidic channels." *Applied Physics Letters*, 85, 1855–1857 (2004).

37. L. Wang, W. Xu, M. Bachman, G. P. Li, and Z. P. Chen. Imaging and quantifying of microflow by phase-resolved optical Doppler tomography." *Optics Communications*, 232, 25–29 (2004).

38. Y. Yasuno, S. Makita, T. Endo, G. Aoki, H. Sumimura, M. Itoh, and T. Yatagai, "One-shot-phase-shifting Fourier domain optical coherence tomography by reference wavefront tilting." *Optics Express*, 12, 6184–6191 (2004).

39. J. I. Youn, T. Akkin, and T. E. Milner. "Electrokinetic measurement of cartilage using differential phase optical coherence tomography," *Physiol. Meas.*, 25, 85–95 (2004).

40. B. J. Vakoc, S. H. Yun, J. F. de Boer, G. J. Tearney, and B. E. Bouma. "Phase-resolved optical frequency domain imaging." *Optics Express*, 13, 5483–5493 (2005).

41. Y. Watanabe, Y. Hayasaka, M. Sato, and N. Tanno. "Full-field optical coherence tomography by achromatic phase shifting with a rotating polarizer." *Applied Optics*, 44, 1387–1392 (2005).

42. J. I. Youn, G. Vargas, B. J. F. Wong, and T. E. Milner. "Depth-resolved phase retardation measurements for laser-assisted non-ablative cartilage reshaping." *Physics in Medicine and Biology*, 50, 1937–1950 (2005).

43. J. Zhang, J. S. Nelson, and Z. P. Chen. "Removal of a mirror image and enhancement of the signal-to-noise ratio in Fourier-domain optical coherence tomography using an electro-optic phase modulator." *Optics Letters*, 30, 147–149 (2005).

44. Y. C. Ahn, W. Y. Jung, and Z. P. Chen. "Turbid two-phase slug flow in a microtube: Simultaneous visualization of structure and velocity field." *Applied Physics Letters*, 89, (2006).

45. M. A. Choma, A. K. Ellerbee, S. Yazdanfar, and J. A. Izatt. "Doppler flow imaging of cytoplasmic streaming using spectral domain phase microscopy." *Journal of Biomedical Optics*, 11, (2006).

46. M. H. De la Torre-Ibarra, P. D. Ruiz, and J. M. Huntley. "Double-shot depth-resolved displacement field measurement using phase-contrast spectral optical coherence tomography." *Optics Express*, 14, 9643–9656 (2006).

47. B. I. Erkmen and J. H. Shapiro. "Phase-conjugate optical coherence tomography." *Physical Review A*, 74, (2006).

48. J. Kim, J. Oh, and T. E. Milner. "Measurement of optical path length change following pulsed laser irradiation using differential phase optical coherence tomography." *Journal of Biomedical Optics*, 11, (2006).

49. Z. H. Ma, R. K. K. Wang, F. Zhang, and J. Q. Yao. "Arbitrary three-phase shifting algorithm for achieving full range spectral optical coherence tomography." *Chinese Physics Letters*, 23, 366–369 (2006).

50. A. Ozcan, M. J. F. Digonnet, and G. S. Kino. "Minimum-phase-function-based processing in frequency-domain optical coherence tomography systems." *Journal of the Optical Society of America a-Optics Image Science and Vision*, 23, 1669–1677 (2006).

51. J. L. Qu and H. B. Niu. "Study of reconstruction algorithms for phase-stepped full-field optical coherence tomography." *Japanese Journal of Applied Physics Part 1-Regular Papers Brief Communications & Review Papers*, 45, 4256–4258 (2006).

52. D. C. Adler, R. Huber, and J. G. Fujimoto. "Phase-sensitive optical coherence tomography at up to 370,000 lines per second using buffered Fourier domain mode-locked lasers." *Optics Letters*, 32, 626–628 (2007).

53. B. Baumann, M. Pircher, E. Gotzinger, and C. K. Hitzenberger. "Full range complex spectral domain optical coherence tomography without additional phase shifters." *Optics Express*, 15, 13375–13387 (2007).
54. P. Bu, X. Z. Wang, and O. Sasaki. "Full-range parallel Fourier-domain optical coherence tomography using sinusoidal phase-modulating interferometry." *Journal of Optics a-Pure and Applied Optics*, 9, 422–426 (2007).
55. J. Fingler, D. Schwartz, C. H. Yang, and S. E. Fraser. "Mobility and transverse flow visualization using phase variance contrast with spectral domain optical coherence tomography." *Optics Express*, 15, 12636–12653 (2007).
56. E. J. McDowell, A. K. Ellerbee, M. A. Choma, B. E. Applegate, and J. A. Izatt. "Spectral domain phase microscopy for local measurements of cytoskeletal rheology in single cells." *Journal of Biomedical Optics*, 12, (2007).
57. E. J. McDowell, M. V. Sarunic, Z. Yaqoob, and C. H. Yang. "SNR enhancement through phase dependent signal reconstruction algorithms for phase separated interferometric signals". *Optics Express*, 15, 10103–10122 (2007).
58. J. Oh, M. D. Feldman, J. Kim, H. W. Kang, P. Sanghi, and T. E. Milner. "Magneto-motive detection of tissue-based macrophages by differential phase optical coherence tomography." *Lasers in Surgery and Medicine*, 39, 266–272 (2007).
59. P. G. Smith, M. N. Patel, J. Kim, K. P. Johnston, and T. E. Milner, "Electrophoretic mobility measurement by differential-phase optical coherence tomography." *Journal of Physical Chemistry C*, 111, 2614–2622 (2007).
60. Y. K. Tao, M. Zhao, and J. A. Izatt. "High-speed complex conjugate resolved retinal spectral domain optical coherence tomography using sinusoidal phase modulation." *Optics Letters*, 32, 2918–2920 (2007).
61. P. H. Tomlins and R. K. Wang. "Digital phase stabilization to improve detection sensitivity for optical coherence tomography." *Measurement Science & Technology*, 18, 3365–3372 (2007).
62. R. K. K. Wang, S. Kirkpatrick, and M. Hinds. "Phase-sensitive optical coherence elastography for mapping tissue microstrains in real time." *Applied Physics Letters*, 90, (2007).
63. D. C. Adler, S. W. Huang, R. Huber, and J. G. Fujimoto. "Photothermal detection of gold nanoparticles using phase-sensitive optical coherence tomography." *Optics Express*, 16, 4376–4393 (2008).
64. J. Fingler, C. Readhead, D. M. Schwartz, and S. E. Fraser. "Phase-Contrast OCT Imaging of Transverse Flows in the Mouse Retina and Choroid." *Investigative Ophthalmology & Visual Science*, 49, 5055–5059 (2008).
65. S. M. R. M. Nezam, C. Joo, G. J. Tearney, and J. F. de Boer. "Application of maximum likelihood estimator in nano-scale optical path length measurement using spectral-domain optical coherence phase microscopy." *Optics Express*, 16, 17186–17195 (2008).
66. J. W. Oh, M. D. Feldman, J. Kim, P. Sanghi, D. Do, J. J. Mancuso, N. Kemp, M. Cilingiroglu, and T. E. Milner. "Detection of macrophages in atherosclerotic tissue using magnetic nanoparticles and differential phase optical coherence tomography." *Journal of Biomedical Optics*, 13, (2008).
67. A. L. Oldenburg, V. Crecea, S. A. Rinne, and S. A. Boppart. "Phase-resolved magnetomotive OCT for imaging nanomolar concentrations of magnetic nanoparticles in tissues." *Optics Express*, 16, 11525–11539 (2008).
68. A. Szkulmowska, M. Szkulmowski, A. Kowalczyk, and M. Wojtkowski. "Phase-resolved Doppler optical coherence tomography—limitations and improvements." *Optics Letters*, 33, 1425–1427 (2008).
69. M. Yamanari, M. Miura, S. Makita, T. Yatagai, and Y. Yasuno. "Phase retardation measurement of retinal nerve fiber layer by polarization-sensitive spectral-domain optical coherence tomography and scanning laser polarimetry." *Journal of Biomedical Optics*, 13, (2008).
70. T. Akkin, D. Landowne, and A. Sivaprakasam. "Optical coherence tomography phase measurement of trans ent changes in squid giant axons during activity." *Journal of Membrane Biology*, 231, 35–46 (2009).
71. H. C. Cheng, J. F. Huang, Y. C. Liu, C. W. Chang, and Y. T. Chang. "Group-delay-based phase-shifting method for Fourier domain optical coherence tomography." *Optical Engineering*, 48, (2009).

72. H. C. Hendargo, M. T. Zhao, N. Shepherd, and J. A. Izatt. "Synthetic wavelength based phase unwrapping in spectral domain optical coherence tomography." *Optics Express*, 17, 5039–5051 (2009).

73. M. H. D. Ibarra, P. D. Ruiz, and J. M. Huntley. "Simultaneous measurement of in-plane and out-of-plane displacement fields in scattering media using phase-contrast spectral optical coherence tomography." *Optics Letters*, 34, 806–808 (2009).

74. M. Lesaffre, S. Farahi, M. Gross, P. Delaye, A. C. Boccara, and F. Ramaz. "Acousto-optical coherence tomography using random phase jumps on ultra-sound and light." *Optics Express*, 17, 18211–18218 (2009).

75. R. K. Manapuram, V. G. R. Manne and K. V. Larin. "Phase-sensitive swept source optical coherence tomography for imaging and quantifying of microbubbles in clear and scattering media." *Journal of Applied Physics*, 105, (2009).

76. M. Pircher, B. Baumann, E. Gotzinger, H. Sattmann, and C. K. Hitzenberger. "Phase contrast coherence microscopy based on transverse scanning." *Optics Letters*, 34, 1750–1752 (2009).

77. B. J. Vakoc, G. J. Tearney, and B. E. Bouma. "Statistical Properties of Phase-Decorrelation in Phase-Resolved Doppler Optical Coherence Tomography." *IEEE Transactions on Medical Imaging*, 28, 814–821 (2009).

78. J. Walther, G. Mueller, H. Morawietz, and E. Koch. "Analysis of in vitro and in vivo bidirectional flow velocities by phase-resolved Doppler Fourier-domain OCT." *Sensors and Actuators a-Physical*, 156, 14–21 (2009).

79. M. Yamanari, Y. Lim, S. Makita, and Y. Yasuno. "Visualization of phase retardation of deep posterior eye by polarization-sensitive swept-source optical coherence tomography with 1-mu m probe." *Optics Express*, 17, 12385–12396 (2009).

80. Z. Yaqoob, W. Choi, S. Oh, N. Lue, Y. Park, C. Fang-Yen, R. R. Dasari, K. Badizadegan, and M. S. Feld. "Improved phase sensitivity in spectral domain phase microscopy using line-field illumination and self phase-referencing." *Optics Express*, 17, 10681–10687 (2009).

81. J. Zhang, B. Rao, L. F. Yu, and Z. P. Chen. "High-dynamic-range quantitative phase imaging with spectral domain phase microscopy." *Optics Letters*, 34, 3442–3444 (2009).

82. W. C. Kuo, C. Y. Chuang, M. Y. Chou, W. H. Huang, and S. T. Cheng. "Phase detection with sub-nanometer sensitivity using polarization quadrature encoding method in optical coherence tomography." *Progress in Electromagnetics Research-Pier*, 104, 297–311 (2010).

83. J. Le Gouet, D. Venkatraman, F. N. C. Wong, and J. H. Shapiro. "Experimental realization of phase-conjugate optical coherence tomography." *Optics Letters*, 35, 1001–1003 (2010).

84. Y. L. Yang, Z. H. Ding, K. Wang, L. Wu, and L. Wu. "Full-field optical coherence tomography by achromatic phase shifting with a rotating half-wave plate." *Journal of Optics*, 12, (2010).

85. C. Fang-Yen, M. C. Chu, H. S. Seung, R. R. Dasari, and M. S. Feld. "Noncontact measurement of nerve displacement during action potential with a dual-beam low-coherence interferometer." *Optics Letters*, 29, 2028–2030 (2004).

86. T. Akkin, D. P. Dave, T. E. Milner, and H. G. Rylander. "Detection of neural activity using phase-sensitive optical low-coherence reflectometry." *Optics Express*, 12, 2377–2386 (2004).

87. J. Park, N. J. Kemp, T. E. Milner, and H. G. Rylander. "Analysis of the phase retardation in the retinal nerve fiber layer of cynomolus monkey by polarization sensitive optical coherence tomography." *Lasers Surg. Med.*, 55–55 (2003).

88. J. I. Youn, T. Akkin, B. J. F. Wong, G. M. Peavy, and T. E. Milner. "Electrokinetic measurements of cartilage using differential phase optical coherence tomography." *Lasers Surg. Med.*, 25, 85–95 (2004).

89. J. Kim, S. A. Telenkov, and T. E. Milner. "Measurement of thermo-refractive and thermo-elastic changes in a tissue phantom using differential phase optical coherence tomography." *Lasers Surg. Med.*, 8–8 (2004).

CHAPTER 8

Principles of Full-Field QPI

8.1 Interferometric Imaging

Quantitative phase imaging deals with measuring the phase shift produced by the specimen at each point within the field of view. Typically, an imaging system is used to create a magnified image of the specimen, as illustrated in Fig. 8.1.

The image field can be expressed in space-time as

$$U_i(x,y;t) = |U_i(x,y)| \cdot e^{-i\left[\langle\omega\rangle t - \langle\mathbf{k}\rangle \cdot \mathbf{r} + \phi(x,y)\right]} \tag{8.1}$$

In Eq. (8.1), $\langle\omega\rangle$ is the mean frequency, $\langle\mathbf{k}\rangle$ the mean wavevector, and ϕ is the phase shift of interest. As described in Chap. 3 when discussing field correlations, for an arbitrary optical field, the frequency spread around $\langle\omega\rangle$ defines temporal coherence and the wavevector spread around $\langle\mathbf{k}\rangle$ characterizes the spatial coherence of the field. Clearly, if the image is recorded by the detector as is, only the *modulus squared* of the field is obtained, $|U_i(x,y)|^2$, and thus, the phase information is lost. However, if the image field is mixed (i.e., interfered) with another (reference) field, U_R, the resulting intensity retains information about the phase,

$$I(x,y) = |U_i(x,y)|^2 + |U_R|^2 + 2|U_R| \cdot |U_i(x,y)|$$
$$\cdot \cos\left[\langle\omega\rangle(t-t_R) - \left(\langle\mathbf{k}\rangle - \mathbf{k}_R\right)\cdot\mathbf{r} + \phi(x,y)\right] \tag{8.2}$$

In Eq. (8.2), we assume that the reference field can have both a delay, t_R, and a different direction of propagation along $\mathbf{k}_R$. It can be seen that measurements at different delays t_R or at different points across the image plane, $\mathbf{r}$, can provide enough information to extract ϕ.

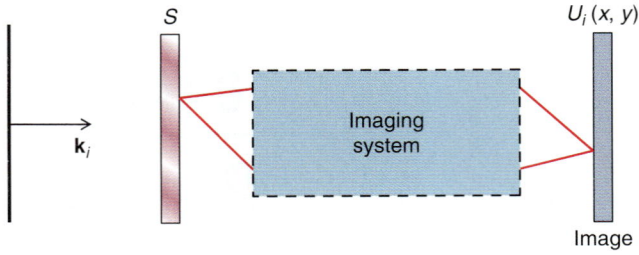

FIGURE 8.1   Generating a complex image field via coherent imaging.

Modulating the time delay is typically referred to as *phase-shifting interferometry*. Using a tilted reference beam is commonly called *off-axis (or shear) interferometry*, as already discussed in the context of holography (Chap. 6). In practice, these methods are not normally used simultaneously, but exceptions exist. Below, we discuss phase retrieval in each case separately.

8.2   Temporal Phase Modulation: Phase-Shifting Interferometry

Phase shifting has been used for a long time in interferometry as an effective means to extract phase information (for a review, see, for example, Ref. 1). The idea is to introduce a control over the phase difference between two interfering fields, such that the intensity of the resulting signal has the form (see Fig. 8.2)

$$I(\delta\phi) = I_1 + I_2 + 2\sqrt{I_1 I_2} \cdot \cos[\phi + \delta\phi] \tag{8.3}$$

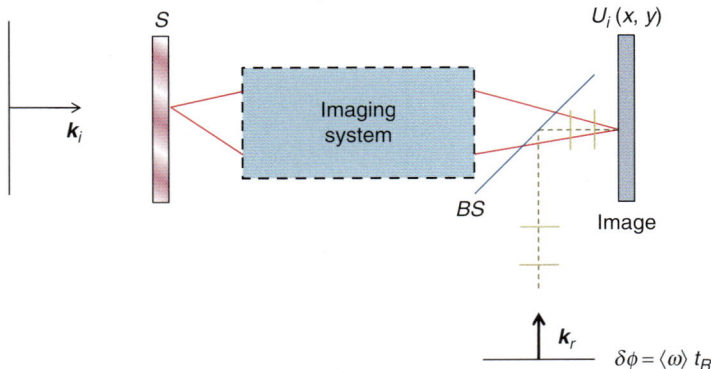

FIGURE 8.2   Schematic for phase shift interferometry: k_i and k_r incident and reference wavevectors, respectively, BS beam splitter.

where ϕ is the phase to be measured and $\delta\phi$ is the phase shift purposely added, $\delta\phi = \langle\omega\rangle(t - t_r)$. Note that, generally, Eq. (8.2) has three unknowns: the intensities of the two fields, I_1 and I_2, and their phase difference, ϕ. Thus, the minimum number of measurements (and, thus, phase shifts) required is three. However, this only provides the value of ϕ over one half of the trigonometric circle. This is due to the fact that the sine and cosine functions are *bijective* over only half the circle, $\left(-\pi/2, \pi/2\right)$ and $(0, \pi)$, respectively. Therefore, most commonly the phase-shifting methods use four phase shifts in increments of $\pi/2$, such that ϕ is retrieved uniquely over the entire circle, i.e., over the $(-\pi, \pi]$ interval, as follows,

$$\phi = \arg\left[I(0) - I(\pi), I\left(3\pi/2\right) - I\left(\pi/2\right)\right] \tag{8.4}$$

where $\arg(x, y)$ is the counterclockwise angle (in radians) from the positive x axis to the line that connects the origin and the point (x, y) (Fig. 8.3). Formally, function $\arg(x, y)$ is defined as

$$\arg(x, y) = \begin{vmatrix} \arctan\left(y/x\right), \text{if } x > 0 \\ \pi/2, \text{if } x = 0, \ y > 0 \\ -\pi/2, \text{if } x = 0, \ y < 0 \\ \pi + \arctan\left(y/x\right), \text{if } x < 0, \ y \geq 0 \\ -\pi + \arctan\left(y/x\right), \text{if } x < 0, \ y < 0 \end{vmatrix} \tag{8.5}$$

In some publications the inverse tangent function, *arctan* or *tan*$^{-1}$, is used instead of *arg*, e.g.. $\phi = \arctan\left\{[I(0) - I(\pi)] / \left[I\left(3\pi/2\right) - I\left(\pi/2\right)\right]\right\}$, but technically this is incorrect because the *arctan* returns values only within $\left(-\pi/2, \pi/2\right)$. In computer programming languages such as Matlab^{TM}, the two-argument function *atan2* is used as a variation of *atan* (short for arctan) that covers the entire trigonometric circle.

Variations on the principle of phase-shifting interferometry have been proposed, including "bucket integrations" and higher number of phase shifts (see, for example, Ref. 2). In Chap. 10, we discuss the main developments that use phase shifting to obtain quantitative phase imaging.

(a)

(b)

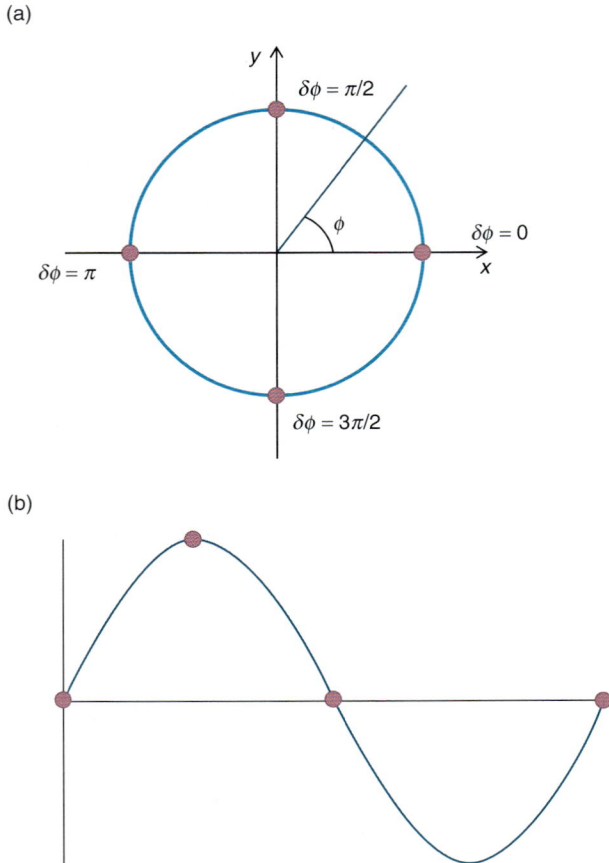

FIGURE 8.3 Phase shifting on the trigonometric circle (*a*) and on a sinusoidal function (*b*).

8.3 Spatial Phase Modulation: Off-Axis Interferometry

Off-axis interferometry takes advantage of the *spatial phase modulation* introduced by the *angularly shifted* reference plane wave and the spatially-resolved measurement allowed by a 2D detector array, such as a CCD (Fig. 8.4).

If the reference field wavevector is in the x-z plane (Fig. 8.4.), the intensity distribution of the interferogram at the detector plane takes the form

$$I(x,y) = |U_i(x,y)|^2 + |U_r|^2 + 2|U_r| \cdot |U_i(x,y)| \cdot \cos[\Delta k \cdot x + \phi(x,y)] \quad (8.6)$$

The goal is to isolate the $\cos[\Delta k \cdot x + \phi(x,y)]$ term from the measurement and then numerically compute its sine counterpart via a Hilbert transform. In order to achieve this, $|U_i|$ and $|U_r|$ can be independently

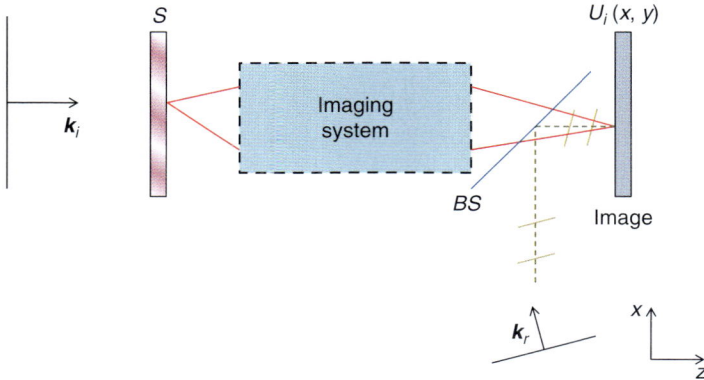

FIGURE 8.4 Schematic for off-axis interferometry: k_i and k_r incident and reference wavevectors, BS beam splitter.

measured by blocking one beam of the interferometer and measuring the resulting intensity of the other. Sometimes, for highly transparent specimens, the spatial dependence of the two amplitudes can be ignored, which eliminates the need for the amplitude measurement. As a result, the *cosine* term is obtained by itself, which can now be interpreted as the real part of a (spatial) complex analytic signal.

The corresponding imaginary part is further obtained via a Hilbert transform, as described in App. A,

$$\sin[\Delta k \cdot x + \phi(x,y)] = P\int \frac{\cos[\Delta k \cdot x' + \phi(x',y)]}{x - x'} dx' \qquad (8.7)$$

where P indicates the principle value integral. Computer programs such as Matlab have a built-in Hilbert transform command that performs this integral in an instant. Finally, the argument of the trigonometric functions is obtained uniquely as

$$\phi(x,y) + \Delta kx = \arg[\cos(\Delta kx + \phi), \sin(\Delta kx + \phi)] \qquad (8.8)$$

where the *arg* function was defined earlier in Eq. (8.5). Importantly, the frequency of modulation, Δk, sets an upper limit on the highest spatial frequency resolvable in an image. As with any other measured signals, the frequency of modulation must exceed that of the desired signal by a factor of at least 2 (Nyquist theorem). This adds constraints on the off-axis angle of the reference beam (i.e., it must be steep enough). Sometimes, this spatial modulation is the final limiting factor in terms of transverse resolution and not diffraction. Of course, this is not ideal as typically the goal of all QPI methods is to preserve the diffraction-limited transverse resolution. We will come back to this point later.

Note that Δkx can have values much higher than 2π, that is to say that the phase $\phi(x,y) + \Delta kx$ is highly *wrapped*. However, the value of Δk is known from the tilt of the reference beam, such that the phase image, $\phi(x,y)$, is obtained by subtracting Δkx from the result in Eq. (8.8). This operation amounts to subtracting a tilted plane, of slope Δk, from the measured surface $\phi(x,y) + \Delta kx$. Phase unwrapping is discussed further in the next section.

8.4   Phase Unwrapping

As already mentioned, the phase measurements yield values, at best, within the $(-\pi, \pi]$ interval and *modulo(2π)*. In other words, without further analysis, the phase measurements cannot distinguish between a certain phase value, say ϕ_0, and a value $\phi_0 + 2\pi$, or $\phi_0 + 4\pi$, etc. It is said that the measurement yields *wrapped phase* information. However, using prior knowledge regarding the continuity of the object under investigation, the measured phase signal can be *numerically* corrected to cover greater intervals, outside $(-\pi, \pi]$. This mathematical operation is called *phase unwrapping*. An illustration of this operation for a 1D signal is shown in Fig. 8.5.

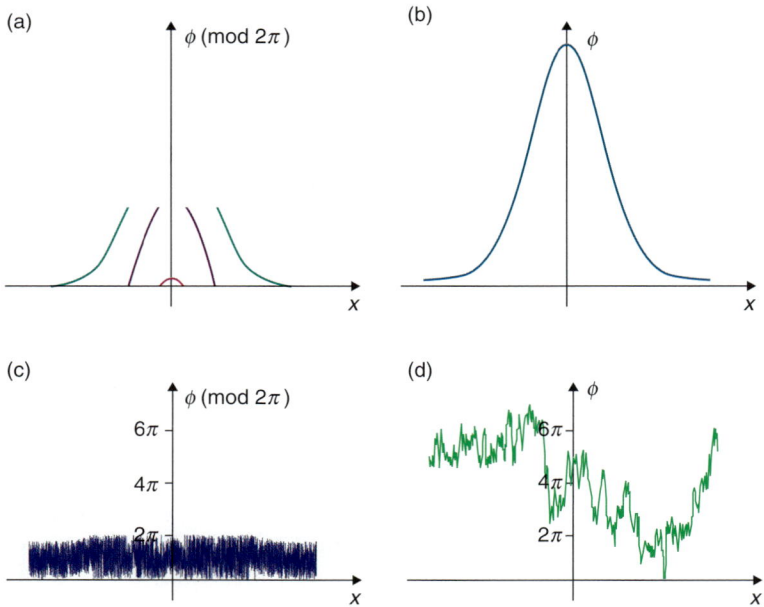

FIGURE 8.5   (*a*) Wrapped phase associated with a noise-free signal. (*b*) Unwrapped phase from (*a*). (*c*) Wrapped phase associated with a noisy signal. (*d*) Unwrapped phase from (*c*); the value may be affected by errors depending on the signal-to-noise ratio.

(a)

$|U_i(x, y)|^2$

(b)

$|U_i(x, y + U_r(x, y)|^2$

(c)

$\cos[\phi(x, y) + \Delta k(x + y)]$

(d)

$[\phi(x, y) + \Delta k(x + y)]_{\text{mod } 2\pi}$

(e)

$\phi(x, y) + \Delta k(x + y)$

(f)

$\phi(x, y)$

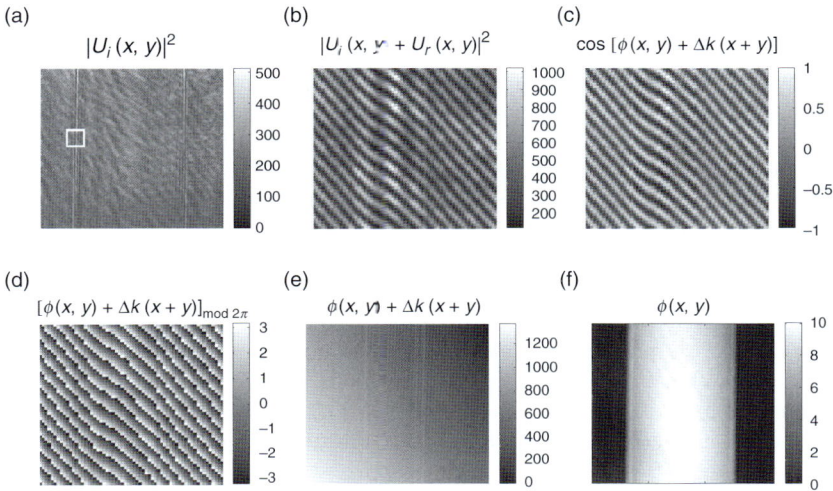

FIGURE 8.6 (a) Transmission intensity image of a transparent object (optical fiber); (b), (c), (d) interferogram, sinusoidal signal, and wrapped phase, respectively, measured from the white box indicated in (a); (e) full-field unwrapped phase, (f) full-field quantitative phase image. [Reproduced with permission from T. Ikeda, G. Popescu, R. R. Dasari and M. S. Feld, Hilbert phase microscopy for investigating fast dynamics in transparent systems Opt. Lett., 30, 1165–1168 (2005). Copyright Optical Society of America 2005.]

Mathematically, the unwrapping operation essentially searches for 2π jumps in the signal and corrects them by adding the 2π values back to the signal. The 1D problem is significantly simpler than the 2D. For example, Matlab™ has a standard command called "unwrap," which is very efficient and fast. However, complications occur with noisy signals, due to the measurement or the roughness of the object itself. Figures 8.5c and d illustrate the effects of the noise on the unwrapping process.

Various algorithms have been proposed for 2D unwrapping, which is a much more difficult problem than the 1D analog and, thus, still the subject of active research.[3] Figure 8.6 illustrates the 2D unwrapping.[4] As mentioned in Sec. 5.3, in off-axis interferometry the spatial-phase modulation introduced by the tilted reference beam, say $\Delta k \cdot x$, adds strong unwrapping (see Fig 8.6e). Again, the noise in the measurement can affect significantly the effectiveness of the unwrapping process.

8.5 Figures of Merit in QPI

Like all instruments, the QPI devices are characterized by certain parameters that quantify their performance. The main figures of merit are: acquisition rate, transverse resolution, temporal phase sensitivity, and spatial phase sensitivity.

8.5.1 Temporal Sampling: Acquisition Rate

Acquisition rate establishes the fastest phenomena that can be studied by a QPI method. According to the Nyquist sampling theorem (or Nyquist-Shannon theorem), the sampling frequency has to be at least twice the frequency of the signal of interest.[5,6] In QPI, acquisition rates vary broadly with the application, from >100 Hz in the case of membrane fluctuations to <1 mHz when studying the cell cycle. The maximum acquisition rate of QPI systems depends on the modality used for phase retrieval. Thus, off-axis interferometry allows for "single shot" measurements, i.e., obtaining the phase map from a single camera exposure. Of course, in this case the acquisition rate is only limited by the camera itself. Today's technology provides access to cameras that can acquire more than 1000 frames per second at megapixel resolution per frame. QPI at several hundred images per second has already been reported in the context of red blood cell dynamics.[7–10] Typically, there is a trade-off between acquisition speed and sensitivity, which the investigator has to consider. Unlike in fluorescence-based imaging, in QPI the image is formed by elastic scattering, which provides stronger signals. The visible or near infrared wavelength illumination in QPI is less damaging to the specimen than the UV typically used to excite fluorescence. This, in principle, allows for higher illumination irradiance, which in turn can shorten the exposure time.

Clearly, phase-shifting techniques are slower. If four exposures are required for each phase image, the overall acquisition rate is at best four times lower than that of the camera. One option to speed up the measurement is to perform an *interlaced* processing, similar to a running average, as illustrated in Fig. 8.7. Thus the four frames A, B, C, D, corresponding to the 0, $\pi/2$, π, $3\pi/2$ phase shifts can be processed either in the fixed order shown in Fig. 8.7a or in a continuously

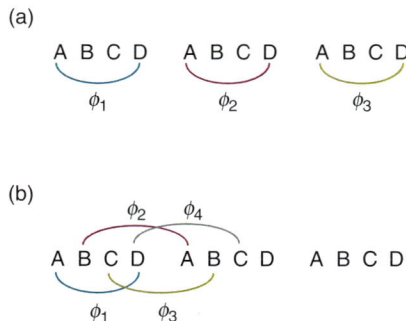

FIGURE 8.7 Grouped (*a*) and interlaced (*b*) processing of the four frames, A, B, C, D, in phase-shifting interferometry.

changing order, as in Fig. 8.7b. The latter type of processing yields four times faster acquisition rates, but the phase information is still smeared over the time necessary to acquire the four frames. Either way, the phase shifter itself (e.g., liquid crystal modulator, piezoelectric transducer, etc.) introduces further delays. It is therefore important that the system is stable over the duration of the four-frame acquisition.

At the other end of the temporal sampling, the longest QPI investigation is limited by the overall stability of the instrument. Typically, the phase noise increases at lower frequencies, which adds constraints on the duration of reliable imaging. The temporal and spatial sensitivity are described in more detail later.

8.5.2 Spatial Sampling: Transverse Resolution

In QPI it is desirable to preserve the diffraction-limited resolution provided by the microscope, as described in Sec. 4.2. We found that the transverse resolution in the case of a circular aperture is given by $\rho_0 = 1.22\lambda/NA$, where λ is the wavelength and NA the numerical aperture, which is defined by the refractive index and the half angle, θ, subtended by the entrance pupil from the sample plane, $NA = n\sin\theta$.

QPI may offer new opportunities in terms of transverse resolution, which were not covered by the intensity-based imaging systems. For example, Goodman gives a clear example of a situation where the definition of transverse resolution is not clear-cut in the case of coherent imaging (see Sec. 6.5.2. in Ref. 11). To review this concept, let us consider two points imaged by a coherent system whereby there exists a phase shift between the two. The intensity profile at the image plane is the result of the coherent superposition of the two fields,[11]

$$I(x) = \left| \frac{J_1[\pi(x-0.61)]}{\pi(x-0.61)} + e^{i\phi} \frac{J_1[\pi(x+0.61)]}{\pi(x+0.61)} \right|^2 \tag{8.9}$$

In Eq. (8.9), x is the normalized coordinate. This intensity distribution is shown in Fig. 8.8 for various values of ϕ and for the incoherent case. Clearly, the phase difference between the two points has a significant effect on the intensity distribution and, specifically, on how well the system "resolves" the two points. Thus, for $\phi = 0$ the points are not distinguishable at all, while for $\phi = \pi/2$ (fields in quadrature) we recover the same profile as for the incoherent case, when the intensities (rather than the fields) add up,

$$I_{incoherent}(x) = \left\{ \frac{J_1[\pi(x-0.61)]}{\pi(x-0.61)} \right\}^2 + \left\{ \frac{J_1[\pi(x+0.61)]}{\pi(x+0.61)} \right\}^2 \tag{8.10}$$

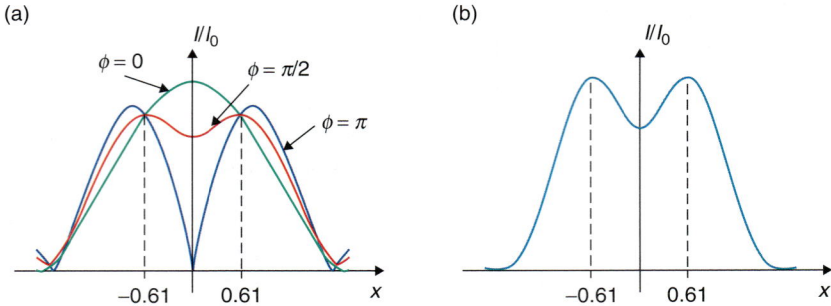

FIGURE 8.8   (a) Intensity profile in the image plane of two coherent point sources separated by the Rayleigh criterion. The vertical lines indicate the locations of the two points. (b) The intensity profile when the two points are incoherent. (Adapted after Ref. 11.)

Finally, when the two fields are out of phase ($\phi = \pi$), the intensity at the origin goes to zero, i.e., the points look most separated, yet shifted in position away from the origin.

This simple example shows that phase values in an image affects the resolving power of the instrument. In other words, when using the common intensity-based definition for resolving power, contrast and resolution appear to be entangled in coherent imaging. Note that this discussion in terms of intensity-based definition fails altogether when imaging phase objects, for which the intensity profile is flat.

Defining a proper measure of transverse resolution in QPI is non-trivial and perhaps worth pursuing by theoretical researchers. Of course, such a definition must take into account that the coherent imaging system is not linear in *phase* (or in *intensity*), but in the *complex field*. For now, when discussing transverse resolution of a QPI system, we mean the definition associated with coherent imaging, as described in Sec. 4.2.

As further discussed in Chaps. 9 and 10, phase-shifting methods are more likely than off-axis methods to preserve the diffraction-limited resolution of the instrument. In off-axis geometries, the issue is complicated by the additional length scale introduced by the spatial modulation frequency (i.e., the fringe period). Again, following the Nyquist sampling theorem, this frequency must be high enough to recover the maximum frequency allowed by the numerical aperture of the objective (Sec. 8.3.). In Chaps. 9 and 10, we show that, with proper precautions, this can be achieved in practice. Still, the spatial filtering in off-axis, involving Fourier transformations back and forth, has at least the detrimental effect of adding noise to the final image. By contrast, in phase shifting, the phase-image recovery involves only simple operations of summation and subtraction (see Sec. 8.2.), which is overall less noisy.

8.5.3   Temporal Stability: Temporal-Phase Sensitivity

Temporal stability is perhaps the most challenging feature to achieve in QPI. Interestingly, in their interferometric studies on the speed of light, Michelson and Morley clearly state the challenges involved in attaining the necessary phase stability: *In the first experiment one of the principal difficulties encountered was [...] its extreme sensitivity to vibration. This was so great that it was impossible to see the interference fringes except at brief intervals when working in the city, even at two o'clock in the morning.*[12]

In studying dynamic phenomena by QPI, the question that often arises is: What is the smallest phase change that can be detected at a given point in the field of view? For instance, studying red blood cell membrane fluctuations requires a displacement sensitivity of the order of 1 nm, which translates roughly to temporal phase sensitivity of a 5–10 mrad, depending on the wavelength. In time-resolved interferometric experiments, uncorrelated noise between the two fields of the interferometer always limits the temporal-phase sensitivity; i.e., the resulting interference signal contains a random phase in the cross-term,

$$I(t) = |U_1|^2 + |U_2|^2 + 2|U_1| \cdot |U_2| \cdot \cos[\phi(t) + \delta\phi(t)] \tag{8.11}$$

where ϕ is the phase under investigation and $\delta\phi(t)$ is the *temporal phase noise*. If $\delta\phi$ fluctuates randomly over the entire interval $(-\pi, \pi]$ during the time scales relevant to the measurement, the information about the quantity of interest, ϕ, is completely lost, i.e., the last term in Eq. (8.11) averages to zero. Sources of phase noise include air fluctuations, mechanical vibrations of optical components, vibrations in the optical table, etc.

A *quantitative* way to assess the phase stability experimentally is to perform successive measurements of a stable sample (or no sample at all). The temporal phase fluctuations, $\delta\phi(t)$, associated with one point can be described globally by the standard deviation (Fig. 8.9)

$$\sigma_t = \sqrt{\left\langle \left[\delta\phi(t) - \langle\delta\phi(t)\rangle_t \right]^2 \right\rangle_t} \tag{8.12}$$

Since the absolute value of the phase is irrelevant and only relative changes are meaningful, we can always subtract a constant from the phase signal and make it into a *zero-average* signal. In this case, the *variance*, σ_0^2, (i.e., square of the standard deviation) is,

$$\sigma_0^2 = \langle\delta\phi^2(t)\rangle_t - \langle\delta\phi(t)\rangle_t^2$$
$$= \langle\delta\phi^2(t)\rangle_t \tag{8.13}$$

(a) (b)

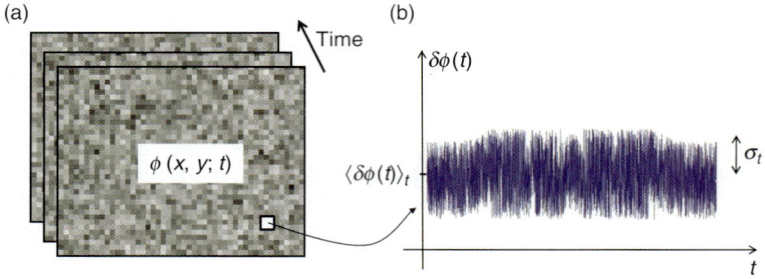

FIGURE 8.9 (*a*) Time-series of quantitative phase images. (*b*) Phase noise at the point shown in (*a*). The average and standard deviation are indicated.

This standard deviation sets the limit to the lowest value of phase change that the instrument can detect at that particular point. In other words, a signal of this standard deviation will correspond to a signal-to-noise ratio of SNR = 1.

A simple yet effective means to reduce the noise in single-shot systems is to reference the phase image to a point in the field of view that is known to be stable (i.e., no dynamic portion of the specimen is present at that point),

$$\phi'(x,y,t) = \phi(x,y,t) - \phi(x_0,y_0,t) \qquad (8.14)$$

where ϕ' is the new, referenced phase map and (x_0,y_0) is the point of reference. What this simple operation does is to remove the *common mode* noise, i.e., the phase fluctuations that are common to the entire field of view. If the measurement is subjected only to this type of noise, say $\Delta\phi(t)$, i.e., if the time dependence at point (x_0,y_0) is only due to noise, such that $\phi(x_0,y_0,t) = \phi_0(x_0,y_0) + \Delta\phi(t)$, this referencing operation eliminates the noise $\Delta\phi$ entirely,

$$\phi'(x,y,t) = \phi(x,y,t) + \Delta\phi(t) - [\phi(x_0,y_0) + \Delta\phi(t)]$$
$$= \phi(x,y,t) - \phi_0(x_0,y_0) \qquad (8.15)$$

With this simple operation, the referenced phase is now noise-free and shifted in value by a constant ϕ_0, which is unimportant. In practice, there is always a noise component that is spatially dependent and, thus, cannot be removed by this procedure, e.g. random rotational modes of optical components.

Sometimes, in the first approximation, we can assume that the noise is the same at all temporal scales covered by the measurement, i.e., assume *white noise*. A fuller descriptor of the temporal phase noise, one that provides information about the *spectral* behavior of the noise, is obtained by computing numerically the power spectrum of the measured signal, $|\delta\phi(\omega)|^2$,

$$|\delta\phi(\omega)|^2 = \left|\int \delta\phi(t) \cdot e^{i\omega t} dt\right|^2 \qquad (8.16)$$

We can normalize this power spectrum to obtain a new spectral function,

$$\Phi(\omega) = A\left|\delta\phi(\omega)\right|^2 \tag{8.17}$$

where the normalization constant, A, is such that the area under Φ gives the *variance* of the signal, i.e.,

$$\int \Phi(\omega)d\omega = \sigma_0^2 \tag{8.18}$$

This new *power spectral density*, Φ, essentially describes the contribution to the variance of each frequency component. For example, if we break the frequency interval associated with the measurement into equal bins of bandwidth $\Delta\omega$, we can write the variance of the noise as the sum of variances within each bin (σ_i^2)

$$\sigma_j^2 = \sum_i \Phi(\omega_i)\Delta\omega$$
$$= \sum_i \sigma_i^2 \tag{8.19}$$

The summation in Eq. (8.18) originates from the fact that noise at different frequencies is uncorrelated and, thus, the variance of their sum is the sum of variances (See Sec. 4.3.).

Function Φ has units of $rad^2/(rad/s)$ and its square-root (in $rad/\sqrt{rad/s}$) is the analog of the *noise equivalent power* (NEP), commonly used as figure of merit for photodetectors. Thus, NEP represents the smallest phase change (in *rad*) that can be measured (with SNR = 1), at a frequency bandwidth of 1 rad/s. As a consequence, much higher phase sensitivity can be achieved by "locking" the measurement onto a narrow band of frequencies around a given frequency, say ω_1 (Fig. 8.10).

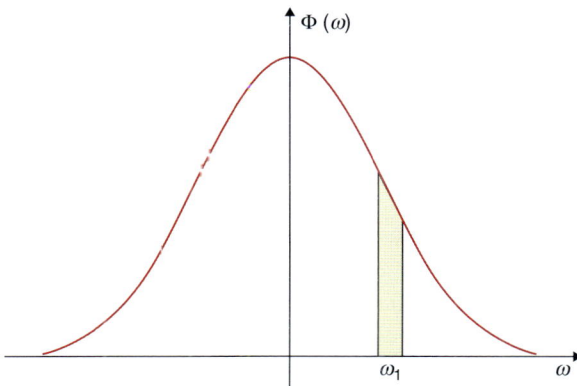

FIGURE 8.10   Noise power spectral density and noise within a narrow band around frequency ω_1. The area of the $\Phi(\omega)$ curve gives the total noise variance. The shaded area represents the contribution to the total variance of the noise of frequency, ω_1.

Example 8.1 Let us consider that the power spectrum in Fig. 8.10 is a Gaussian of r.m.s. bandwidth 10 rad/s and that the zero-average signal has a standard deviation $\sigma_0 = 1$ rad (a huge value for all practical purposes),

$$\Phi(\omega) = \sigma_0^2 \cdot \frac{\exp\left(-\dfrac{\omega^2}{2\Delta\omega^2}\right)}{\displaystyle\int_0^\infty \exp\left(-\dfrac{\omega^2}{2\Delta\omega}\right) d\omega} \tag{8.20}$$

where $\sigma_0^2 = 1$ rad^2 and $\Delta\omega = 10$ rad/s. If the measurement is performed around a frequency $\omega_1 = 10$ rad/s, with a bandwidth of 1 rad/s, the resulting phase noise is only

$$\sigma_1^2 = \int_{10-0.5}^{10+0.5} \Phi(\omega) d\omega$$

$$= 0.24 \text{ rad}^2 \tag{8.21}$$

$$\sigma_1 = 0.16 \text{ rad}$$

Further, if the measurement is performed at the same frequency $\omega_1 = 10$ rad/s but a bandwidth of 0.1 rad/s, the phase noise drops to $\sigma_2^2 = 2.4 \times 10^{-3}$ rad^2 ($\sigma_2 = 0.05$ rad$=\sigma_0/20$). This example illustrates that measuring within a narrow band around a particular frequency can enhance the sensitivity of the measurement by orders of magnitude. Of course, the price paid for the increased signal to noise is longer acquisition time.

    In order to improve experimentally the stability of QPI systems, there are several approaches typically pursued: (i) passive stabilization, (ii) active stabilization, (iii) differential measurements, and (iv) common path interferometry.

(i) *Passive stabilization.* Includes damping mechanical oscillations from the system (e.g., from the optical table), placing the interferometer in vacuum-sealed enclosures, etc. To some extent, most QPI systems incorporate some degree of passive stabilization; floating the optical table is one such example. Interestingly, in their experiments, Michelson and Morley used an optical table made of stone that floated in a bath of mercury (see Fig. 8.11).[12] This also allowed them to seamlessly rotate the interferometer and measure the speed of light along different directions. Unfortunately, these procedures are often insufficient to ensure sensitive phase measurements of biological relevance.

(ii) *Active stabilization* involves the continuous cancellation of noise via a feedback loop and an active element (e.g., a piezoelectric transducer) that tunes the pathlength difference in the interferometer. This principle has been implemented in various geometries in the past with some success. Such active stabilization drastically complicates the measurement by

(a)   (b)

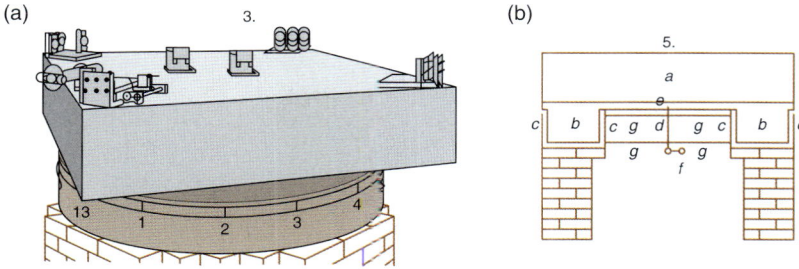

FIGURE 8.11   Experimental setup used by Michelson and Morley: (*a*) overview, (*b*) cross section. The stone (*a*) is 1.5 m square and 0.8 m thick, resting on an annular wooden float. The float rests on mercury, approximately 1.5 cm thick. [Reproduced with permission from A. A. Michelson and E. W. Morley, On the relative motion of the luminiferous ether, *American Journal of Science*, 34, 333 (1887).]

adding dedicated electronics and optical components (for an example, see Sec. 9.2.2.1).

(iii)   *Differential measurements* can also be used effectively to increase QPI sensitivity. The main idea is to perform two noisy measurements whereby the noise in the two signals is correlated and, thus, can be subtracted. We discussed this idea for point-scanning methods in Sec. 7.6. Another particular example was described above, when the quantitative phase image was referenced to a "fixed" point in the field of view (Eq. 8.14)

(iv)   *Common path interferometry* refers to QPI geometries where the two fields travel along paths that are physically very close. In this case, the noise in both fields is very similar and hence automatically cancels in the interference (cross) term. A good illustration is to observe a live phase-contrast microscope image. We have known since Abbe[13] that an image is, in fact, a (complicated) interferogram. Yet, the field distribution in a phase-contrast image is extremely stable, unlike, say, the fringe pattern generated by a common Michelson interferometer. This stability is precisely due to the fact that the interfering fields travel along a *common path* and through the same set of optics. As discussed already in Sec. 6.1, Gabor's in-line holography benefits from the stability of common path geometry.

Examples of common path QPI methods that proved their potential for biological studies will be described in more detail in Chaps. 11 and 12 (Chap. 12 describes white light methods that happen to also be common path).

8.5.4   Spatial Uniformity: Spatial Phase Sensitivity

Analog to the "frame-to-frame" phase noise discussed in the previous section, there is a "point-to-point" (*spatial*) phase noise that affects

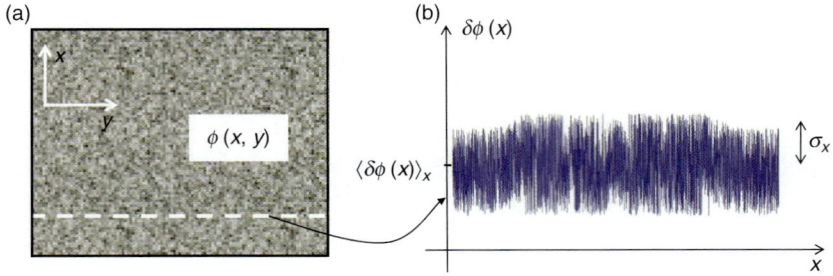

FIGURE 8.12   (a) Quantitative phase image. (b) Phase noise along the profile shown in (a). The average and standard deviation are indicated.

the QPI measurement. This spatial phase sensitivity limits the smallest *topographic* or refractive index change that the QPI system can detect. Again, we can define an overall spatial standard deviation by measuring a flat background (see Fig. 8.12). The standard deviation for the entire field of view, following an analog definition to the time domain (Eq. 8.12), is

$$\sigma_r = \sqrt{\left\langle \left[\delta\phi(x,y) - \langle \delta\phi(x,y) \rangle_{x,y} \right]^2 \right\rangle_{x,y}} \tag{8.22}$$

Quantitatively, the spatial phase noise is best described in the (spatial) frequency domain, in complete analogy with the temporal noise (Eq. 8.15). Thus, the normalized spatial power spectrum density (NEP squared) is defined as

$$\Phi(k_x, k_y) \propto \left| \int_A \int \delta\phi(x,y) \cdot e^{i(k_x \cdot x + k_y \cdot y)} dx\, dy \right|^2 \tag{8.23}$$

where $\Phi(k_x, k_y)$ has units of $\mathrm{rad}^2/(\mathrm{rad}/\mu m)^2$ and is normalized such that its integral over the frequencies gives the (spatial) variance,

$$\int_{A_k} \int \Phi(k_x, k_y) dk_x\, dk_y = \sigma_0^2 \tag{8.24}$$

Again, phase sensitivity can be increased significantly if the measurement is band-passed around a certain spatial frequency (k_{x1}, k_{y1}).

Unlike with temporal noise, there are no clear cut solutions to improve spatial sensitivity besides keeping the optics pristine and decreasing the coherence length of the illumination light. The spatial nonuniformities in the phase background are mainly due to the random interference pattern (i.e., *speckle*) produced by fields scattered from impurities on optics, specular reflections from the various surfaces in the system, etc. This spatial noise is worst when using highly coherent sources, i.e., *lasers*. Using white light as illumination drastically reduces the effects of speckle, as detailed in Chap. 9, while preserving the requirement of a coherence area that is at least as large as the field of view. In post-processing, sometimes subtracting the constant phase background (no sample QPI) helps.

Finally, the *spatial and temporal power spectrum* can be combined into a single function,

$$\Phi(\mathbf{k},\omega) \propto \left| \int_A \int \int_{-\infty}^{\infty} \delta\phi(\mathbf{r},t) \cdot e^{i(\omega t - \mathbf{k}\cdot\mathbf{r})} \, dt \, d^2\mathbf{r} \right|^2 \tag{8.25}$$

the units of which are $\mathrm{rad}^2 / [(\mathrm{rad}/\mathrm{s}) \cdot (\mathrm{rad}/\mu m)^2]$, $\mathbf{r} = (x,y)$, $\mathbf{k} = (k_x, k_y)$.

Example 8.2   Revisiting Ex. 8.1 in the previous section, let us assume an overall (*spatiotemporal*) standard deviation $\sigma_0 = 1$ rad (very high), which is due to a spectral function that has a Gaussian shape in both ω and $\mathbf{k}$,

$$\Phi(\omega,\mathbf{k}) = \sigma_c^2 \cdot \frac{\exp\left(-\dfrac{\omega^2}{2\Delta\omega^2}\right)}{A_\omega} \cdot \frac{\exp\left[-\dfrac{(k_x^2 + k_y^2)}{2\Delta k^2}\right]}{A_{k_x} \cdot A_{k_y}} \tag{8.26}$$

where A_ω, A_{k_x} and A_{k_y} are the integrals of the Gaussians over the ω and $\mathbf{k}$ intervals of interest.

Note that now band-passing can take place in all three dimensions (ω, k_x, k_y), which essentially elevates the effect to the power 3! Previously we found that band-passing the signal with a bandwidth 100 times smaller than the total bandwidth (0.1 vs. 10 rad/s) yields a standard deviation $\sigma_2 = 0.05$ rad, i.e., 20 smaller than the original σ_0. If a similar procedure is applied spatially, the band-passed standard deviation becomes a factor of $20^3 = 8000$ times smaller than σ_0. This procedure can render extreme sensitivities provided that the phenomenon under investigation reveals itself within such narrow bands of *spatiotemporal frequencies*.

As a side note, the MIT-Caltech Laser Interferometer Gravitational-Wave Observatory (LIGO) is an instrument for detecting gravitational waves produced by a pair of colliding neutron stars. Interestingly, the measurable quantity that is expected to eventually report on the space-time ripples is precisely a phase shift between the two arms of several kilometer-long interferometers.[14] The necessary pathlength sensitivity at a reasonable frequency is in the order of $10^{-9} - 10^{-12}$ nm!

8.6   Summary of QPI Approaches and Figures of Merit

The discussion in this chapter perhaps made it apparent that there is no perfect QPI method, i.e., there is no technique that performs optimally with respect to all figures of merit identified in the last section. We can summarize the QPI approaches and their performances in Table 8.1.

	Acquisition rate	Transverse resolution	Temporal sensitivity	Spatial sensitivity
Off-axis	X			
Phase-shifting		X		
Common-path			X	
White light				X

TABLE 8.1

Clearly, the *off-axis* methods are fast as they are single shot, *phase-shifting* preserves the diffraction-limited transverse resolution without special measures, *common-path* methods are stable and *white light* illumination suffers less from speckle and, thus, images are spatially more uniform. As the diagonal table above suggests, we can think of these four figures of merit as the "normal modes" of categorizing QPI techniques. However, as we will see in the chapters to follow, there are methods that combine these four approaches, seeking to add the respective individual benefits. Thus, there are $C_4^2 = 6$ possible combinations of two geometries, as follows,

- Off-axis and phase-shifting (discussed in Sec. 9.1.2)
- Phase shifting and white light (Sec. 10.1)
- Phase shifting and common path (Sec. 11.1.)
- Off-axis and common path (Sec. 11.2)
- Common path and white light (Chap. 12)
- Off-axis and white light (Sec. 14.2.2., this method is also common path)

More recently, even three of these approaches have been combined. The possible number of combinations is $C_4^3 = 4$:

- Phase shifting, common path, white light (Chap. 12)
- Off-axis, common path, white light (Sec. 14.2.2.)
- Off-axis, phase shifting, common path (see, for instance, Refs. 15 and 16)
- Off-axis, phase shifting, white light (perhaps to be developed)

References

1. K. Creath. "Phase-measurement interferometry techniques." *Prog. Opt.*, 26, 349–393 (1988).
2. Y. Y. Cheng and J. C. Wyant. "Phase shifter calibration in phase-shifting interferometry." *Applied Optics*, 24, 3049–3052 (1985).

3. D. C. Ghiglia and M. D. Pritt. *Two-Dimensional Phase Unwrapping: Theory, Algorithms, and Software.* (Wiley, New York, 1998).

4. T. Ikeda, G. Popescu, R. R. Dasari, and M. S. Feld. "Hilbert phase microscopy for investigating fast dynamics in transparent systems." *Optics Letters*, 30, 1165–1168 (2005).

5. H. Nyquist. "Certain Topics in Telegraph Transmission Theory." *Transactions of the A. I. E. E.*, 617 (1928).

6. C. E. Shannon. "Communication in the Presence of Noise." *Proceedings of the IRE*, 37, 10 (1949).

7. G. Popescu, T. Ikeda, C. A. Best, K. Badizadegan, R. R. Dasari, and M. S. Feld. "Erythrocyte structure and dynamics quantified by Hilbert phase microscopy." *J. Biomed. Opt. Lett.*, 10, 060503 (2005).

8. G. Popescu, T. Ikeda, R. R. Dasari, and M. S. Feld. "Diffraction phase microscopy for quantifying cell structure and dynamics." *Optics Letters*, 31, 775–777 (2006).

9. Y. K. Park, M. Diez-Silva, G. Popescu, G. Lykotrafitis, W. Choi, M. S. Feld, and S. Suresh. "Refractive index maps and membrane dynamics of human red blood cells parasitized by *Plasmodium falciparum*." *Proceedings of the National Academy of Sciences*, 105, 13730 (2008).

10. Y. K. Park, C. A. Best, K. Badizadegan, R. R. Dasari, M. S. Feld, T. Kuriabova, M. L. Henle, A. J. Levine, and G. Popescu. *Measurement of red blood cell mechanics during morphological changes, Proceedings of the National Academy of Sciences* (2010).

11. J. W. Goodman. *Introduction to Fourier Optics.* (McGraw-Hill, New York, 1996).

12. A. A. Michelson and E. W. Morley. "On the relative motion of the luminiferous ether." *American Journal of Science*, 34, 333 (1887).

13. E. Abbe. "Beiträge zur Theorie des Mikroskops und der mikroskopischen Wahrnehmung." *Arch. Mikrosk. Anat.*, 9, 431 (1873).

14. A. Abramovici, W. E. Althouse, R. W. P. Drever, Y. Gursel, S. Kawamura, F. J. Raab, D. Shoemaker, L. Sievers, R. E. Spero, K. S. Thorne, R. E. Vogt, R. Weiss, S. E. Whitcomb, and M. E. Zucker. "Ligo - the Laser-Interferometer-Gravitational-Wave-Observatory." *Science*, 256, 325–333 (1992).

15. V. Mico, Z. Zalevsky, and J. Garcia. "Common-path phase-shifting digital holographic microscopy: A way to quantitative phase imaging and superresolution." *Optics Communications*, 281, 4273–4281 (2008).

16. P. Gao, I. Harder, V. Nercissian, K. Mantel, and B. Yao. "Common path phase shifting microscopy based on grating diffraction." *Optics Letters*, 35, 712 (2010).

CHAPTER **9**

Off-Axis Methods

9.1 Digital Holographic Microscopy (DHM)

As described in Chap. 6, the principle of digital holography was demonstrated in the 1960s.[1,2] Later, in 1982, Takeda et al. reported a "Fourier transform method of fringe pattern analysis for computer-based topography and interferometry," where off-axis interferometry and FFT (fast Fourier transform) processing were combined to study topography of structures.[3] However, the impact of digital holography to microscopy became significant much later, in the 1990s, when implemented with charged coupled devices (CCDs) as recording media. Thus, in 1994, Schnars and Juptner, with the Bremen Institute in Germany, reported, "Direct recording of holograms by a CCD target and numerical reconstruction."[4] Essentially this study demonstrated "lensless" off-axis holography using a CCD as detector. The authors recognized the advantages of using CCDs, i.e., bypassing the chemical process of developing and potential for high acquisition rates. However, it was pointed out that their resolution at the time (~100 lines/mm) was significantly lower than in standard films, which limited their applicability to objects that subtend small angles. On the other hand, Schnars and Juptner correctly anticipated that "future CCD chips, with higher resolutions, will improve the quality of the image."

Soon after, the benefits of digital holography have been exploited in microscopy by several different groups. In particular, very productive research on *digital holographic microscopy* (DHM) has been conducted at the laboratory directed by Christian Depeursinge at Ecole Polytechnique Federal de Lausanne. The principle and applications of this method are reviewed below.

9.1.1 Principle

In 1999, the Depeursinge group reported high-resolution quantitative phase imaging using DHM, both with a lensless configuration[5] and by using a microscopy objective.[6] Here, we describe the latter setup, which allowed for higher transverse resolution and became the standard configuration in the biological studies that followed.

141

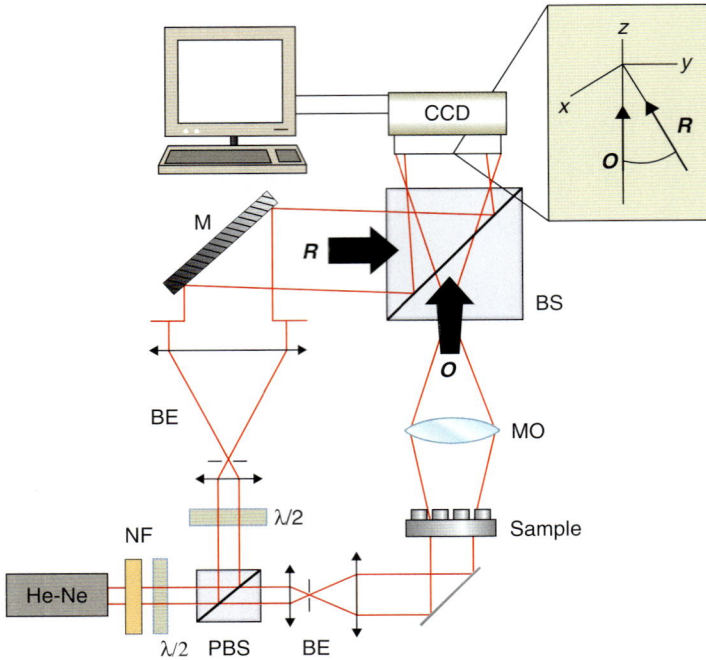

FIGURE 9.1   Schematic of DHM for transmission imaging. NF, neutral density filter; PBS, polarizing beam splitter; BE, beam expander with spatial filter; $\lambda/2$, half-wave plate; M, mirror; BS, beam splitter; O, object wave; R, reference wave. Inset: detail showing the off-axis geometry at the CCD plane. [(From Ref. 5.) The Optical Society, used with permission, *Optics Letters*, E. Cuche, F. Bevilacqua & C. Depeursinge, "Digital holography for quantitative phase-contrast imaging," 24, 291–293 (1999).]

Both the transmission and reflection configurations have been demonstrated, but here we focus on the transmission geometry, which is more commonly used for biomedical applications.

The experimental arrangement uses a Mach-Zender interferometer, as illustrated in Fig. 9.1.[6] A He-Ne laser beam is spatially filtered and expanded to approach a plane wave and further split into two paths. Combinations of $\lambda/2$ waveplates and polarization beam splitters provide independent control over the intensities of the two fields.

The specimen is placed at a distance d_s in front of the microscope objective, as shown in Fig. 9.2. The CCD is positioned a distance d in front of the image plane, IP, where the objective forms the image of the sample. Thus, the CCD records the interference between this out-of-focus sample field, U_S, and the off-axis reference field, U_R.

The sample field can be calculated using the Fresnel propagation from the sample plane to the objective, applying the lens transformation and, finally Fresnel-propagating one more time to the CCD plane

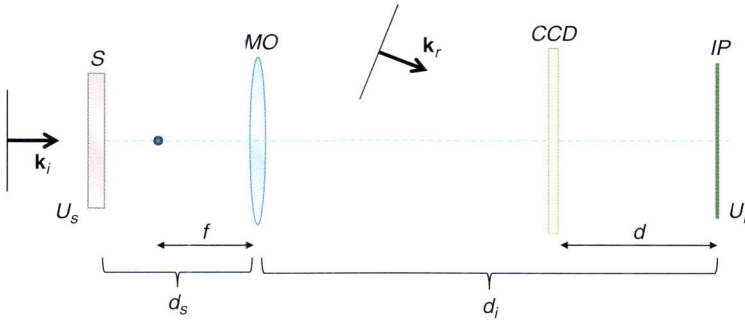

FIGURE 9.2 The sample arm in DHM: k_i, k_r, incident and reference wavevectors; S, specimen; MO, microscope objective; CCD, charged coupled device; IP, image plane; f, objective focal distance; d_i and d_s, image distance and sample distance, respectively; d, distance between image and CCD.

(see Chap. 2 for a review on Fresnel diffraction integral). However, a more practical way to perform the calculation is to start with the image field at plane IP, U_i, and back propagate it via the Fresnel transformation to the CCD plane. The image field, U_i, is the magnified replica of the sample field.

$$U_i(x,y) = \frac{1}{M^2} U_S\left(\frac{x}{M}, \frac{y}{M}\right)$$ (9.1)

In Eq. (9.1), the transverse magnification, M, equals the ratio between the sample and image distances of the

$$M = \frac{d_i}{d_0}$$ (9.2)

Thus, the field at the CCD plane is

$$U_F(x,y) = U_i(x,y) \otimes e^{-\frac{ik_0}{2d}(x^2+y^2)}$$ (9.3)

In Eq. (9.3), the Fresnel wavelet differs from that in Eq. (2.21) by its sign in the exponent. Here the negative sign indicates a *convergent* field that propagates backwards from the image plane, or, equivalently, a field that propagates over a distance, $-d$.

The total (hologram) field at the CCD becomes

$$U_h(x,y) = U_F(x,y) + |U_r| \cdot e^{i\mathbf{k}\cdot\mathbf{r}}$$

$$= U_F(x,y) + |U_r| \cdot e^{i(k_{rx}\cdot x + k_{rz}\cdot z)}$$ (9.4)

where $k_{rx} = k \cdot \sin\theta$, $k_{rz} = k \cdot \cos\theta$, and θ is the off-axis angle of the reference field. The intensity distribution of the hologram is

$$I_h(x,y) = |U_h(x,y)|^2$$

$$= |U_F(x,y)|^2 + |U_r|^2 +$$

$$+ U_F(x,y) \cdot |U_r| \cdot e^{-ik_{rx} \cdot x} + U_F(x,y)^* \cdot |U_r| \cdot e^{ik_{rx} \cdot x} \tag{9.5}$$

The digital reconstruction requires Fourier transforming the recorded CCD image followed by a *numerical* Fresnel propagation. Thus, taking the Fourier transform of Eq. (9.5), we obtain

$$\tilde{I}_H(k_x, k_y) = I_0(k_x, k_y) + I_{+1}(k_x, k_y) + I_{-1}(k_x, k_y) \tag{9.6}$$

where I_0 is the zeroth order (DC component) and $I_{\pm1}$ are the two diffraction orders of interest as shown

$$\tilde{I}_0(k_x, k_y) = \Im\left[|U_F(x,y)|^2 + |U_r|^2 \right]$$

$$\tilde{I}_{+1}(k_x, k_y) = |U_r| \cdot \tilde{U}_F(k_x - k_{rx}, k_y) \tag{9.7}$$

$$\tilde{I}_{-1}(k_x, k_y) = |U_r| \cdot \tilde{U}_F(k_x + k_{rx}, k_y)$$

Schematically, the three orders are shown in Fig. 9.3.[6]

Using *spatial filtering*, one of the diffraction orders, say I_{+1}, can be isolated and shifted back by k_{rx} to the origin. This term can be further expressed via the Fresnel convolution in Eq. (9.3) and using that its Fourier transform yields a product (see App. B for the convolution theorem),

$$\tilde{I}_1(k_x, k_y) = |U_r| \cdot \tilde{U}_i(k_x, k_y) \cdot e^{i\frac{d}{2k_0}(k_x^2 + k_y^2)} \tag{9.8}$$

In Eq. (9.8), we used the Fourier transform properties of a Gaussian,

$$e^{-i\frac{k_0}{2d}(x^2 + y^2)} \leftrightarrow e^{i\frac{d}{2k_0}(k_x^2 + k_y^2)} \tag{9.9}$$

Finally, the complex field associated with the original sample is obtained by Fourier transforming $\tilde{U}_i(k_x, k_y)$ back to the spatial domain,

$$U_i(x,y) \propto \Im\left[I_1(k_x, k_y) \cdot e^{-i\frac{d}{2k_0}(k_x^2 + k_y^2)} \right] \tag{9.10}$$

Note that the operation in Eq. (9.10) is nothing more than a *deconvolution*, which essentially undoes the Fresnel propagation over the distance d. Figure 9.4 illustrates how the complex (amplitude and phase) image plane is retrieved from a recorded hologram.[6]

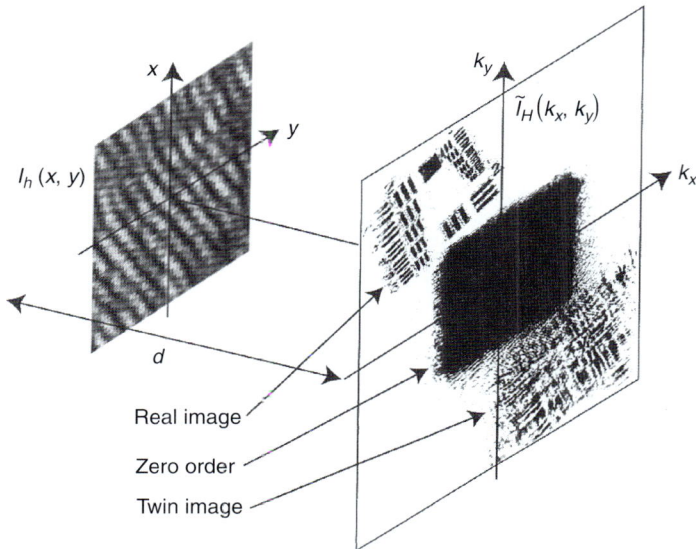

FIGURE 9.3 Geometry for hologram reconstruction: *x-y*, hologram plane; *kx-ky*, observation plane; *d*, reconstruction distance. [(Fig. 4, *From Ref. 6.*) The Optical Society, used with permission, *Applied Optics*, E. Cuche, P. Marquet, & C. Depeursinge, "Simultaneous amplitude-contrast & quantitative phase-contrast microscopy by numerical reconstruction of Fresnel off-axis holograms," 38, 6994–7001(1999).]

The need for the numerical deconvolution in Eq. (9.10) can be eliminated if the CCD is placed at the image plane of the microscope. This was pointed out by Cuche et al.:[6] *"Classical microscopy can be achieved with this arrangement by translation of the object or the hologram plane such that the image is found on the CCD."* However, the authors conclude that this geometry is not compatible with off-axis holography: *"Such a disposition is not appropriate for off-axis holography but is used in interference microscopy where phase shifting interferometry is applied to obtain the phase information."* It turns out that, when imaging transparent specimens, the image plane measurement is not only possible, but also provides some advantages over the out-of-focus (Fresnel) measurements. This approach will be discussed in detail in Secs. 9.2 and 11.2.

Of course, the CCD pixel size and number of pixels are important parameters to keep in mind when recording and reconstructing the hologram. Thus, the CCD size limits the frequency coverage and the pixel dimension constrains the dimensions of the object that can be recorded by DHM.[5] Further, there is a minimum number of pixels necessary to sample the fringe pattern. The Nyquist theorem requires that at least two pixels are used for a given sinusoidal signal. However, in digital holography, because the signal is proportional to the field of interest and not intensity, the minimum number of pixels turns out to be three.[7] Typically, in practice, a higher number of pixels

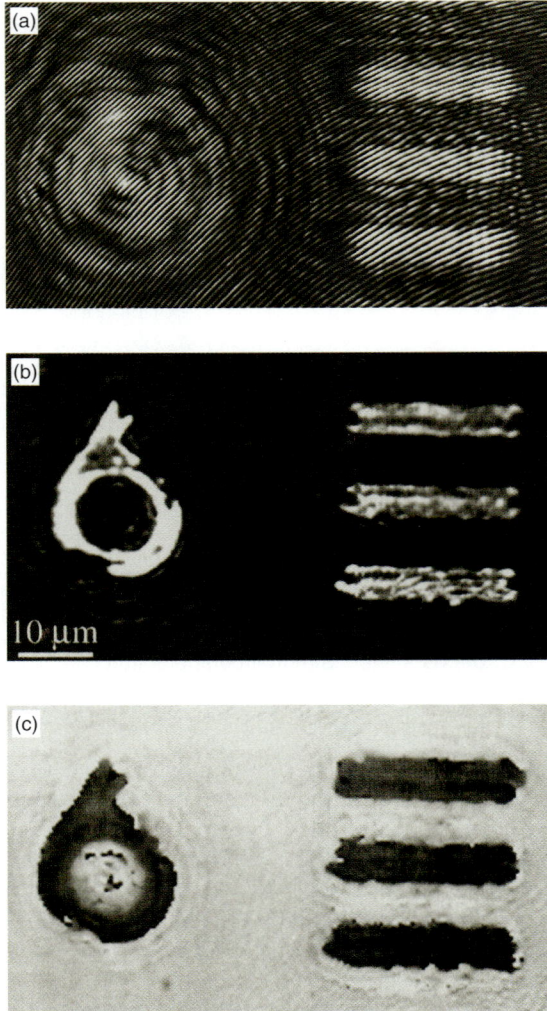

FIGURE 9.4   (*a*) Digital hologram recorded by the CCD camera. (*b*) Amplitude image. (*c*) Phase image. [(Adapted from Figs. 6 and 7, Ref. [6]). The Optical Society, used with permission, *Applied Optics,* E. Cuche, P. Marquet, & C. Depeursinge, "Simultaneous amplitude-contrast & quantitative phase-contrast microscopy by numerical reconstruction of Fresnel off-axis holograms," 38, 6994–7001(1999).]

per fringe are used for safety. We will come back to this point later in Secs. 9.2 and 11.2.

9.1.2   Further Developments

Many different digital holographic microscopy techniques have been developed on a similar principle to that outlined above. In

1997, Yamaguchi and Zhang developed a *phase shifting* version of digital holography,[8] which was later applied in a microscopy setup.[9] Another implementation of phase-shifting digital holography was later developed by Guo and Devaney in a Mach-Zender geometry.[10] A related approach in digital holography, *in-line* or *on-axis* digital holographic microscopy,[11–13] is closer to Gabor's original method and carries the benefit of stability at the expense of spatial modulation. We will discuss this compromise in further detail in Chap. 11.

Kim proposed a digital holographic method that employs wavelength-scanning for achieving modulation.[14] Digital holography with *spatiotemporal* modulation was reported by Indebetouw and Klysubun as a method to relax constraints on spatial coherence.[15]

An important application of the complex field information is to detect the plane of best focus during imaging thin specimens, otherwise a challenging task. This was proposed by Dubois et al. in 2006.[16] Further numerical refocusing based on digital holography data was employed by Langehanenberg et al.[17]

Further effort has been devoted by several groups towards achieving the high transverse resolution necessary for studying microscopic objects such as live cells.[13–23] Ferraro et al. showed that the phase information can be used to compensate for wavefront curvature introduced by the microscope objective.[24]

Parshall and Kim employed a dual-wavelength method for phase unwrapping, which improved the quality of cell imaging by digital holography.[25] Dual-wavelength imaging was used also in a single measurement by Kuhn et al.[26] In 2009, Ash et al. demonstrated the principle of combining digital holography and total internal reflection microscopy, which allows for imaging in near field.[27]

In order to improve transverse resolution, Alexandrov et al. reported a synthetic aperture approach to microscopy, in which the complex Fourier field was measured via digital holography.[28] A number of groups have discussed the problem of improving resolution beyond the Rayleigh limit by using the complex field information provided by digital holography.[29–33]

The Wax group proposed a combination of off-axis and phase-shifting techniques[34] and also investigated the effects of different off-axis angles.[13,35]

There is no question that technology refinement in DHM will continue to develop. However, in recent years we have been witnessing a significant increase in biological applications targeted by DHM methods. These are discussed in more detail in the next section.

9.1.3 Biological Applications

With the progress in transverse resolution and signal to noise, DHM started to demonstrate potential as a *label-free* method for cell imaging.[17,19,21–23,27,36–51]

Cell Imaging

In 2005, Marquet et al. from EPFL Switzerland claimed the first DHM images of cells in culture.[52] The results illustrating high-quality images of live neurons demonstrated the potential of DHM to become a useful tool in cell biology (Fig. 9.5). Later, by measuring phase

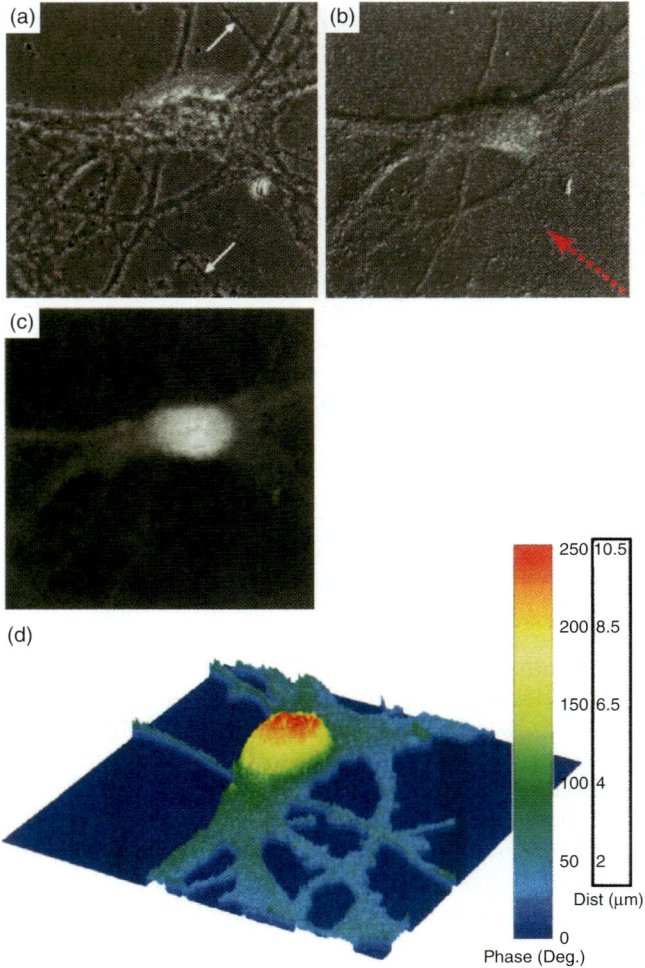

FIGURE 9.5 Images of a living mouse cortical neuron in culture: (*a*) dark phase contrast image; (*b*) DIC image; (*c*) phase image; (*d*) perspective image in false colors of the phase distribution obtained with the DHM. [(From Fig. 2, Ref. 52.) The Optical Society, used with permission, *Optics Letters*, P. Marquet, B. Rappaz, P.J. Magistretti, E. Cuche, Y. Emery, T. Colomb, & C. Depeursinge, "Digital holographic microscopy: a noninvasive contrast imaging technique allowing quantitative visualization of living cells with subwavelength axial accuracy," 30, 460–470(2005).]

(a)　　　　　　　(b)　　　　　　　(c)　　　　　　　(d)

(e)

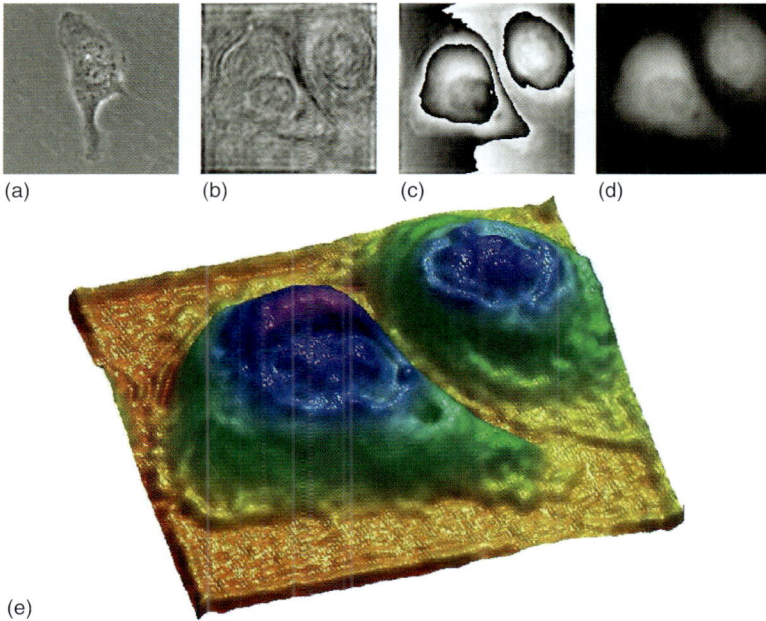

FIGURE **9.6**　Holography of non-confluent SKOV-3 cells. The image area is 60 × 60 μm^2 (404 × 404 pixels) and the image is at $z = 5$ μm from the hologram: (a) Zernike phase contrast image; (b) holographic amplitude and (c) phase images; (d) unwrapped phase image; (e) 3D pseudocolor rendering of (d). [(From Fig. 4, Ref. 19.) The Optical Society, used with permission, *Optics Express*, C.J. Mann, L.F. Yu, C.M. Lo, & M.K. Kim, "High-resolution quantitative phase-contrast microscopy by digital holography," 13, 8693–8698 (2005).]

images of the same cells in two media of different refractive index, the authors estimated the mean cell refractive index.[53]

Figure 9.6 shows images of ovarian cancer cells reported in 2005 by the Kim group at University of South Florida.[19] The images are diffraction limited in terms of transverse resolution and provide 30-nm accuracy in terms of pathlength, which allows intracellular structures to be observed without staining.

The von Bally group at University Medical Center of Muenster, Germany, demonstrated the autofocus capability based on DHM, which allowed dynamic imaging of live cells. Figure 9.7 illustrates the imaging of a spreading keratinocyte cell and the effectiveness of the autofocus procedure.[17]

Cell Growth

In 2009, Rappaz et al. studied the *dry mass* production during the cell cycle in wild type yeast cells. Exploiting the relationship between dry mass and optical phase shift (see also Chap. 1 for a review of this

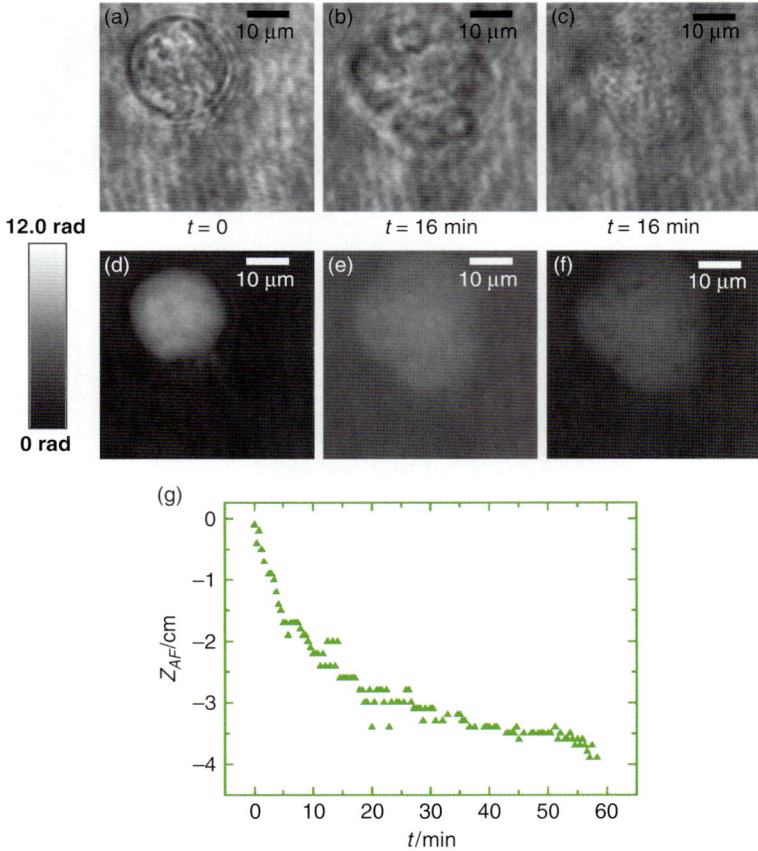

FIGURE 9.7   Digital holographic investigations of spreading keratinocyte cell that has been recorded with a fixed focus in the CCD sensor (image) plane (40× microscope lens, NA = 0.6): (a)–(c) reconstructed amplitude and (d)–(f) phase distributions at (a); (d) t = 0 and (b), (e) t = 16 min with fixed focus; (c), (f) t = 16 min with applied numerical digital holographic autofocus. (g) Temporal dependence of the autofocus position z_AF on the measurement time, t, with fixed mechanical focus of the microscope lens. [(From Fig. 5, Ref. 17.) The Optical Society, used with permission, Applied Optics: P. Langehanenberg, B. Kemper, D. Dirksen, and G. Von Bally, "Autofocusing in digital holographic phase contrast microscopy on pure phase objects for live cell imaging," 47, D176–D182 (2008).]

relationship), the authors demonstrated that DHM can provide a non-invasive and quantitative measure of cell growth (Fig. 9.8; Ref. [44]).

These proof-of-principle applications and others, not mentioned here, occurred roughly within the past five years. It is anticipated that these early developments will set the stage for many more biological applications for DHM in future.

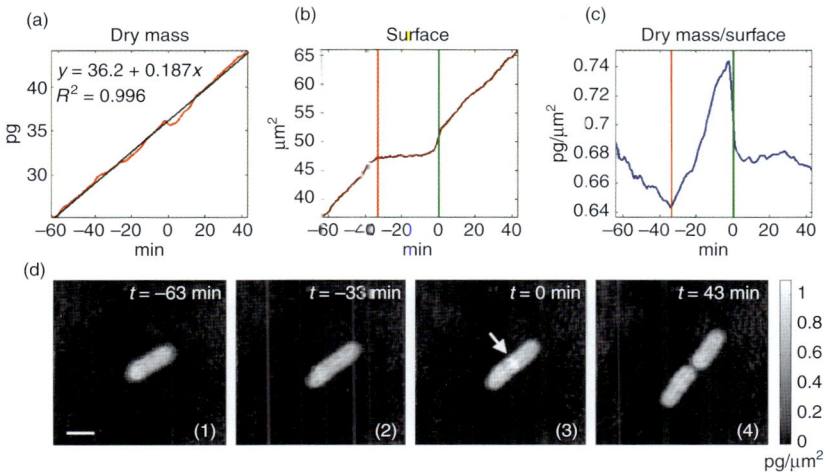

FIGURE 9.8 Time evolution of: (a) dry mass fit: linear regression, (b) projected cell surface, and (c) dry mass (DM) concentration. Vertical lines: red, end of surface growth; green, cytokinesis. (d) Representative DM surface density images recorded at the beginning (1), at the end of cell growth (2), just before cytokinesis (arrow: septum) (3), and at the end of the recording period 4. Scale bar 5 μm. The results are representative of five wild-type cells. [(From Fig. 1, Ref. 44.) SPIE, used with permission, *J. Biomed. Opt.*, B. Rappaz, E. Cano, T. Colomb, J. Kuhn, C. Depeursinge, V. Simanis, P.J. Magistretti, and P. Marquet, "Noninvasive characterization of the fission yeast cell cycle by monitoring dry mass with digital holographic microscopy," 14, 034049 (2009).]

9.2 Hilbert Phase Microscopy (HPM)

Hilbert phase microscopy (HPM) is a QPI method related to digital holographic microscopy discussed in the previous section, as it also uses off-axis interferometry. However, HPM is specifically geared towards imaging optically thin specimens such as cells, and, as such, employs a geometry in which the interferogram is recorded in the image plane, rather than in an out-of-focus (Fresnel) plane. As already discussed in the introduction (Chap. 1), ideally the phase measurement should take place in the plane where the field is the smoothest, such that issues caused by spatial sampling and phase discontinuities are avoided. In the case of live cells, this plane is always the image plane. The HPM principle and applications are described below.

9.2.1 Principle

In 2005, our group lead by Michael Feld at Massachusetts Institute of Technology reported HPM as a new method of measuring quantitative phase images from only one spatial interferogram recording.[54,55] The experimental setup is shown in Fig. 9.9a. A He-Ne laser ($\lambda = 632$ nm) is coupled into a 1×2 single-mode fiber-optic coupler and collimated

(a)

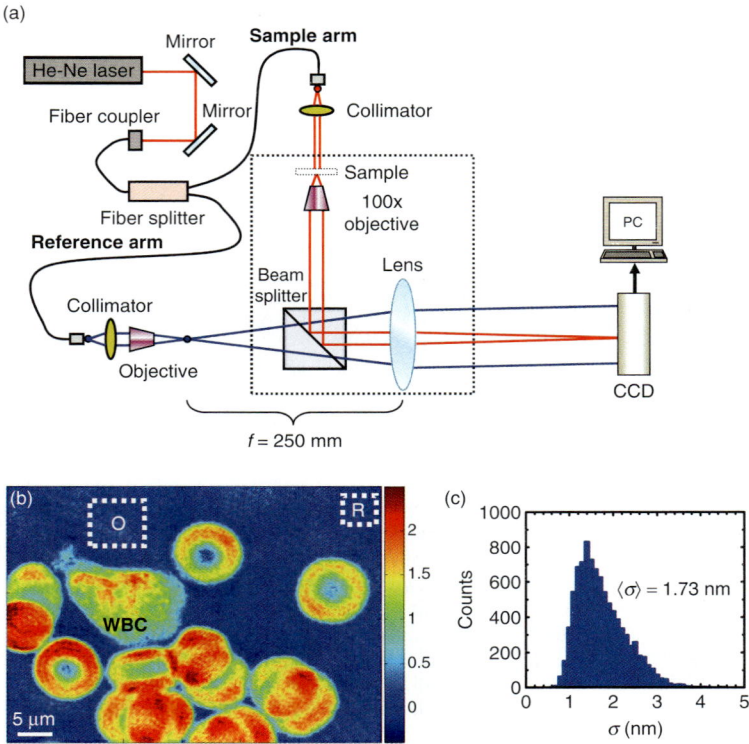

FIGURE 9.9   (*a*) HPM experimental setup. (*b*) HPM image of a droplet of blood. (*c*) The histogram of standard deviations associated with a region in the field of view containing no cells. [(From Figs. 1 and 2, Ref. 55.) SPIE, used with permission, J. of Biomedical Optics Letters, G. Popescu, T. Ikeda, C.A. Best, K. Badizadegan, R.R. Desari, and M.S. Feld, "Erythrocyte structure and dynamics quantified by Hilbert phase microscopy," 10, 060503 (2005).]

on each of the two outputs. One output field acts as the illumination field for an inverted microscope equipped with a 100X objective. The tube lens is such that the image field, U_i, associated with the sample is formed at the CCD plane via the beam splitter cube. The reference field, U_r, can be approximated by a plane wave and is tilted with respect to the sample field such that uniform fringes are along, say, the x-axis. The CCD had an acquisition rate of 291 frames/s at the full resolution of 640×480 pixels, at 1 to 1.5 ms exposure time. The fringes are typically sampled by 5 to 6 pixels per period. The spatial irradiance associated with the interferogram across one direction is given by

$$I(x,y) = U_i(x,y)^2 + U_r^2 +$$
$$+ U_i(x,y) \cdot U_r \cdot e^{-ik_{rx} x} + U_i(x,y)^* \cdot U_r \cdot e^{ik_{rx} x}$$

(9.11)

where we use the notations from Sec. 9.1. Note that the main difference with respect to the DHM measurement (Eq. 9.4) is that now U_i itself and not its Fresnel transform (U_F) is part of the detected intensity. In other words, the intensity distribution contains the phase information of interest directly.

$$I(x,y) = I_r + I_i(x,y) + 2\sqrt{I_r I_i(x,y)} \cos[k_{rx}x + \phi(x,y)] \qquad (9.12)$$

Using high-pass spatial filtering and the Hilbert transformation (in honor of whom the method was named), the quantity of interest, ϕ, is retrieved in each point as described in Sec. 8.3 (see Refs. [54,55] for further details).

The ability of HPM to perform live cell dynamic measurements at the millisecond and nanometer scales, we obtained time-resolved HPM images of red blood cells (RBCs). Droplets of whole blood were simply sandwiched between cover slips, with no additional preparation. Figure 9.9b shows a quantitative phase image of live blood cells; both isolated and agglomerated erythrocytes are easily identifiable. A white blood cell (WBC) is also present in the field of view. Using the refractive index of the cell and surrounding plasma of 1.40 and 1.34, respectively,[56] the phase information associated with the RBCs is translated into nanometer scale image of the cell topography. The assumption of optical homogeneity of RBC is commonly used[57,58] and justified by the knowledge that cellular content consists mainly of hemoglobin solution. In order to eliminate the longitudinal noise between successive frames, each phase image was referenced to the average value across an area in the field of view containing no cells (denoted in Fig. 9.9b by R). To quantify the residual noise of the instrument in a spatially relevant way, we recorded sets of 1,000 images, at 10.3 ms/frame and analyzed the path-length fluctuations of individual points within a 100×100 pixel area (denoted in Fig. 9.9b by O). The path-length associated with each point in O was averaged over 5×5 pixels, which approximately corresponds to the dimensions of the diffraction limit spot. The histogram of the standard deviations associated with all the spots within region O is shown in Fig. 9.9c. The average value of this histogram is indicated. This noise assessment demonstrates that the HPM instrument is capable of providing quantitative information about structure and dynamics of biological systems, such as RBCs, at the nanometer scale.

9.2.2   Further Developments

Several subsequent extensions of HPM have been proposed. Here we describe two techniques: one for improving the HPM stability and the other for combining it with confocal microscopy.

Actively Stabilized HPM (sHPM)

Note that, like DHM, HPM is not common path, and, thus, is subject to phase noise (see Sec. 8.5). In many situations, referencing the frame

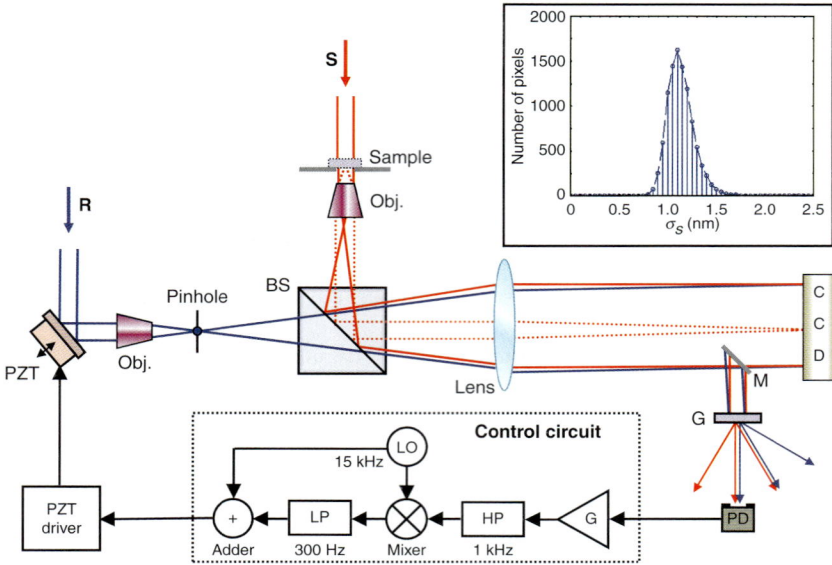

FIGURE 9.10 Stabilized Hilbert phase microscope. PZT, piezo-electric transducer; Obj, microscope objective; BS, beam splitter cube; M, mirror; PD, photodetector; G, amplifier; HP, high-pass filter; LP, low-pass filter. [(From Fig. 1, Ref. 59.) APS, used with permission, *Physical Review Letters*, G. Popescu, T. Ikeda, K. Goda, C.A. Best-Popescu, M. Laposata, S. Manley, R.R. Dasari, K. Badizadegan & M.S. Feld, "Optical measurement of cell membrane tension," 97, 218101(2006).]

to a stable point in the field of view helps remove most of the frame-to-frame instability. However, studying dynamics in live cell environments sometimes does not provide the luxury of having a fixed point in the field of view (i.e., all points may fluctuate). Therefore, in 2006, we developed an active feedback loop that was interfaced with the HPM system, as shown in Fig. 9.10 (Ref. [59]). The core of the instrument is the same as in Fig. 9.9, except that now a portion of the interfering fields is extracted from the interferometer and sent into the feedback loop. Thus, in order to suppress the inherent optical path-length noise present in the interferometer, we used an electronic feedback system that locks the interferometer on an interference fringe. The feedback circuit actively operates via a piezo-electric transducer (PZT) in the reference arm, using a combination of spatial and temporal modulation, as follows. A small mirror, M, deflects a portion of the interfering beams before they reach the CCD. At a plane conjugate to the image (CCD) plane, we place an amplitude grating, which has the same period as the interferogram. Due to this spatial matching, each diffraction order n produced by the sample beam overlaps with the order $n - 1$ of the reference beam. The two interfering beams propagating on the axis are spatially isolated by an aperture and detected by the photodiode PD. The feedback loop operates on a principle similar to that described in Ref. [60]. The PD signal is mixed with the local oscillator (LC) that modulates the reference arm length, and then is low-pass filtered, yielding the error signal. The frequency of the LO is chosen to be $\Omega/(2\pi) = 15$ kHz, such that the displacements of the sample can be measured up to about 1 kHz. The control signal that corrects the reference arm length with respect to the sample arm against the fringe fluctuations is combined with the LO modulation signal and fed to the PZT. The stability of the instrument against the residual noise was assessed by acquiring successive phase images of the field of view with no sample. The optical path-length standard deviation σ_s calculated from 128 phase images at 10 ms/frame was calculated for each pixel. The inset of Fig. 9.1 shows the histogram of the standard deviation obtained from an area of 100×100 pixels. The average standard deviation had a value of 1.2 nm, which demonstrates the efficacy of the active stabilization and suitability of the instrument for quantifying membrane motions.

This instrument allowed nanometer sensitive and quantitative measurements of red blood cell membrane fluctuations, as described in Sec. 9.2.3.

HPM and Confocal Reflectance Microscopy

In 2009, Lue at al., from MIT, demonstrated the experimental combination of HPM and confocal reflectance microscopy.[61] The benefit of this composite instrument is that the confocal microscope can provide the physical thickness of the specimen independently from the phase map rendered by HPM, which, in turn, yields the refractive index. Essentially, the two methods used in combination on the same field of view provide a *decoupling* between thickness and refractive index. The potential of the instrument for measuring refractive index of live cells has been demonstrated with experiments of He-La cells.

9.2.3   Biological Applications

Since its development, HPM has been employed successfully in a number of biological applications, as described below.

Red Blood Cell Morphology

In 2005, we showed that the newly developed HPM can quantify morphology of red blood cells (RBCs) with nanoscale accuracy and dynamically.[55] Mature erythrocytes represent a very particular type of structure; they lack nuclei and organelles and thus can be modeled as optically homogeneous objects, i.e., they produce local optical phase shifts that are proportional to their thickness. Therefore, measuring quantitative phase images of red blood cells provides cell-thickness profiles with an accuracy that corresponds to a very small fraction of the optical wavelength. An example of nanometer-scale topography of red blood cell quantified by Hilbert phase microscopy is shown in Fig. 9.11. The thickness profile of the cell, $h(x, y)$, relates to the measured phase, $\phi(x, y)$, as $h(x, y) = (\lambda/2\pi\Delta n)\phi(x, y)$, where Δn is the refractive index contrast between the hemoglobin contained in the cell and the surrounding fluid (plasma). The cell thickness profile obtained by this method is shown in Figs. 9.11b and 9.11d.

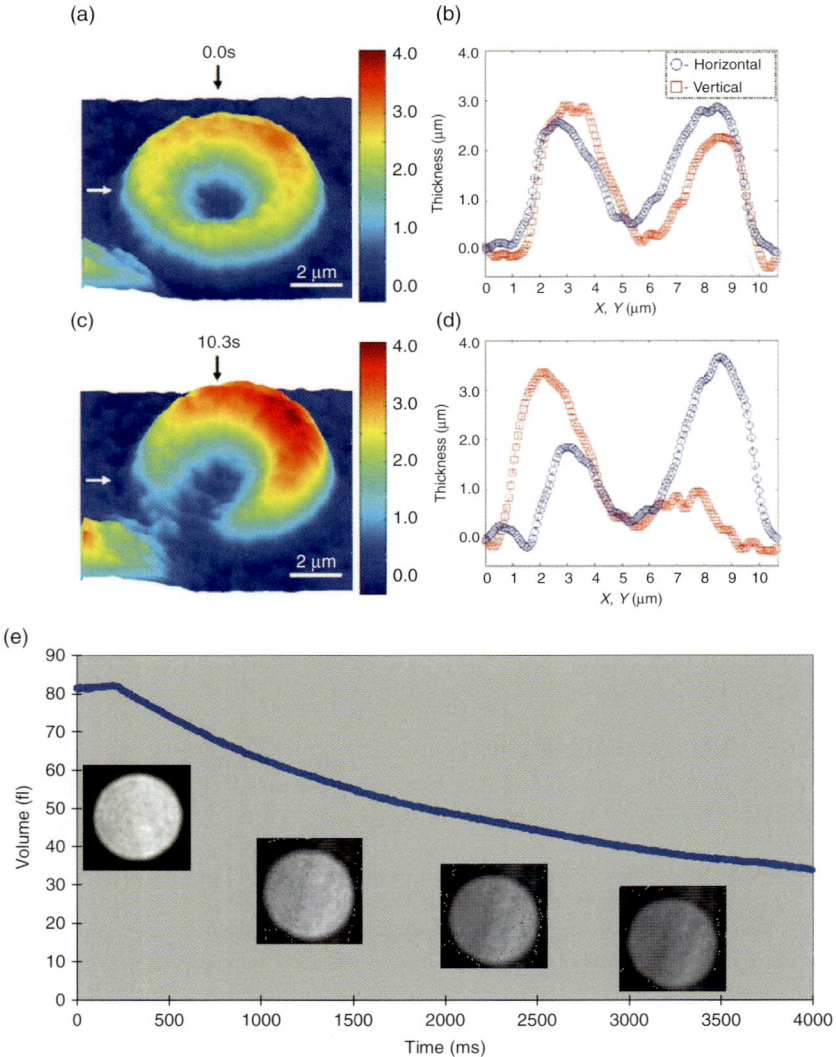

FIGURE 9.11   Quantitative assessment of the shape transformation associate with a RBC during a 10-s period. The profiles in (b) and (d) are measured along the profiles indicated by the arrows in (a) and (c). (e) Cell volume as a function of time during hemolysis that starts at approximately t = 300 ms. (Adapted from Ref. 55.)

The volume of individual cells is measured from the HPM data as $V = \int h(x,y)dxdy$. Figure 9.11e depicts the volume of RBC measured by HPM during spontaneous hemolysis, i.e., after the membrane ruptured and the cell started to lose hemoglobin. This result demonstrates the ability of quantitative-phase imaging to provide RBC volumetry without the need for preparation. It represents a significant advance with respect to current techniques that require the cells to be prepared such that they assume spherical shapes.[62]

Cell Refractometry in Microfluidic Channels

In 2006, HPM was employed to quantify the average refractive index of live cells flowing in microfluidic devices.[63] Interaction of optical fields with tissues is largely determined by the three-dimensional refractive index distribution associated with the biological structure.[64–67] Yet, direct measurements of cell and tissue refractive index are not readily available today.

In order to extract the refractive index information independently from the cell thickness, we placed live He-La cells (a human epithelial carcinoma cell line) in microchannels of fixed dimension that confine the cell in vertical direction.[63] Single input and single output microchannels of rectangular cross sections were prepared by molding elastomer on microstructures fabricated on a silicon wafer (for details see Ref. [68]). First, the refractive index of the culture medium (CM) was determined by acquiring successive phase images of microchannels filled both with CM and water. Thus, we measured $n_{CM} = 1.337$ with a standard deviation of 0.0015.

He-La cells suspended in CM were then introduced into the microchannels, which deformed the cells and confined them in the vertical direction. Figure 9.12a shows an example of a quantitative phase image of such a sample. Because the microchannel thickness exhibits some variability due to the fabrication process, we quantified the phase shift of both the cell and CM with respect to the microchannel, which eliminated the need for *a priori* knowledge of the microchannel height. This procedure was applied for measuring the refractive index of 17 cells and the results summarized in Fig. 9.12b. This histogram shows that the measured refractive index values are characterized by a small standard deviation, $n_{cell} = 1.384 \pm 0.0018$. This value for the average refractive index is comparable to other results published in the literature[69–71]. We performed also independent measurements of refractive index on cells in suspension. In this case, we used the fact that cells in suspension assume a spherical shape, which enables extraction of the refractive index from the quantitative phase image. The value obtained, $n_{cell} = 1.3846 \pm 0.0049$, agrees very well with the result obtained from the microchannel experiments.

Red Blood Cell Membrane Fluctuations

In 2006, our group at MIT applied the stabilized HPM system to study RBC membrane fluctuations. By far, the study of RBC dynamics

(a)

(b)

Mean = 1.3843
SD = 0.00177

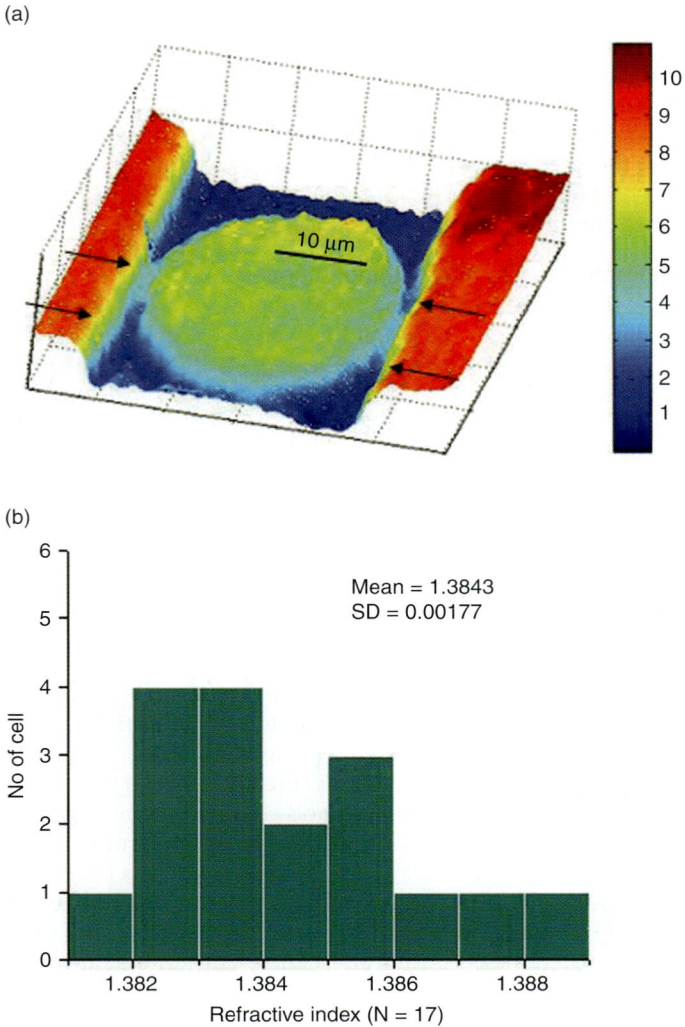

FIGURE 9.12 (*a*) He La cell confined in microchannel. Colorbar indicates path-length in radian. (*b*) Histogram of cell refractive index. (Adapted from Ref. [63]).

has been the most impactful application of QPI methods in biology.[43,54,55,59,72–85] This area of study will be revisited later in the context of other QPI techniques. In this section, we restrict our discussion to measurements by (stabilized) HPM.

Because RBCs have a relatively simple structure,[86,87] they represent a convenient model for studying cell membranes. On the other hand, understanding cell membranes has broad applications in both science and technology.[88,89] The lipid bilayer is 4 to 5 nm thick, and exhibits fluidlike behavior, characterized by a finite bending modulus, κ, and a vanishing shear modulus, $\mu \approx 0$. The resistance to shear, crucial for RBC function, is provided by the spectrin network, which has a mesh size of ~80 nm.

Spontaneous membrane fluctuations, or "flickering," have been modeled theoretically under both static and dynamic conditions in an attempt to connect the statistical properties of the membrane displacements to relevant mechanical properties of the cell.[57,90–93] These thermally-induced membrane motions exhibit 100-nm scale amplitudes at frequencies of tens of hertz. In past studies, measurements of the membrane mean squared displacement vs. spatial wave number, $\Delta u^2(q)$, revealed a q^{-4} dependence predicted by the equipartition theorem, which is indicative of *fluidlike behavior,* i.e., no shear resistance.[57,58,94–96] These results conflicted with the static deformation measurements provided by micropipette aspiration,[97,98] high-frequency electric fields,[99,100] and, more recently, optical tweezers,[101] which indicate an average value for the shear elasticity of the order of $\mu \sim 10^{-6}$ J/m^2. Gov et al. predicted that the cytoskeleton pinning of the membrane has an overall effect of confining the fluctuations and, thus, gives rise to superficial tension-like term much larger than in the case of free bilayers.[90]

Prior optical methods for studying RBC dynamics, including phase contrast microscopy (PCM),[57] reflection interference contrast microscopy (RICM),[58] and fluorescence interference contrast (FLIC)[102] are limited in their ability to measure cell membrane displacements. It is well known that PCM provides phase shifts quantitatively only for samples that are optically much thinner than the wavelength of light, which is a condition hardly satisfied by any cell type. Similarly, a single RICM measurement cannot provide the absolute cell thickness unless additional measurements or approximations are made.[103] FLIC relies on inferring the absolute position of fluorescent dye molecules attached to the membrane from the absolute fluorescence intensity, which may limit both the sensitivity and acquisition rate of the technique.[102]

We performed highly sensitive experimental measurements of thermal fluctuations associated with RBCs under different morphological conditions. The results reveal the effect of the cytoskeleton on the RBC fluctuations and support the model proposed by Gov et al.

In order to quantify membrane fluctuations at the nanometer and millisecond scales with high transverse resolution, we used the stabilized Hilbert phase microscopy (HPM),[54,55] referred to as the stabilized Hilbert phase microscopy (sHPM) (see Fig. 9.10).[59]

Our samples were primarily composed of RBCs with typical discocytic shapes, but also contained cells with abnormal morphology which formed spontaneously in the suspension, such as echinocytes, with a spiculated shape, and spherocytes, approaching a spherical shape. By taking into account the free energy contributions of both the bilayer and cytoskeleton, these morphological changes have been successfully modeled.[104]

Figures 9.13a-c show typical sHPM images of cells in these three groups. For comparison, we also analyzed the motions of RBCs fixed with 40 μM gluteraldehyde, using a standard procedure.[105] The resultant mean squared displacements, $\Delta u^2(q)$, for each group of 4-5 cells are summarized in Fig. 9.13d. The fixed cells show significantly diminished fluctuations, as expected. The curves associated with the three untreated RBC groups exhibit a power law behavior with an exponent $\alpha = 2$. As in the case of vesicles, this dependence is an indication of tension; however, the RBC tension is determined by the confinement of the bilayer by the cytoskeleton.[90,106] Based on this model, we fitted the data to extract the tension coefficient for each individual cell. The average values obtained for the discocytes, echinocytes, and spherocytes are, respectively, $\sigma = (1.5 \pm 0.2) \cdot 10^{-6}$ J/m^2, $\sigma = (4.05 \pm 1.1) \cdot 10^{-6}$ J/m^2, and $\sigma = (8.25 \pm 1.6) \cdot 10^{-6}$ J/m^2. The tension coefficient of red blood cells is 4-24 times larger than what we measured for vesicles, which suggests that the contribution of the cytoskeleton might be responsible for this enhancement. Further, it is known that the cytoskeleton plays a role in the transitions from a normal red blood cell shape to abnormal morphology, such as echinocyte and spherocyte.[104] Therefore, the consistent increase in tension we measured for the discocyte-echinocyte-spherocyte transition can be explained by changes in the cytoskeleton, which pins the bilayer. Compared to other optical techniques used for studying membrane fluctuations, the sHPM technique used here is quantitative in terms of membrane topography and displacements, highly sensitive to the nanoscale membrane motions, and provides high transverse resolution.

Very recently, new theoretical advances have shown that the tension-like behavior can be explained by taking into account the curvature of the cell.[73,107] We will discuss this in more detail in Chap. 11.

Tissue Refractometry

In 2007, we used HPM to measure the refractive index distribution in tissue slices.[108] HPM has been applied for the first time to image unstained 5-μm thick tissue slices of mouse brain, spleen, and liver.[108] The refractive properties of the tissue are retrieved in terms of the average refractive index and its spatial variation (Fig. 9.14). It was found that the average refractive index varies significantly with tissue type, such that the brain is characterized by the lowest value and the liver by the highest. This approach opens a new possibility for "stain-free" characterization of tissues, where the diagnostic power is provided by the intrinsic refractive properties of the biological structure. Results obtained on liver tissue affected by a lysosomal storage

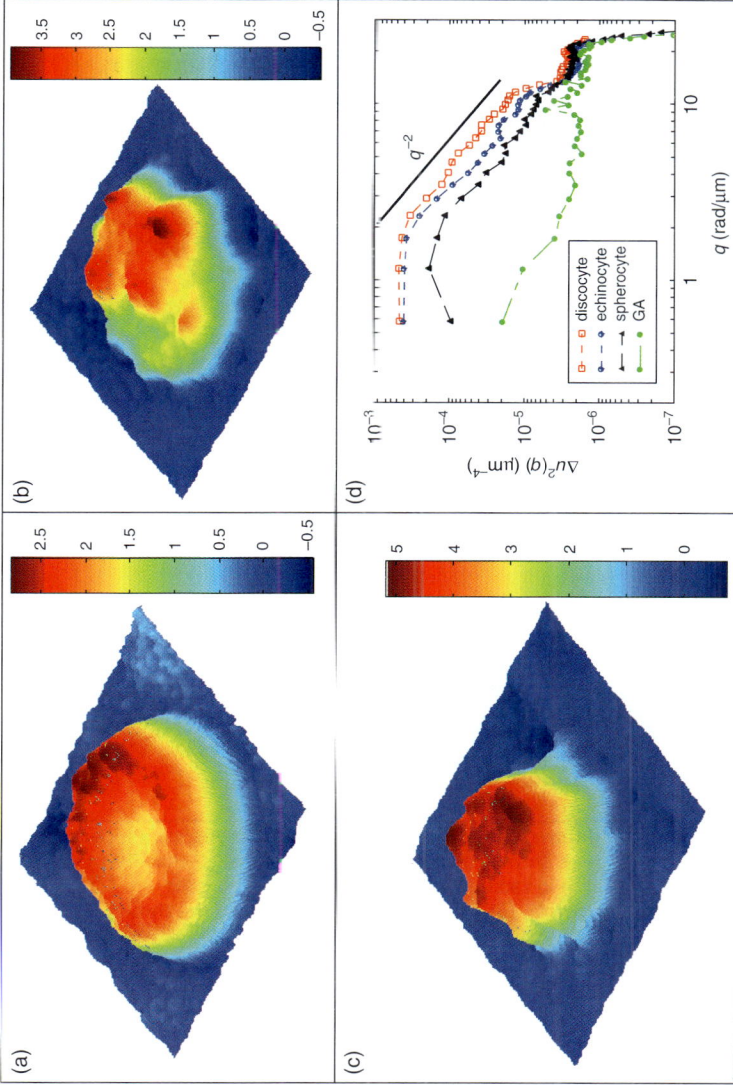

FIGURE 9.13 (a-c) sHPM images of a discocyte (a), echinocyte (b), and spherocyte (c). The colorbar shows thickness in microns. (d) Mean squared displacements for the three RBC groups and for the gluteraldehyde (GA) fixed cells (Fig. 3, Ref. [59]). APS, used with permission, Physical Review Letters, G. Popescu, T. Ikeda, K. Goda, C.A. Best-Popescu, M. Laposata, S. Manley, R.R. Dasari, K. Badizadegan & M.S. Feld, "Optical measurement of cell membrane tension," 97, 218101 (2006).

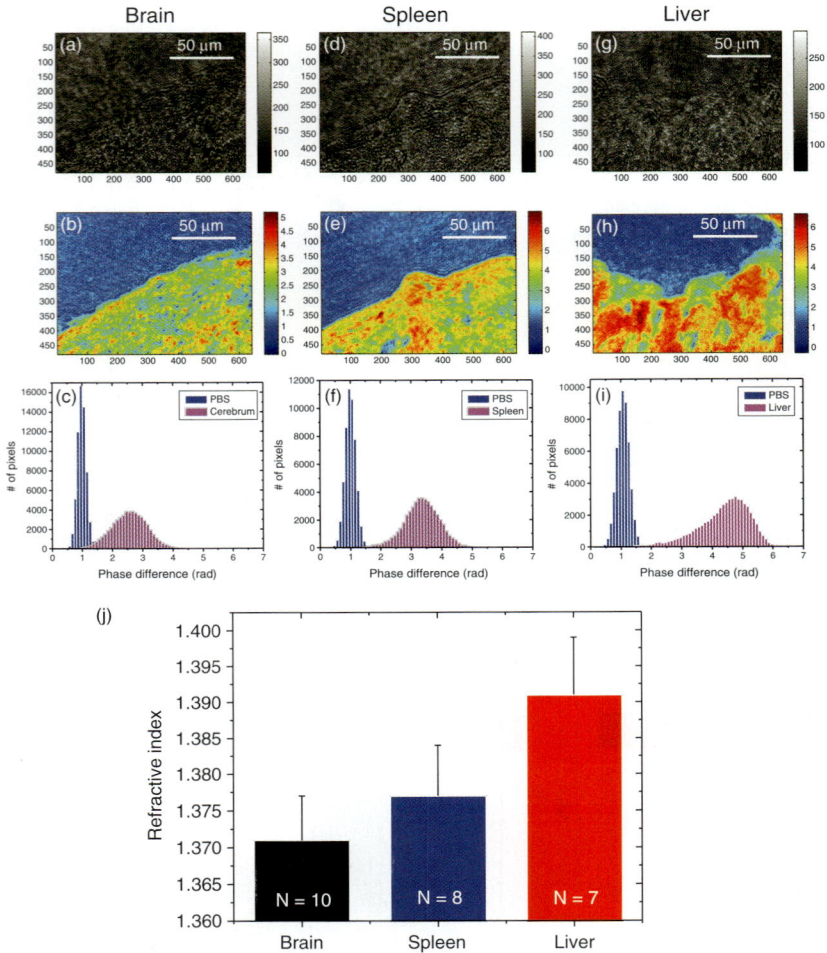

FIGURE 9.14 Examples of quantitative phase imaging studies of tissues slices: (*a-c*) brain, (*d-f*) spleen, (*g-j*) liver. Row 1 shows the bright field images, row 2 the HPM images (colorbars represent phase in radian), and row 3 the corresponding histogram of the phase shifts in row 2. (*k*) Average refractive index for the three sample groups, as indicated. The error bars indicate the sample to sample variation (N samples group, as shown) (*Adapted from Ref 108.*).

disease showed that QPI can quantify structural changes during this disease development.

We will discuss further the ability of QPI to measure scattering properties of tissues in Chap. 13, in the context of Fourier transform light scattering, as well as in Chap. 15, in the context of QPI prospects for clinical applications.

References

1. B. R. Brown and A. W. Lohmann. "Complex spatial filtering with binary masks." *Applied Opics,* 5, 967 (1966).
2. J. W. Goodman and R. W. Lawrence. "Digital image formation from electronically detected holograms." *Applied Physics Letters,* 11, 77 (1967).
3. M. Takeda, H. Ina, and S. Kobayashi. "Fourier-transform method of fringe-pattern analysis for computer-based topography and interferometry." *Journal of the Optical Society of America,* 72, 156–160 (1982).
4. U. Schnars and W. Jüptner. "Direct recording of holograms by a CCD target and numerical reconstruction." *Applied Optics,* 33, 179–181 (1994).
5. E. Cuche, F. Bevilacqua, and C. Depeursinge. "Digital holography for quantitative phase-contrast imaging." *Optics Letters,* 24, 291–293 (1999).
6. E. Cuche, P. Marquet, and C. Depeursinge. "Simultaneous amplitude-contrast and quantitative phase-contrast microscopy by numerical reconstruction of Fresnel off-axis holograms." *Applied Optics,* 38, 6994–7001 (1999).
7. J. W. Goodman. *Introduction to Fourier Optics* (McGraw-Hill, New York, 1996).
8. I. Yamaguchi and T. Zhang. "Phase-shifting digital holography." *Optics Letters,* 22, 1268–1270 (1997).
9. T. Zhang and I. Yamaguchi. "Three-dimensional microscopy with phase-shifting digital holography." *Optics Letters,* 23, 1221–1223 (1998).
10. P. Y. Guo and A. J. Devaney. "Digital microscopy using phase-shifting digital holography with two reference waves." *Optics Letters,* 29, 857–859 (2004).
11. J. Garcia-Sucerquia, W. B. Xu, S. K. Jericho, P. Klages, M. H. Jericho, and H. J. Kreuzer. "Digital in-line holographic microscopy." *Applied Optics,* 45, 836–850 (2006).
12. W. B. Xu, M. H. Jericho, I. A. Meinertzhagen, and H. J. Kreuzer. "Digital in-line holography for biological applications." *Proceedings of the National Academy of Sciences of the United States of America,* 98, 11301–11305 (2001).
13. N. T. Shaked, T. M. Newpher, M. D. Ehlers, and A. Wax. "Parallel on-axis holographic phase microscopy of biological cells and unicellular microorganism dynamics." *Applied Optics,* 49, 2872–2878 (2010).
14. M. K. Kim. "Wavelength-scanning digital interference holography for optical section imaging." *Optics Letters,* 24, 1693–1695 (1999).
15. G. Indebetouw and P. Klysubun. "Spatiotemporal digital microholography." *Journal of the Optical Society of America a-Optics Image Science and Vision,* 18, 319–325 (2001).
16. F. Dubois, C. Schockaert, N. Callens, and C. Yourassowsky. "Focus plane detection criteria in digital holography microscopy by amplitude analysis." *Optics Express,* 14, 5895–5908 (2006).
17. P. Langehanenberg, B. Kemper, D. Dirksen, and G. von Bally. "Autofocusing in digital holographic phase contrast microscopy on pure phase objects for live cell imaging." *Applied Optics,* 47, D176–D182 (2008).
18. P. Ferraro, S. Grilli, D. Alfieri, S. De Nicola, A. Finizio, G. Pierattini, B. Javidi, G. Coppola, and V. Striano. "Extended focused image in microscopy by digital holography." *Optics Express,* 13, 6738–6749 (2005).
19. C. J. Mann, L. F. Yu, C. M. Lo, and M. K. Kim. "High-resolution quantitative phase-contrast microscopy by digital holography." *Optics Express,* 13, 8693–8698 (2005).
20. F. Palacios, J. Ricardo, D. Palacios, E. Goncalves, J. L. Valin, and R. De Souza. "3D image reconstruction of transparent microscopic objects using digital holography." *Optics Communications,* 248, 41–50 (2005).
21. P. Ferraro, D. Alferi, S. De Nicola, L. De Petrocellis, A. Finizio, and G. Pierattini. "Quantitative phase-contrast microscopy by a lateral shear approach to digital holographic image reconstruction." *Optics Letters,* 31, 1405–1407 (2006).
22. T. R. Hillman, S. A. Alexandrov, T. Gutzler, and D. D. Sampson. "Microscopic particle discrimination using spatially-resolved Fourier-holographic light scattering angular spectroscopy." *Optics Express,* 14, 11088–11102 (2006).

23. C. J. Mann, L. F. Yu, and M. K. Kim. "Movies of cellular and sub-cellular motion by digital holographic microscopy." *Biomedical Engineering Online*, 5, 21 (2006).
24. P. Ferraro, S. De Nicola, A. Finizio, G. Coppola, S. Grilli, C. Magro, and G. Pierattini. "Compensation of the inherent wave front curvature in digital holographic coherent microscopy for quantitative phase-contrast imaging." *Applied Optics*, 42, 1938–1946 (2003).
25. D. Parshall and M. K. Kim. "Digital holographic microscopy with dual-wavelength phase unwrapping." *Applied Optics*, 45, 451–459 (2006).
26. J. Kuhn, T. Colomb, F. Montfort, F. Charriere, Y. Emery, E. Cuche, P. Marquet, and C. Depeursinge. "Real-time dual-wavelength digital holographic microscopy with a single hologram acquisition." *Optics Express*, 15, 7231–7242 (2007).
27. W. M. Ash, L. Krzewina, and M. K. Kim. "Quantitative imaging of cellular adhesion by total internal reflection holographic microscopy." *Applied Optics*, 48, H144–H152 (2009).
28. S. A. Alexandrov, T. R. Hillman, T. Gutzler, and D. D. Sampson. "Synthetic aperture fourier holographic optical microscopy." *Physical Review Letters*, 97, 168102 (2006).
29. C. Liu, Z. G. Liu, F. Bo, Y. Wang, and J. Q. Zhu. "Super-resolution digital holographic imaging method." *Applied Physics Letters*, 81, 3143–3145 (2002).
30. V. Mico, Z. Zalevsky, P. Garcia-Martinez, and J. Garcia. "Synthetic aperture superresolution with multiple off-axis holograms." *Journal of the Optical Society of America a-Optics Image Science and Vision*, 23, 3162–3170 (2006).
31. G. Indebetouw, Y. Tada, J. Rosen, and G. Brooker. "Scanning holographic microscopy with resolution exceeding the Rayleigh limit of the objective by superposition of off-axis holograms," *Applied Optics*, 46, 993–1000 (2007).
32. J. R. Price, P. R. Bingham, and C. E. Thomas. "Improving resolution in microscopic holography by computationally fusing multiple, obliquely illuminated object waves in the Fourier domain." *Applied Optics*, 46, 827–833 (2007).
33. M. Paturzo, F. Merola, S. Grilli, S. De Nicola, A. Finizio, and P. Ferraro. "Super-resolution in digital holography by a two-dimensional dynamic phase grating." *Optics Express*, 16, 17107–17118 (2008).
34. K. J. Chalut, W. J. Brown, and A. Wax. "Quantitative phase microscopy with asynchronous digital holography." *Optics Express*, 15, 3047–3052 (2007).
35. N. T. Shaked, Y. Z. Zhu, M. T. Rinehart, and A. Wax. "Two-step-only phase-shifting interferometry with optimized detector bandwidth for microscopy of live cells." *Optics Express*, 17, 15585–15591 (2009).
36. M. Kemmler, M. Fratz, D. Giel, N. Saum, A. Brandenburg, and C. Hoffmann. "Noninvasive time-dependent cytometry monitoring by digital holography." *Journal of Biomedical Optics*, 12, 064002-1 (2007).
37. I. Moon and B. Javidi. "Three-dimensional identification of stem cells by computational holographic imaging." *Journal of the Royal Society Interface*, 4, 305–313 (2007).
38. B. Kemper and G. von Bally. "Digital holographic microscopy for live cell applications and technical inspection." *Applied Optics*, 47, A52–A61 (2008).
39. H. Y. Sun, B. Song, H. P. Dong, B. Reid, M. A. Player, J. Watson, and M. Zhao. "Visualization of fast-moving cells in vivo using digital holographic video microscopy." *Journal of Biomedical Optics*, 13, 014007-1 (2008).
40. W. C. Warger, J. A. Newmark, C. M. Warner, and C. A. DiMarzio. "Phase-subtraction cell-counting method for live mouse embryos beyond the eight-cell stage." *Journal of Biomedical Optics*, 13, 034005-1 (2008).
41. L. F. Yu, S. Mohanty, G. J. Liu, S. Genc, Z. P. Chen, and M. W. Berns. "Quantitative phase evaluation of dynamic changes on cell membrane during laser microsurgery." *Journal of Biomedical Optics*, 13, 050508-1 (2008).
42. Y. S. Choi and S. J. Lee. "Three-dimensional volumetric measurement of red blood cell motion using digital holographic microscopy." *Applied Optics*, 48, 2983–2990 (2009).

43. B. Rappaz, A. Barbul, A. Hoffmann, D. Boss, R. Korenstein, C. Depeursinge, P. J. Magistretti, and P. Marquet. "Spatial analysis of erythrocyte membrane fluctuations by digital holographic microscopy." *Blood Cells Molecules and Diseases,* 42, 228–232 (2009).

44. B. Rappaz, E. Cano, T. Colomb, J. Kuhn, C. Depeursinge, V. Simanis, P. J. Magistretti, and P. Marquet. "Noninvasive characterization of the fission yeast cell cycle by monitoring dry mass with digital holographic microscopy." *Journal of Biomedical Optics.* 14, 034049 (2009).

45. C. Remmersmann, S. Sturwald, B. Kemper, P. Langehanenberg, and G. von Bally, "Phase noise optimization in temporal phase-shifting digital holography with partial coherence light sources and its application in quantitative cell imaging." *Applied Optics,* 48, 1463–1472 (2009).

46. A. Rosenhahn, F. Staier, T. Nisius, D. Schafer, R. Barth, C. Christophis, L. M. Stadler, S. Streit-Nierobisch, C. Gutt, A. Mancuso, A. Schropp, J. Gulden, B. Reime, J. Feldhaus, E. Weckert, B. Pfau, C. M. Gunther, R. Konnecke, S. Eisebitt, M. Martins, B. Faatz, N. Guerassimova, K. Honkavaara, R. Treusch, E. Saldin, S. Schreiber, E. A. Schneidmiller, M. V. Yurkov, I. Vartanyants, G. Grubel, M. Grunze, and T. Wilhein. "Digital In-line Holography with femtosecond VUV radiation provided by the free-electron laser FLASH." *Optics Express,* 17, 8220–8228 (2009).

47. M. K. Kim. "Applications of digital holography in biomedical microscopy." *Journal of the Optical Society of Korea,* 14, 77–89 (2010).

48. A. Ligresti, L. De Petrocellis, D. H. P. de la Ossa, R. Aberturas, L. Cristino, A. S. Moriello, A. Finizio, M. E. Gil, A. I. Torres, J. Molpeceres, and V. Di Marzo. "Exploiting nanotechnologies and TRPV1 channels to investigate the putative anandamide membrane transporter." *PLoS ONE,* 5, e10239 (2010).

49. O. Mudanyali, D. Tseng, C. Oh, S. O. Isikman, I. Sencan, W. Bishara, C. Oztoprak, S. K. Seo, B. Khademhosseini, and A. Ozcan. "Compact, light-weight and cost-effective microscope based on lensless incoherent holography for telemedicine applications." *Lab on a Chip,* 10, 1417–1428 (2010).

50. C. Pache, J. Kuhn, K. Westphal, M. F. Toy, J. Parent, O. Buchi, A. Franco-Obregon. C. Depeursinge, and M. Egli, "Digital holographic microscopy real-time monitoring of cytoarchitectural alterations during simulated microgravity." *Journal of Biomedical Optics,* 15, 026021-1 (2010).

51. N. Warnasooriya, F. Joud, F. Bun, G. Tessier, M. Coppey-Moisan, P. Desbiolles, M. Atlan, M. Abboud, and M. Gross. "Imaging gold nanoparticles in living cell environments using heterodyne digital holographic microscopy." *Optics Express,* 18, 3264–3273 (2010).

52. P. Marquet, B. Rappaz, P. J. Magistretti, E. Cuche, Y. Emery, T. Colomb, and C. Depeursinge. "Digital holographic microscopy: a noninvasive contrast imaging technique allowing quantitative visualization of living cells with subwavelength axial accuracy." *Optics Letters,* 30, 468–470 (2005).

53. B. Rappaz, P. Marquet, E. Cuche, Y. Emery, C. Depeursinge, and P. J. Magistretti. "Measurement of the integral refractive index and dynamic cell morphometry of living cells with digital holographic microscopy." *Optics Express,* 13, 9361–9373 (2005).

54. T. Ikeda, G. Popescu, R. R. Dasari, and M. S. Feld. "Hilbert phase microscopy for investigating fast dynamics in transparent systems." *Optics Letters,* 30, 1165–1168 (2005).

55. G. Popescu, T. Ikeda, C. A. Best, K. Badizadegan, R. R. Dasari, and M. S. Feld. "Erythrocyte structure and dynamics quantified by Hilbert phase microscopy." *Journal of Biomedical Optics Letters,* 10, 060503 (2005).

56. M. Hammer, D. Schweitzer, B. Michel, E. Thamm, and A. Kolb. "Single scattering by red blood cells." *Applied Optics,* 37, 7410–7418 (1998).

57. F. Brochard and J. F. Lennon. "Frequency spectrum of the flicker phenomenon in erythrocytes." *Journal de Physique,* 36, 1035–1047 (1975).

58. A. Zilker, H. Engelhardt, and E. Sackmann. "Dynamic reflection interference contrast (ric-) microscopy—a new method to study surface excitations of cells and to measure membrane bending elastic-moduli." *Journal de Physique,* 48, 2139–2151 (1987).

59. G. Popescu, T. Ikeda, K. Goda, C. A. Best-Popescu, M. Laposata, S. Manley, R. R. Dasari, K. Badizadegan, and M. S. Feld. "Optical measurement of cell membrane tension." *Physical Review Letters*, 97, 218101 (2006).
60. R. W. P. Drever, J. L. Hall, F. V. Kowalski, J. Hough, G. M. Ford, A. J. Munley, and H. Ward. "Laser phase and frequency stabilization using an optical-resonator." *Applied Physics B-Photophysics and Laser Chemistry*, 31, 97–105 (1983).
61. N. Lue, W. Choi, G. Popescu, Z. Yaqoob, K. Badizadegan, R. R. Dasari, and M. S. Feld. "Live cell refractometry using Hilbert phase microscopy and confocal reflectance microscopy." *Journal of Physical Chemistry A*, 113, 13327–13330 (2009).
62. C. A. Best. *Fatty acid ethyl esters and erythrocytes: Metabolism and membrane effects, Ph.D. Thesis*, (Northeastern University, Boston, MA, 2005).
63. N. Lue, G. Popescu, T. Ikeda, R. R. Dasari, K. Badizadegan, and M. S. Feld. "Live cell refractometry using microfluidic devices." *Optics Letters*, 31, 2759 (2006).
64. V. Backman, M. B. Wallace, L. T. Perelman, J. T. Arendt, R. Gurjar, M. G. Muller, Q. Zhang, G. Zonios, E. Kline, J. A. McGilligan, S. Shapshay, T. Valdez, K. Badizadegan, J. M. Crawford, M. Fitzmaurice, S. Kabani, H. S. Levin, M. Seiler, R. R. Dasari, I. Itzkan, J. Van Dam, and M. S. Feld. "Detection of preinvasive cancer cells." *Nature*, 406, 35–36 (2000).
65. R. Drezek, A. Dunn and R. Richards-Kortum. "Light scattering from cells: finite-difference time-domain simulations and goniometric measurements." *Applied Optics*, 38, 3651–3661 (1999).
66. J. R. Mourant, I. Canpolat, C. Brocker, O. Esponda-Ramos, T. M. Johnson, A. Matanock, K. Stetter, and J. P. Freyer. "Light scattering from cells: The contribution of the nucleus and the effects of proliferative status." *Journal of Biomedical Optics*, 5, 131–137 (2000).
67. V. V. Tuchin. *Tissue optics* (SPIE–The International Society for Optical Engineering, 2000).
68. G. M. Whitesides, E. Ostuni, S. Takayama, X. Jiang, and D. E. Ingber. "Soft lithography in biology and biochemistry." *Annual Review of Biomedical Engineering*, 3, 335–373 (2001).
69. J. S. Maier, S. A. Walker, S. Fantini, M. A. Franceschini, and E. Gratton. "Possible Correlation between blood-glucose concentration and the reduced scattering coefficient of tissues in the near-infrared." *Optics Letters*, 19, 2062–2064 (1994).
70. J. Beuthan, O. Minet, J. Helfmann, M. Herrig, and G. Muller. "The spatial variation of the refractive index in biological cells." *Physics in Medicine and Biology*, 41, 369–382 (1996).
71. H. Liu, B. Beauvoit, M. Kimura, and B. Chance. "Dependence of tissue optical properties on solute-induced changes in refractive index and osmolarity." *Journal of Biomedical Optics*, 1, 200–211 (1996).
72. Y. K. Park, C. A. Best, T. Kuriabova, M. L. Henle, M. S. Feld, A. J. Levine, and G. Popescu. "Measurement of nonlinear microrheology of red blood cells." *Physical Review Letters*, (under review).
73. Y. K. Park, C. A. Best, K. Badizadegan, R. R. Dasari, M. S. Feld, T. Kuriabova, M. L. Henle, A. J. Levine and G. Popescu. "Measurement of red blood cell mechanics during morphological changes." *Proc. Nat. Acad. Sci.*, (2010).
74. Y. K. Park, C. A. Best, T. Auth, N. Gov, S. A. Safran, G. Popescu, S. Suresh and M. S. Feld. "Metabolic remodeling of the human red blood cell membranes." *Proc. Nat. Acad. Sci.*, 107, 1289 (2010).
75. Y. Park, M. Diez-Silva, D. Fu, G. Popescu, W. Choi, I. Barman, S. Suresh, and M. S. Feld. "Static and dynamic light scattering of healthy and malaria-parasite invaded red blood cells." *Journal of Biomedical Optics*, 15, 020506 (2010).
76. H. F. Ding and G. Popescu. "Instantaneous spatial light interference microscopy." *Optics Express*, 18, 1569–1575 (2010).
77. G. Popescu, Y. Park, W. Choi, R. R. Dasari, M. S. Feld, and K. Badizadegan. "Imaging red blood cell dynamics by quantitative phase microscopy." *Blood Cells Molecules and Diseases*, 41, 10–16 (2008).
78. Y. K. Park, M. Diez-Silva, G. Popescu, G. Lykotrafitis, W. Choi, M. S. Feld, and S. Suresh. "Refractive index maps and membrane dynamics of human

red blood cells parasitized by Plasmodium falciparum." *Proc. Nat. Acad. Sci.,* 105, 13730 (2008).

79. G. Popescu, Y. K. Park, R. R Dasari, K. Badizadegan, and M. S. Feld. "Coherence properties of red blood cell membrane motions." *Physical Review E,* 76, 031902 (2007).

80. M. S. Amin, Y. K. Park, N. Lue, R. R. Dasari, K. Badizadegan, M. S. Feld, and G. Popescu. "Microrheology of red blood cell membranes using dynamic scattering microscopy." *Optics Express,* 15, 17001 (2007).

81. G. Popescu, T. Ikeda, R. R Dasari, and M. S. Feld. "Diffraction phase microscopy for quantifying cell structure and dynamics." *Optics Letters,* 31, 775–777 (2006).

82. G. Popescu, K. Badizadegan, R. R. Dasari, and M. S. Feld. "Observation of dynamic subdomains in red blood cells." *Journal of Biomedical Optics Letters,* 11, 040503 (2006).

83. Y. K. Park, G. Popescu, K. Badizadegan, R. R. Dasari, and M. S. Feld. "Diffraction phase and fluorescence microscopy." *Optics Express,* 14, 8263 (2006).

84. G. Popescu, L. P. Deflores, J C. Vaughan, K. Badizadegan, H. Iwai, R. R. Dasari, and M. S. Feld. "Fourier phase microscopy for investigation of biological structures and dynamics." *Optics Letters,* 29, 2503–2505 (2004).

85. B. Rappaz, A. Barbul, Y. Emery, R. Korenstein, C. Depeursinge, P. J. Magistretti, and P. Marquet. "Comparative study of human erythrocytes by digital holographic microscopy, confocal microscopy, and impedance volume analyzer." *Cytometry Part A,* 73A, 895–903 (2008).

86. D. Boal. *Mechanics of the cell* (Cambridge University Press, 2002).

87. R. M. Hochmuth and R. E. Waugh. "Erythrocyte membrane elasticity and viscocity." *Annual Review of Physiology.* 49, 209–219 (1987).

88. R. Lipowsky. "The conformation of membranes." *Nature,* 349, 475–481 (1991).

89. E. Sackmann. "Supported membranes: Scientific and practical applications." *Science,* 271, 43–48 (1996).

90. N. Gov, A. G. Zilman, and S. Safran. "Cytoskeleton confinement and tension of red blood cell membranes." *Physical Review Letters,* 90, 228101 (2003).

91. N. Gov. "Membrane undulations driven by force fluctuations of active proteins." *Physical Review Letters,* 93, 268104 (2004).

92. N. S. Gov and S. A. Safran. "Red blood cell membrane fluctuations and shape controlled by ATP-induced cytoskeletal defects." *Biophysics Journal,* 88, 1859–1874 (2005).

93. R. Lipowski and M. Girardet. "Shape fluctuations of polymerized or solidlike membranes." *Physical Review Letters ,* 65, 2893–2896 (1990).

94. K. Zeman, E. H. and E. Sackman. *Eur. Biophys. J.,* 18, 203 (1990).

95. A. Zilker, M. Ziegler and E. Sackmann. "Spectral-analysis of erythrocyte flickering in the 0.3-4-Mu-M-1 regime by microinterferometry combined with fast image-processing." *Physics Review A,* 46, 7998–8002 (1992).

96. H. Strey, M. Peterson, and E. Sackmann. "Measurement of erythrocyte membrane elasticity by flicker eigenmode decomposition." *Biophysics Journal,* 69, 478–488 (1995).

97. D. E. Discher, N. Mohandas, and E. A. Evans. "Molecular maps of red cell deformation: Hidden elasticity and in situ connectivity." *Science,* 266, 1032–1035 (1994).

98. R. M. Hochmuth, P. R. Worthy, and E. A. Evans. "Red cell extensional recovery and the determination of membrane viscosity." *Biophysics Journal,* 26, 101–114 (1979).

99. H. Engelhardt and E. Sackmann. "On the measurement of shear elastic moduli and viscosities of erythrozyte plasma membranes by transient deformation in high frequency electric fields." *Biophysics Journal,* 54, 495–508 (1988).

100. H. Engelhardt, H. Gaub, and E. Sackmann, "Viscoelastic properties of erythrocyte membranes in high-frequency electric fields." *Nature,* 307, 378–380 (1984).

101. S. Suresh, J. Spatz, J. P. Mills, A. Micoulet, M. Dao, C. T. Lim, M. Beil, and T. Seufferlein. "Connections between single-cell biomechanics and human disease states: gastrointestinal cancer and malaria." *Acta Biomaterialia,* 1, 15–30 (2005).

102. Y. Kaizuka and J. T. Groves. "Hydrodynamic damping of membrane thermal fluctuations near surfaces imaged by fluorescence interference microscopy." *Physics Review Letters*, 96, 118101 (2006).
103. A. Zidovska and E. Sackmann. "Brownian motion of nucleated cell envelopes impedes adhesion." *Physics Review Letters*, 96, 048103 (2006).
104. H. W. G. Lim, M. Wortis, and R. Mukhopadhyay. "Stomatocyte-discocyte-echinocyte sequence of the human red blood cell: Evidence for the bilayer-couple hypothesis from membrane mechanics. *Proc. Nat. Acad. Sci.* 99, 16766–16769 (2002).
105. C. A. Best, J. E. Cluette-Brown, M. Teruya, A. Teruya, and M. Laposata. "Red blood cell fatty acid ethyl esters: A significant component of fatty acid ethyl esters in the blood." *Journal of Lipid Ressearch*, 44, 612–620 (2003).
106. N. Gov, A. Zilman and S. Safran. "Cytoskeleton confinement of red blood cell membrane fluctuations." *Biophysics Journal*, 84, 486A–486A (2003).
107. T. Kuriabova and A. J. Levine. "Nanorheology of viscoelastic shells: Applications to viral capsids." *Physical Review E*, 77, - (2008).
108. N. Lue, J. Bewersdorf, M. D. Lessard, K. Badizadegan, K. Dasari, M. S. Feld, and G. Popescu. "Tissue refractometry using Hilbert phase microscopy." *Optics Letters*, 32, 3522 (2007).

CHAPTER **10**

Phase-Shifting
Methods

Phase-shifting interferometry has been used in metrology for decades (see, for example, Refs. 1 to 57). Of course, the implementation of this principle in QPI methods dedicated to biological investigations is more recent.[58–72] The principle of phase-shifting QPI has been already discussed in Sec. 8.2. The main advantage of phase-shifting over off-axis techniques is in the ability to render quantitative phase maps via simple mathematical operations, i.e., subtractions and divisions, which overall reduces the spatial noise. In this chapter, we present two phase-shifting QPI methods that have already demonstrated their capability for biological studies: Digitally Recorded Interference Microscopy with Automatic Phase Shifting and Optical Quadrature Microscopy.

10.1 Digitally Recorded Interference Microscopy with Automatic Phase Shifting (DRIMAPS)

The group lead by Graham Dunn at King's College, London, has used QPI very early on for in-depth cell biology studies (see, for example, Refs. 62, 63, and 71 to 75) The principle of DRIMAPS and the main results in cell biology enabled by this method are discussed below.

10.1.1 Principle

In 1995, Zicha and Dunn reported, "An image-processing system for cell behavior studies in subconfluent cultures."[71] DRIMAPS is a phase-shifting QPI method that is implemented on an existing *Horn* microscope. This experimental arrangement is shown in Fig. 10.1. The details of operating the microscope for optimal performance are presented in Ref. 74. Briefly, the illumination light is separated into two paths by the beam splitter to form Mach-Zender interferometer. The reference arm contains a sample compensator ("dummy specimen"), such that the two interferometer arms are optically similar. Before the detector, the two beams are recombined via a second beam

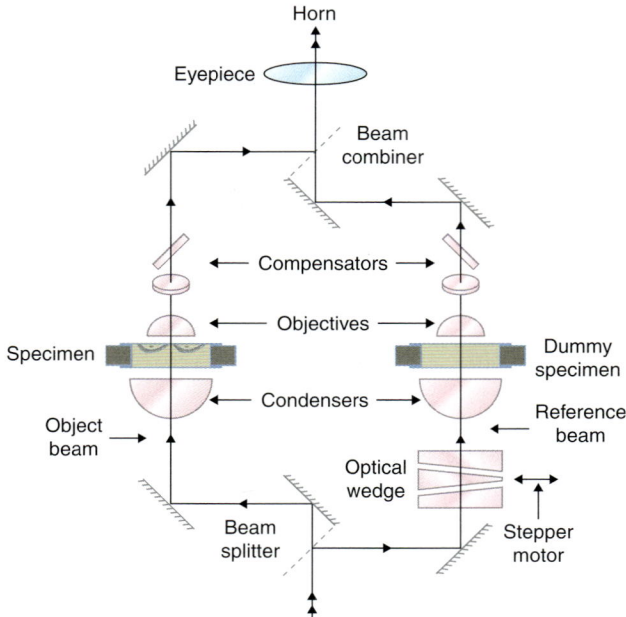

FIGURE 10.1   DRIMAPS system implemented with a Horn microscope. (*From Fig. 1, Ref. 74.*) Academic Press, San Diego, used with permission, G. A. Dunn and D. Zicha, in *Cell Biology: A Laboratory Handbook*, J. E. Celis (ed.), 1998.

splitter. The phase shifting in increments of $\pi/2$ is achieved by sliding the optical wedge horizontally. From the four recorded intensity frames, the quantitative phase image is obtained as in typical phase-shifting interferometry (see Sec. 8.2),

$$\phi(x,y) = \arg[I(x,y;0) - I(x,y;\pi), I(x,y;3\pi/2) - I(x,y;\pi/2)] \quad (10.1)$$

10.1.2   Further Developments

The same group employed a similar QPI method based on a different microscope platform, the *Jamin-Lebedeff* microscope, as shown in Fig. 10.2.[73] The main difference with respect to the Horn microscope is that in the Jamin-Lebedeff the two light paths of the interferometer are separated through polarizing optics (for details, see Ref. 74). Thus the illumination field is separated into two beams via a polarizing beam splitter, such that the resulting fields have orthogonal polarizations. The phase shifting is achieved by controlling the rotation angle of the analyzer with respect to the half waveplate ($\lambda/2$ in Fig. 10.2). The main feature of this arrangement is its stability. However, one drawback may arise if the sample exhibits birefringence.[74] Note that a similar limitation was encountered when discussing differential phase-contrast using OCT (see Sec. 7.5.1).

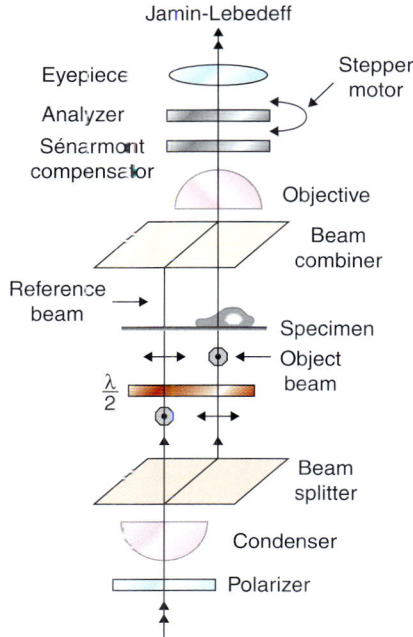

FIGURE 10.2   DRIMAPS system implemented with a Jamin-Lebedeff microscope. The Mach-Zehnder interferometer is obtained via polarization optics (*From Fig. 1, Ref. 74*). Academic Press, San Diego, used with permission, G. A. Dunn and D. Zicha, in *Cell Biology: A Laboratory Handbook*, J. E. Celis (ed.), 1998.

In the early days of interference microscopy, it was understood that there is a linear relationship between the phase map of a live cell and its non-aqueous, or *dry mass* content (see pioneering work by Barer and Davies in Refs. [77,78] and further discussion in Secs. 11.1.3.2 and 15.2.1). Much of the biological research involving DRIMAPS used this principle to generate maps of dry mass ("dry maps") of cells in cultures. These applications of the DRIMAPS system are reviewed below.

10.1.3   Applications

In 1995, Dunn and Zicha applied DRIMAPS to study "Dynamics of fibroblast spreading."[63] Figure 10.3 shows the dry mass vs. time curves for seven-day primary cells.[63] These measurements performed on chick heart fibroblasts indicated largely monotonous growth in dry mass. Breaks in the individual trajectories correspond to periods when the cells are in contact with others, although they may also result from cells leaving the field of view or dividing. The noise associated with some of these growth trajectories is due to measurement errors caused by rapid cell movement. Larger spikes and dips on the traces are caused by cells colliding with debris or leaving small fragments attached to the substrate.[63]

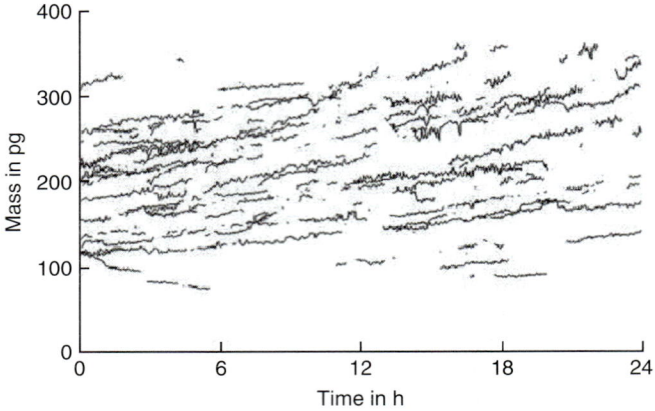

FIGURE 10.3   Trajectories of dry mass in pg vs. time in hours for chick fibroblast cultures (*From Fig. 2, Ref. 63*). *J. Cell Sci.*, used with permission, G. A. Dunn & D. Zicha, "Dynamics of Fibroblast Spreading," 108, 1239–1249 (1995).

The QPI data provided valuable information regarding the relationship between cell mass and projected area. Figure 10.4 shows a summary of these measurements, where Dunn and Zicha found clear upper limits of area per mass ratios, of the order of 7 to 8 μm^2/pg. Note that the vertical trajectories correspond to cells rounding up during mitosis. The authors point out that the ratio of daughter cell masses is not necessarily 1 and, in fact, it was observed to be as high as 1.35. It is worth emphasizing that QPI is perhaps the only

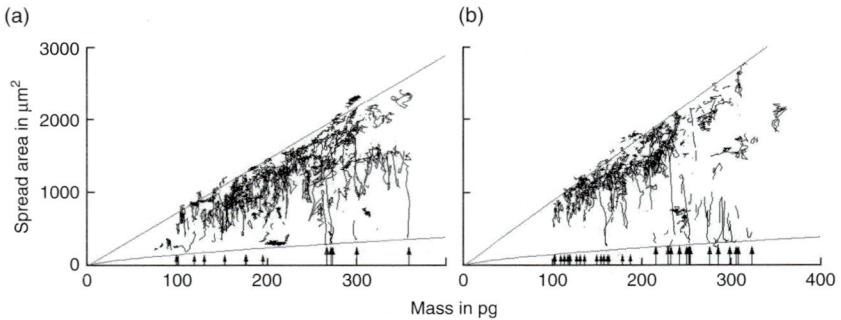

FIGURE 10.4   Trajectories of spread area in μm^2 vs. dry mass in pg for the seven-day cells (*a*) and for the eight-day cells (*b*). The straight inclined line in each plot indicates the approximate upper limit on spreading per unit mass, which is 7.2 μm^2/pg (*a*) and 8.8 μm^2/pg (*b*). The lower curve in each plot is the equatorial, cross-sectional area of a sphere of density 0.07 pg mm-3 plotted as a function of its mass. Long arrows and short arrows indicate the masses of cells measured within 1 hour before and after cytokinesis, respectively (*From Fig. 3, Ref. 63*). *J. Cell Sci.*, used with permission, G. A. Dunn & D. Zicha, "Dynamics of Fibroblast Spreading," 108, 1239–1249 (1995).

Phase-Shifting Methods 173

approach to date that can generate such growth data quantitatively, dynamically, and non-invasively. We will reiterate the importance of QPI methods for studying cell growth, in Secs. 11.1.3.2, 12.2.3.2, and 15.1.4.

DRIMAPS was also employed to study the dynamic fluctuations of cell margins, where the phase measurement provided quantitative information about protrusion and retraction rates.[75] In 1999, the effect of type 1 transforming growth factor-beta (TGFβ1) upon cell motility was investigated quantitatively using DRIMAPS.[72] These measurements are illustrated in Fig. 10.5. Outlines of cells at time 0 minutes are reproduced by white curves on images of cells taken 30 minutes later. During the 30-minute interval the TGFβ1-treated cell moved over a distance of 26 μm whereas the control cell only moved 7 μm. The pseudocolor scale represents mass density, and it can be seen that this reaches a higher level in the control cell, which indicates that the treated cell is more thinly spread. In the right-hand column, the protrusion regions are represented by those parts of the cells that lie outside the white outlines, whereas retraction regions are the black areas within each white outline.[72] These controlled experiments resulted in new findings regarding TGFβ1: (1) it does not alter the rate of increase in cell mass; (2) it increases the cell cycles and, thus, according to (1), causes a progressive increase in cell size; (3) it induces cell-cycle-dependent increase in motility; and (4) it induces an increase in cell spreading.

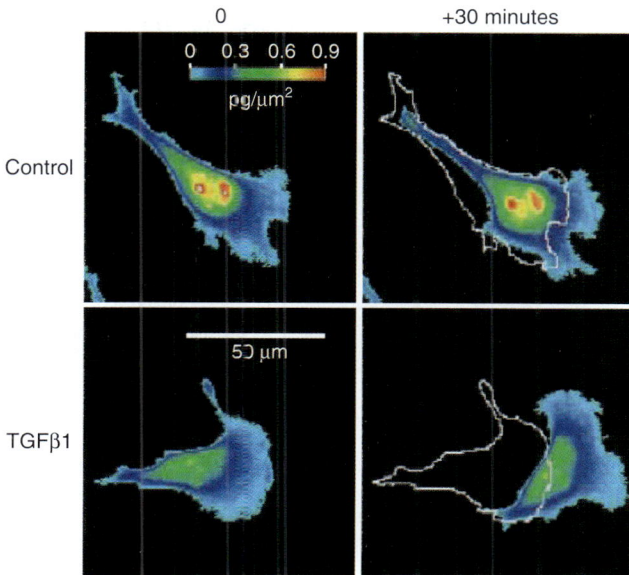

FIGURE 10.5 DRIMAPS recordings demonstrating cell displacement over a period of 30 minutes (From Fig. 1, Ref. 72). J. Cell Sci., used with permission, D. Zicha, E. Genot, G.A. Dunn & I.M. Kramer, "TFG beta 1 induces a cell-cycle-dependent increase in motility of epithelial cells," 112, 447–454 (1999).

While not an intrinsically stable method, DRIMAPS benefited a great deal from the advantages of an existing microscope environment, which made it suitable for in-depth biological studies. Thus, the cell biology studies by DRIMAPS took the concept of quantitative imaging pioneered by Barer and Davies (Refs. 77,78) to an entirely new level.

10.2   Optical Quadrature Microscopy (OQM)

10.2.1   Principle

OQM was developed at Northeastern University, and its biological applications were mainly pursued by the group lead by Charles DiMarzio.[79–88] This quadrature technique was adapted from laser radar, where it was applied to determine the sign of the Doppler velocity.[88] As shown in Fig. 10.6, the principle relies on using

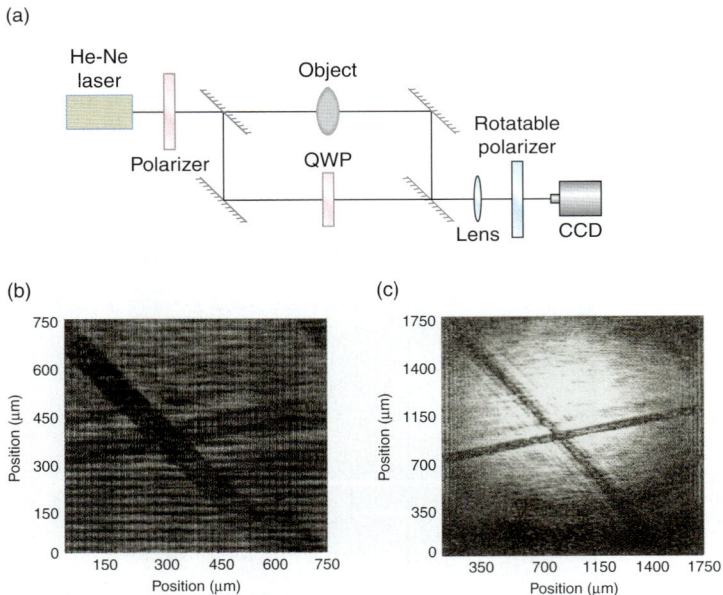

(a)

(b)   (c)

FIGURE 10.6   (*a*) Quadrature interferometer. (*b-c*) Reconstructed images of two crossed hairs separated by 6 mm. Each hair is 80 mm in diameter. (*a*) The image distance is set to 4.5 mm. The front hair is in focus. (*b*) The image distance is 10.4 mm, and the back hair is in focus. Note the scale change, which corresponds to the larger field of view at larger distances from the image plane (*From Figs. 1 and 5, Ref. 87*). The Optical Society, used with permission, *Optics Letters*, D.O. Hogenboom, C.A. DiMarzio, T.J. Gaudette, A.J. Devaney & S.C. Lindbert, "Three-dimensional images generated by quadrature interferometry," 23, 783–785 (1998).

polarization waveplates for adding phase shifts between the two fields in a Mach-Zender interferometer.[87] The quarter waveplate on the reference arm transform the initial linear polarization into circular polarization. After combining the two fields via the second beam splitter, the analyzer accomplishes the desired phase-shift difference by rotations with controlled angles. Thus, with two measurements in *quadrature* (i.e., $\pi/2$ out of phase), both the cosine and sine components of the phase image of interest are obtained. If the intensities of the reference and sample beams are measured separately, then the phase information is obtained uniquely.[87]

10.2.2 Further Developments

Later this principle was generalized to produce simultaneously all the four necessary phase shifts, i.e., $0, \pi/2, \pi, 3\pi/2$ and thus remove the need for independent intensity measurements. The measurement is performed by four cameras simultaneously, as shown in Fig. 10.7 (this OQM experimental setup is described in more detail in Ref. 79). A linearly polarized He-Ne laser is coupled into a single-mode fiber and further collimated. A non-polarizing beam splitter separates the light into a reference and sample beam of a Michelson interferometer. The sample arm of the interferometer contains an existing microscope

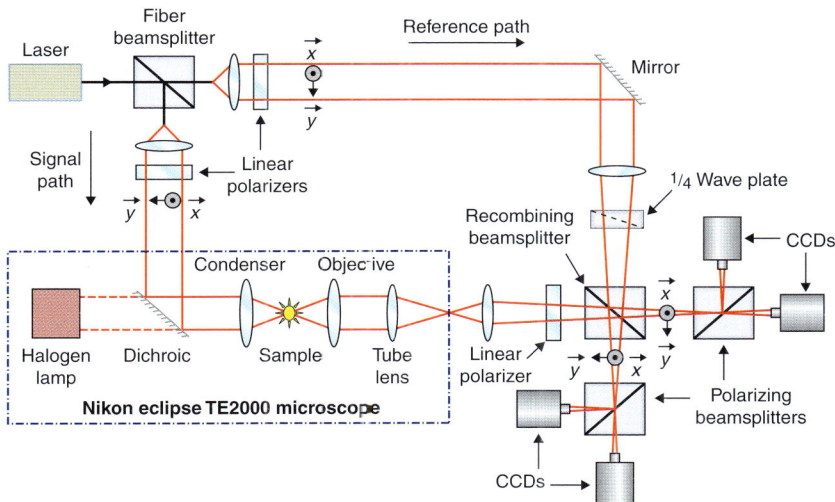

FIGURE 10.7 Optical layout for OQM. The *x*- and *y*-basis vectors are labeled along the optical path within the individual arms of the interferometer and after recombination. The unlabeled lenses are single element lenses (*From Fig. 2, Ref. 79*). The Optical Society, used with permission, *Optics Express*, W.C. Warger & C.A. DiMarzio, "Computational signal-to-noise ratio analysis for optical quadrature microscopy," 17, 2400–2422 (2009).

that contains its own illumination (halogen lamp) and, thus, provides bright field and differential interference contrast images. The sample field is recombined with the reference field via another non-polarizing beam splitter. The reference field passes through a lens, which introduces a wavefront curvature that matches that of the sample field. Importantly, the reference field passes through a *quarter waveplate* that, again, changes the field from linearly to circularly polarized.

Note that the two fields at the output of the interferometer are $\pi/2$ out of phase such that the total energy is conserved (essentially conservation of energy can be proved from the trigonometric relationship $\sin^2(x) + \cos^2(x) = 1$). Thus, the two polarizing beam splitters generate four output intensity distributions that correspond to the four frames in common phase- shifting interferometry (see Sec. 8.2.),

$$I_n = |U_R|^2 + |U_S|^2 + 2|U_R \cdot U_S| \cdot \cos\left(\phi + {n\pi}/{2}\right)$$ (10.2)

where U_R and U_S are the reference and sample field, respectively, and $n = 0, 1, 2, 3$. A thorough SNR analysis that incorporates noise terms for fluctuations in the laser, aberrations within the individual paths of the Mach-Zehnder interferometer, and imperfections within the beam splitters and CCD cameras is performed in Ref. 79.

Of course, this arrangement requires that the images recorded by all four CCDs are perfectly *registered*, that is, the recorded images overlap with subpixel accuracy, which is a nontrivial task.[82] Nevertheless, compared to typical phase-shifting interferometry, OQM enables simultaneous measurements of the required four frames, which can potentially provide high acquisition rates. The main biological applications of OQM are illustrated below.

10.2.3   Applications

One interesting application of OQM is in counting the number of cells in embryos.[81,85] The number of cells in a preimplantation embryo is directly correlated to the health and viability of the embryo.[84] In 2007, Newmark et al. used OQM to count the number of cells in mouse preimplantation embryos noninvasively. Figure 10.8 illustrates the procedure of the phase subtraction method (PSM) by which individual cells are digitally segmented and subtracted one by one from the image. This method exploits the multimodal imaging capability obtained by multiplexing OQM and DIC microscopy, as detailed in Ref. 84.

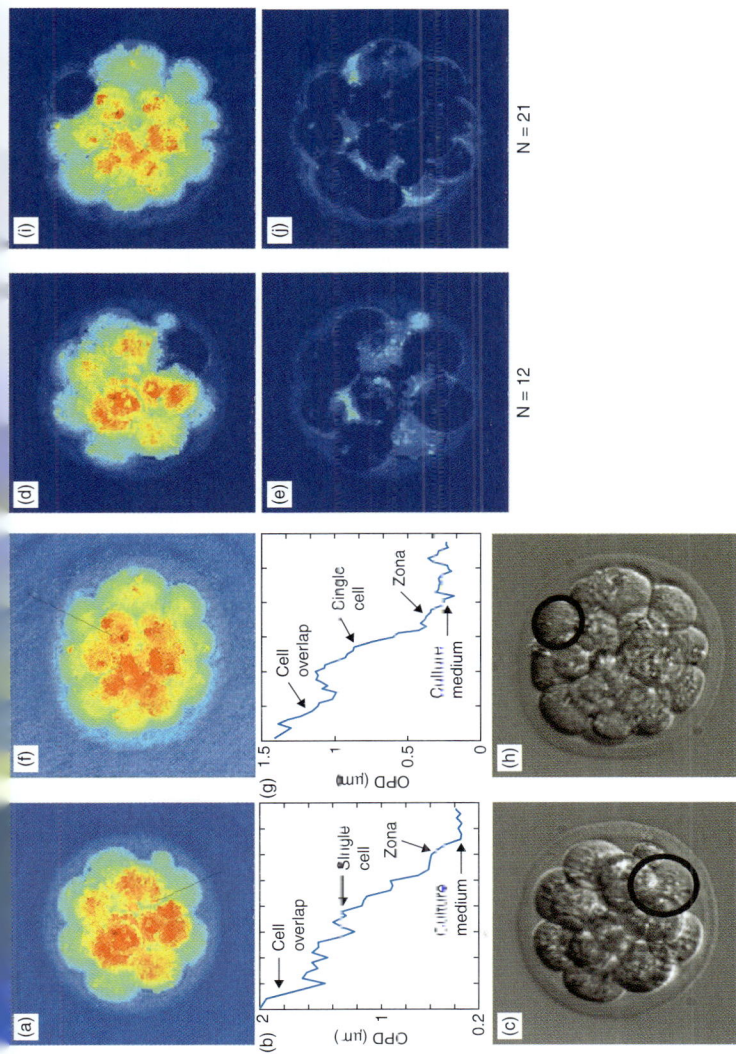

Figure 10.8 Counting the number of cells in embryos by using the phase subtraction method (PSM). Images (*a*) to (*e*) are from the 12-cell embryo and images (*f*) to (*j*) are from the 21-cell embryo. Lines were drawn through a single cell on the OQM images (*a*) and (*f*), which were used to generate plots of distance along the line vs. OPD (B, G). The plots were used in conjunction with the DIC images to generate elliptical boundaries for the cells (*c*) and (*h*). The ellipses were used to sequentially subtract cells from the OQM images (*e*) and (*i*). At the end of the process, the polar body was the only remaining cell in the OQM images (*e*) and (*j*). The resulting number of cells, by using the PSM, is shown at the bottom of each column. (*From Fig. 3, Ref. 85.*) *Microscopy & Microanalysis*, used with permission, J.A. Newmark, W.C. Warger, C. Chang, G.E. Herrera, D.H. Brooks, C.A. DiMarzio & C.M. Warner, "Determination of the number of cells in preimplantation embryos by using noninvasive optical quadrature microscopy in conjunction with differential interference contrast microscopy," 13, 118–127 (2007).

References

1. X. Zhang, E. Y. Lam and T. C. Poon, Reconstruction of sectional images in holography using inverse imaging, *Optics Express*, 16, 17215–17226 (2008).
2. K. Yamamoto, Y. Sugawara, M. R. McCartney and D. J. Smith, Phase-shifting electron holography for atomic image reconstruction, *J Electron Microsc (Tokyo)*, 59, S81–S88 (2010).
3. T. Zhang and I. Yamaguchi, Three-dimensional microscopy with phase-shifting digital holography, *Opt. Lett.*, 23, 1221–1223 (1998).
4. I. Yamaguchi and T. Zhang, Phase-shifting digital holography, *Opt. Lett.*, 22, 1268–1270 (1997).
5. I. Yamaguchi, T. Ida, M. Yokota and K. Yamashita, Surface shape measurement by phase-shifting digital holography with a wavelength shift, *Applied Optics*, 45, 7610–7616 (2006).
6. I. Yamaguchi, K. Yamamoto, G. A. Mills and M. Yokota, Image reconstruction only by phase data in phase-shifting digital holography, *Appl. Opt.*, 45, 975–983 (2006).
7. X. F. Xu, L. Z. Cai, X. F. Meng, G. Y. Dong and X. X. Shen, Fast blind extraction of arbitrary unknown phase shifts by an iterative tangent approach in generalized phase-shifting interferometry, *Opt. Lett.*, 31, 1966–1968 (2006).
8. N. Warnasooriya and M. K. Kim, LED-based multi-wavelength phase imaging interference microscopy, *Optics Express*, 15, 9239–9247 (2007).
9. K. Wang, Z. H. Ding, Y. Zeng, J. Meng and M. H. Chen, Sinusoidal B-M method based spectral domain optical coherence tomography for the elimination of complex-conjugate artifact, *Optics Express*, 17, 16820–16833 (2009).
10. X. G. Wang, D. M. Zhao, F. Jing and X. F. Wei, Information synthesis (complex amplitude addition and subtraction) and encryption with digital holography and virtual optics, *Opt. Exp.*, 14, 1476–1486 (2006).
11. H. H. Wahba and T. Kreis, Characterization of graded index optical fibers by digital holographic interferometry, *Applied Optics*, 48, 1573–1582 (2009).
12. P. H. Tomlins and R. K. Wang, Simultaneous analysis of refractive index and physical thickness by Fourier domain optical coherence tomography, IEE Proceedings-Optoelectronics, 153, 222–228 (2006).
13. S. Tamano, Y. Hayasaki and N. Nishida, Phase-shifting digital holography with a low-coherence light source for reconstruction of a digital relief object hidden behind a light-scattering medium, *Appl. Opt.*, 45, 953–959 (2006).
14. A. Stern and B. Javidi, Space-bandwith conditions for efficient phase-shifting digital holographic microscopy, *Journal of the Optical Society of America a-Optics Image Science and Vision*, 25, 736–741 (2008).
15. M. D. Stenner and M. A. Neifeld, Motion compensation and noise tolerance in phase-shifting digital in-line holography, *Opt. Exp.*, 14, 4286–4299 (2006).
16. G. Situ, J. P. Ryle, U. Gopinathan and J. T. Sheridan, Generalized in-line digital holographic technique based on intensity measurements at two different planes, *Applied Optics*, 47, 711–717 (2008).
17. H. Sasaki, K. Yamamoto, T. Hirayama, S. Ootomo, T. Matsuda, F. Iwase, R. Nakasaki and H. Ishii, Mapping of dopant concentration in a GaAs semiconductor by off-axis phase-shifting electron holography, *Applied Physics Letters*, 89, (2006).
18. M. V. Sarunic, B. E. Applegate and J. A. Izatt, Real-time quadrature projection complex conjugate resolved Fourier domain optical coherence tomography, *Optics Letters*, 31, 2426–2428 (2006).
19. V. Lauer, New approach to optical diffraction tomography yielding a vector equation of diffraction tomography and a novel tomographic microscope, *Journal of Microscopy-Oxford*, 205, 165–176 (2002).
20. J. T. Oh and B. M. Kim, Artifact removal in complex frequency domain optical coherence tomography with an iterative least-squares phase-shifting algorithm, *Applied Optics*, 45, 4157–4164 (2006).
21. N. A. Ochoa, M. Mora-Gonzalez and F. M. Santoyo, Flatness measurement by a grazing Ronchi test, *Optics Express*, 11, 2177–2182 (2003).

22. M. B. North-Morris, J. VanDelden and J. C. Wyant, Phase-shifting birefringent scatterplate interferometer, *Applied Optics*, 41, 668–677 (2002).
23. T. Nomura, S. Murata, E. Nitanai and T. Numata, Phase-shifting digital holography with a phase difference between orthogonal polarizations, *Appl. Opt.*, 45, 4873–4877 (2006).
24. R. M. Neal and J. C. Wyant, Polarization phase-shifting point-diffraction interferometer, *Applied Optics*, 45, 3463–3476 (2006).
25. G. A. Mills and I. Yamaguchi, Effects of quantization in phase-shifting digital holography, *Appl. Opt.*, 44, 1216–1225 (2005).
26. X. F. Meng, L. Z. Cai, X. F. Xu, X. L. Yang, X. X. Shen, G. Y. Dong and Y. R. Wang, Two-step phase-shifting interferometry and its application in image encryption, *Opt. Lett.*, 31, 1414–1416 (2006).
27. L. Martinez-Leon, M. Araiza, B. Javidi, P. Andres, V. Climent, J. Lancis and E. Tajahuerce, Single-shot digital holography by use of the fractional Talbot effect, *Optics Express*, 17, 12900–12909 (2009).
28. X. X. Lu, Y. M. Zhang, L. Y. Zhong, Y. L. Luo and C. L. She, Fourier algorithm method for reconstruction of large-aperture digital holograms based on phase compensation, *Opt. Lett.*, 29, 614–616 (2004).
29. J. P. Liu and T. C. Poon, Two-step-only quadrature phase-shifting digital holography, *Optics Letters*, 34, 250–252 (2009).
30. R. A. Leitgeb, R. Michaely, T. Lasser and S. C. Sekhar, Complex ambiguity-free Fourier domain optical coherence tomography through transverse scanning, *Optics Letters*, 32, 3453–3455 (2007).
31. H. H. Lee, J. H. You and S. H. Park, Phase-shifting lateral shearing interferometer with two pairs of wedge plates, *Optics Letters*, 28, 2243–2245 (2003).
32. E. S. Lee, J. Y. Lee and Y. S. Yoo, Nonlinear optical interference of two successive coherent anti-Stokes Raman scattering signals for biological imaging applications, *Journal of Biomedical Optics*, 12, 024010 (2007).
33. G. M. Lai, Q. X. Ru, K. Aoyama and A. Tonomura, Electron-Wave Phase-Shifting Interferometry in Transmission Electron-Microscopy, *Journal of Applied Physics*, 76, 39–45 (1994).
34. K. Khare and N. George, Direct coarse sampling of electronic holograms, *Opt. Lett.*, 28, 1004–1006 (2003).
35. J. W. Kang and C. K. Hong, Phase-contrast microscopy by in-line phase-shifting digital holography: shape measurement of a titanium pattern with nanometer axial resolution, *Optical Engineering*, 46, 040506 (2007).
36. T. Kakue, Y. Moritani, K. Ito, Y. Shimozato, Y. Awatsuji, K. Nishio, S. Ura, T. Kubota and O. Matoba, Image quality improvement of parallel four-step phase-shifting digital holography by using the algorithm of parallel two-step phase-shifting digital holography, *Optics Express*, 18, 9555–9560 (2010).
37. B. Javidi and D. Kim, Three-dimensional-object recognition by use of single-exposure on-axis digital holography, *Opt. Lett.*, 30, 236–238 (2005).
38. M. Z. He, L. Z. Cai, Q. Liu and X. L. Yang, Phase-only encryption and watermarking based on phase-shifting interferometry, *Applied Optics*, 44, 2600–2606 (2005).
39. P. Y. Guo and A. J. Devaney, Digital microscopy using phase-shifting digital holography with two reference waves, *Opt. Lett.*, 29, 857–859 (2004).
40. C. S. Guo, Z. Y. Rong, H. T. Wang, Y. R. Wang and L. Z. Cai, Phase-shifting with computer-generated holograms written on a spatial light modulator, *Appl. Opt.*, 42, 6975–6979 (2003).
41. M. Guizar-Sicairos and J. R. Fienup, Measurement of coherent x-ray focused beams by phase retrieval with transverse translation diversity, *Optics Express*, 17, 2670–2685 (2009).
42. M. Gross, M. Atlan and E. Absil, Noise and aliases in off-axis and phase-shifting holography, *Applied Optics*, 47, 1757–1766 (2008).
43. L. Granero, V. Mico, Z. Zalevsky and J. Garcia, Superresolution imaging method using phase-shifting digital lensless Fourier holography, *Optics Express*, 17, 15008–15022 (2009).
44. B. Gombkoto, A. Kornis, Z. Fuzessy, M. Kiss and P. Kovacs, Difference displacement measurement by digital holography by use of simulated wave fronts, *Appl. Opt.*, 43, 1621–1624 (2004).

45. S. Gioux, A. Mazhar, D. J. Cuccia, A. J. Durkin, B. J. Tromberg and J. V. Frangioni, Three-dimensional surface profile intensity correction for spatially modulated imaging, *Journal of Biomedical Optics*, 14, 034045 (2009).
46. E. Darakis and J. J. Soraghan, Compression of interference patterns with application to phase-shifting digital holography, *Appl. Opt.*, 45, 2437–2443 (2006).
47. H. C. Cheng, J. F. Huang, Y. C. Liu, C. W. Chang and Y. T. Chang, Group-delay-based phase-shifting method for Fourier domain optical coherence tomography, *Optical Engineering*, 48, 075004 (2009).
48. L. Z. Cai, Q. Liu, Y. R. Wang, X. F. Meng and M. Z. He, Experimental demonstrations of the digital correction of complex wave errors caused by arbitrary phase-shift errors in phase-shifting interferometry, *Appl. Opt.*, 45, 1193–1202 (2006).
49. L. Z. Cai, M. Z. He, Q. Liu and X. L. Yang, Digital image encryption and watermarking by phase-shifting interferometry, *Applied Optics*, 43, 3078–3084 (2004).
50. L. Z. Cai, M. Z. He and Q. Liu, Correction of wave-front errors caused by the slight tilt of a reference beam in phase-shifting interferometry, *Appl. Opt.*, 43, 3466–3471 (2004).
51. L. Z. Cai, Q. Liu and X. L. Yang, Phase-shift extraction and wave-front reconstruction in phase-shifting interferometry with arbitrary phase steps, *Optics Letters*, 28, 1808–1810 (2003).
52. Y. Bitou, Digital phase-shifting interferometer with an electrically addressed liquid-crystal spatial light modulator, *Optics Letters*, 28, 1576–1578 (2003).
53. Y. Bitou, Two-wavelength phase-shifting interferometry with a superimposed grating displayed on an electrically addressed spatial light modulator, *Applied Optics*, 44, 1577–1581 (2005).
54. Y. Bitou, H. Inaba, F. L. Hong, T. Takatsuji and A. Onae, Phase-shifting interferometry with equal phase steps by use of a frequency-tunable diode laser and a Fabry-Perot cavity, *Applied Optics*, 44, 5403–5407 (2005).
55. Y. Awatsuji, T. Koyama, T. Tahara, K. Ito, Y. Shimozato, A. Kaneko, K. Nishio, S. Ura, T. Kubota and O. Matoba, Parallel optical-path-length-shifting digital holography, *Applied Optics*, 48, H160–H167 (2009).
56. Y. Awatsuji, T. Tahara, A. Kaneko, T. Koyama, K. Nishio, S. Ura, T. Kubota and O. Matoba, Parallel two-step phase-shifting digital holography, *Applied Optics*, 47, D183–D189 (2008).
57. Y. Awatsuji, A. Fujii, T. Kubota and O. Matoba, Parallel three-step phase-shifting digital holography, *Appl. Opt.*, 45, 2995–3002 (2006).
58. Y. Awatsuji, M. Sasada, A. Fujii and T. Kubota, Scheme to improve the reconstructed image in parallel quasi-phase-shifting digital holography, *Appl. Opt.*, 45, 968–974 (2006).
59. K. J. Chalut, W. J. Brown and A. Wax, Quantitative phase microscopy with asynchronous digital holography, *Optics Express*, 15, 3047–3052 (2007).
60. W. Chen, C. Quan, C. J. Tay and Y. Fu, Quantitative detection and compensation of phase-shifting error in two-step phase-shifting digital holography, *Optics Communications*, 282, 2800–2805 (2009).
61. W. Choi, C. Fang-Yen, K. Badizadegan, S. Oh, N. Lue, R. R. Dasari and M. S. Feld, Tomographic phase microscopy, *Nature Methods*, 4, 717–719 (2007).
62. G. A. Dunn and D. Zicha, Phase-shifting interference microscopy applied to the analysis of cell behaviour, *Symp Soc Exp Biol*, 47, 91–106 (1993).
63. G. A. Dunn and D. Zicha, Dynamics Of Fibroblast Spreading, *J. Cell Sci.*, 108, 1239–1249 (1995).
64. H. Iwai, C. Fang-Yen, G. Popescu, A. Wax, K. Badizadegan, R. R. Dasari and M. S. Feld, Quantitative phase imaging using actively stabilized phase-shifting low-coherence interferometry, *Opt Lett*, 29, 2399–2401 (2004).
65. V. Mico, Z. Zalevsky and J. Garcia, Common-path phase-shifting digital holographic microscopy: A way to quantitative phase imaging and superresolution, *Optics Communications*, 281, 4273–4281 (2008).
66. M. Peckham, C. Wells, P. Taylor-Harris, D. Coles, D. Zicha and G. A. Dunn, Using molecular genetics as a tool in understanding crawling cell locomotion in myoblasts, *Biochemical Society Symposium*, 65, 281–299 (1999).
67. C. Remmersmann, S. Sturwald, B. Kemper, P. Langehanenberg and G. von Bally, Phase noise optimization in temporal phase-shifting digital holography

with partial coherence light sources and its application in quantitative cell imaging, *Applied Optics*, 48, 1463–1472 (2009).

68. N. T. Shaked, T. M. Newpher, M. D. Ehlers and A. Wax, Parallel on-axis holographic phase microscopy of biological cells and unicellular microorganism dynamics, *Applied Optics*, 49, 2872–2878 (2010).

69. N. T. Shaked, Y. Z. Zhu, M. T. Rinehart and A. Wax, Two-step-only phase-shifting interferometry with optimized detector bandwidth for microscopy of live cells, *Optics Express*, 17, 15585–15591 (2009).

70. N. Warnasooriya and M. Kim, Quantitative phase imaging using three-wavelength optical phase unwrapping, *Journal of Modern Optics*, 56, 67–74 (2009).

71. D. Zicha and G. A. Dunn, An Image-Processing System For Cell Behavior Studies In Subconfluent Cultures, *J. Microscopy*, 179, 11–21 (1995).

72. D. Zicha, E. Genot, G. A. Dunn and I. M. Kramer, TGF beta 1 induces a cell-cycle-dependent increase in motility of epithelial cells, *J. Cell Sci.*, 112, 447–454 (1999).

73. A. F. Brown and G. A. Dunn, Microinterferometry Of The Movement Of Dry-Matter In Fibroblasts, *J. Cell Sci.*, 92, 379–389 (1989).

74. G. A. Dunn and D. Zicha, in *Cell Biology: A Laboratory Handbook*, Celis, J. E., ed.), (Academic press, San Diego, CA, 1998).

75. G. A. Dunn, D. Zicha and P. E. Fraylich, Rapid, microtubule-dependent fluctuations of the cell margin, *J. Cell Sci.*, 110, 3091–3098 (1997).

76. J. P. Heath and G. A. Dunn, Cell to substratum contacts of chick fibroblasts and their relation to the microfilament system. A correlated interference-reflexion and high-voltage electron-microscope study, *J Cell Sci*, 29, 197–212 (1978).

77. R. Barer, Interference microscopy and mass determination, *Nature*, 169, 366–367 (1952).

78. H. G. Davies and M. H. Wilkins, Interference microscopy and mass determination, *Nature*, 169, 541 (1952).

79. W. C. Warger and C. A. DiMarzio, Computational signal-to-noise ratio analysis for optical quadrature microscopy, *Optics Express*, 17, 2400–2422 (2009).

80. H. Sierra, C. A. DiMarzio and D. H. Brooks, Modeling phase microscopy of transparent three-dimensional objects: a product-of-convolutions approach, *Journal of the Optical Society of America a-Optics Image Science and Vision*, 26, 1268–1276 (2009).

81. W. C. Warger, J. A. Newmark, C. M. Warner and C. A. DiMarzio, Phase-subtraction cell-counting method for live mouse embryos beyond the eight-cell stage, *Journal of Biomedical Optics*, 13, 034005-1 (2008).

82. C. L. Tsai, W. C. Warger, G. S. Laevsky and C. A. Dimarzio, Alignment with sub-pixel accuracy for images of multi-modality microscopes using automatic calibration, *Journal of Microscopy-Oxford*, 232, 164–176 (2008).

83. W. S. Rockward, A. L. Thomas, B. Zhao and C. A. DiMarzio, Quantitative phase measurements using optical quadrature microscopy, *Applied Optics*, 47, 1684–1696 (2008).

84. W. C. Warger, G. S. Laevsky, D. J. Townsend, M. Rajadhyaksha and C. A. DiMarzio, Multimodal optical microscope for detecting viability of mouse embryos in vitro, *Journal of Biomedical Optics*, 12, 044006 (2007).

85. J. A. Newmark, W. C. Warger, C. Chang, G. E. Herrera, D. H. Brooks, C. A. DiMarzio and C. M. Warner, Determination of the number of cells in preimplantation embryos by using noninvasive optical quadrature microscopy in conjunction with differential interference contrast microscopy, *Microscopy and Microanalysis*, 13, 118–127 (2007).

86. J. A. Newmark, D. J. Townsend, G. E. Herrera, C. A. DiMarzio and C. M. Warner, New imaging techniques for the evaluation of the health and viability of preimplantation embryos., *Biology of Reproduction*, 68, 251–252 (2003).

87. D. O. Hogenboom, C. A. DiMarzio, T. J. Gaudette, A. J. Devaney and S. C. Lindberg, Three-dimensional images generated by quadrature interferometry, *Optics Letters*, 23, 783–785 (1998).

88. D. O. Hogenboom and C. A. DiMarzio, Quadrature detection of a Doppler signal, *Applied Optics*, 37, 2569–2572 (1998).

CHAPTER **11**

Common-Path Methods

Temporal-phase stability was identified as one the most important figures of merit in QPI. This chapter presents *common-path interferometry* as a valuable solution to achieve stable-phase measurements. In essence, the methods presented in this chapter operate by the physical principle rooted in the following observation: A microscope image (e.g., bright field, phase contrast), although an *interferogram*, is extremely stable. This stability occurs because all the interfering fields making up the image (that is, all Fourier components scattered from the specimen) travel in close proximity to each other, passing through the same optical components. Therefore, the noise caused by air fluctuations and vibrations in the components is common to all fields and largely cancels out in the resulting interferogram. In this chapter, we present *Fourier-Phase microscopy*, a common-path and phase-shifting method, and *diffraction-phase microscopy*, a common-path and off-axis method.

11.1 Fourier Phase Microscopy (FPM)

As described in Chap. 5, Abbe understood image formation as an interference phenomenon. In particular, the microscope image can be described as the interference between the scattered light from the specimen and the average, unscattered light, which acts as the reference field. This picture later opened the door for phase-contrast microscopy, in which Zernike enhanced the contrast of this interferogram by having the two fields in quadrature and matching their amplitudes. *Fourier-phase microscopy* is a method that uses the same principles to extract the phase across the specimen quantitatively. Using the *Fourier* (hence the name) decomposition of a low-coherence optical image field into two spatial components that can be controllably shifted in phase with respect to each other, a high-transverse resolution quantitative-phase image can be obtained. Its main technical implementations and biological applications are described below.

11.1.1   Principle

In 2004, our group at MIT reported "Fourier phase microscopy for investigation of biological structures and dynamics."[1] FPM combines the principles of phase-contrast microscopy and phase-shifting interferometry, whereby the scattered and unscattered light from a sample are used as the object and reference fields of an interferometer. The experimental setup is shown in Fig. 11.1. The collimated low-coherence field from a superluminescent diode (SLD, center wavelength 809 nm and bandwidth 20 nm) is used as the illumination source for a typical inverted microscope. Through the output port, the microscope produces a magnified image positioned at the image plane IP. The lens, L_1, is positioned at the same plane IP and has a focal length such that it collimates the zero-spatial frequency field. The Fourier transform of the image field is projected by the lens, L_2 (50-cm focal distance), onto the surface of a programmable-phase

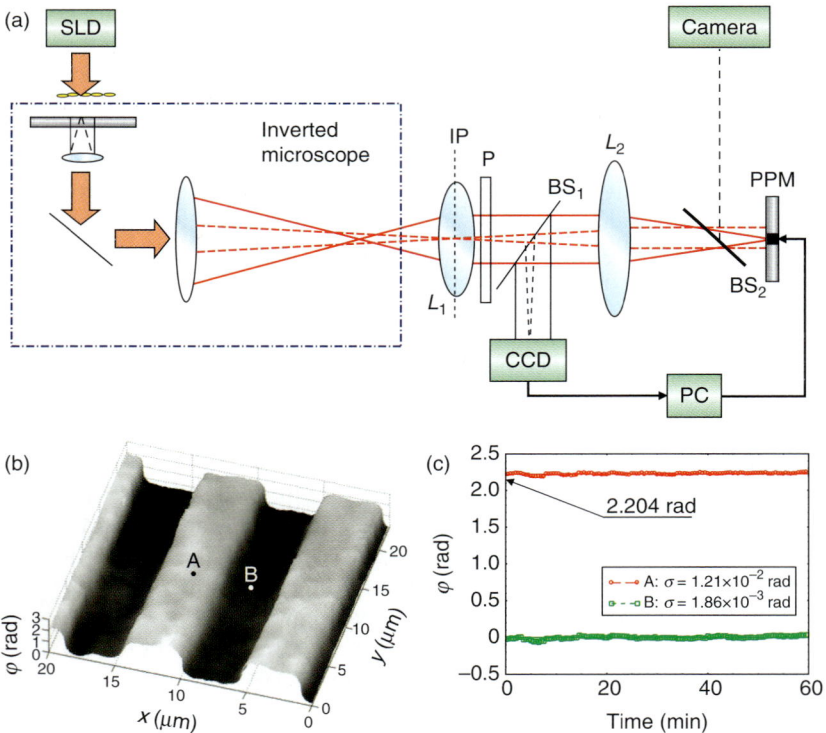

FIGURE 11.1   (a) FPM experimental setup. (b) Quantitative-phase image of a phase grating. (c) Temporal fluctuations of the path lengths associated with points A and B on the grating in (b).

modulator, essentially a spatial light modulator used in "phase mode." This PPM consists of an optically addressed, two-dimensional liquid crystal array with 768×768 active pixels. The polarizer, P, adjusts the field polarization in a direction parallel to the axis of the liquid crystal. In this configuration, the PPM produces precise control over the phase of the light reflected by its surface. The PPM pixel size is 26×26 µm^2, whereas the dynamic range of the phase control is 8 bits over 2π. In the absence of PPM modulation, an exact phase and amplitude replica of the image field is formed at the CCD plane, via the beam splitter BS$_1$. For alignment purposes, a camera is used to image the surface of the PPM via the beam splitter BS$_2$.

The PPM is used to controllably shift the phase of the scattered field component, U_1, (dotted line) in four successive increments of $\pi/2$ with respect to the average field, U_0 (solid line), as in typical phase-shifting interferometry measurements (see Sec. 8.2.). The phase difference between U_1 and U_0 is obtained by combining four recorded interferograms as follows[1]

$$\Delta\varphi(x,y) = \tan^{-1}\left[\frac{I(x,y;3\pi/2) - I(x,y;\pi/2)}{I(x,y;0) - I(x,y;\pi)}\right] \tag{11.1}$$

where $I(x, y; \alpha)$ represents the irradiance distribution of the interferogram corresponding to the phase shift, α. If we define $\beta(x,y) = |U_1(x,y)|/|U_0|$, then the phase associated with the image field $U(x,y)$ can be determined

$$\varphi(x,y) = \tan^{-1}\left[\frac{\beta(x,y)\sin(\Delta\varphi(x,y))}{1 + \beta(x,y)\cos(\Delta\varphi(x,y))}\right] \tag{11.2}$$

The amplitude ratio, β, contained in Eq. (11.2) can be obtained from the four frames, using that $\beta_{\varphi\to0} = 0$.[1] Kadono et al. developed a phase-shifting interferometer based on the similar interference between the scattered and unscattered light, but the phase-image reconstruction required the separate measurement of the unscattered field amplitude.[2] A similar system was later implemented by Ng et al.[3]

The phase image retrieval rate is limited by the refresh rate of the liquid crystal PPM, which in our case is 8 Hz. However, this acquisition rate can be further improved using a faster phase shifter. In fact, we later improved the data acquisition by approximately two orders of magnitude,[4] as discussed in the next section.

The phase accuracy and stability are illustrated in Fig. 11.1b and c. Figure 11.1b shows an example of an FPM measurement, obtained for a transmission phase grating. Using a 40X (NA = 0.65) microscope objective, we retrieved the spatially varying phase delay induced by this grating, which is made of glass with the refractive index, $n = 1.51$.

The profile of the grating was measured by stylus profilometry and the height was found to be 570 ± 10 nm while its pitch had a value of 4 μm. This corresponds to a phase profile of height $\varphi = 2.217 \pm 0.039$ rad. As can be seen in Fig. 11.1b, the measurement correctly recovers the expected phase distribution. Figure 11.1c shows the values of the reconstructed phase associated with the point A and B indicated in Fig. 11.1b, as a function of time. The phase values are averaged over an area that corresponds to 0.6×0.6 μm^2 in the sample plane, which is approximately the diffraction limit of the microscope. The values of the standard deviation associated with the two points are 18 and 12 mrad, respectively, which demonstrate the significant stability of the technique in the absence of active stabilization. Interestingly, the phase stability of the measurement is actually better when wet samples are studied.[5]

This first implementation of FPM was limited in acquisition speed to several frames per minute. This limitation was addressed in a next iteration, as described in the following section.

11.1.2   Further Developments

In 2007, we developed a faster version of FPM based on a liquid crystal modulator that operated in transmission and with a faster response time.[4] The new instrument, referred to as the *fast* Fourier-phase microscope (f-FPM), provides a factor of 100 higher acquisition rates compared to the Fourier-phase microscope described previously. The resulting quantitative phase images are characterized by diffraction-limited transverse resolution and path-length stability better than 2 nm at acquisition rates of 10 fps or more.

The experimental setup is shown in Fig. 11.2. The second harmonic of the CW Nd:YAG laser (wavelength $\lambda = 532$ nm, 500 mW) is used as an illumination source for a typical inverted microscope (Axiovert 100, Carl Zeiss, Inc). To ensure spatial coherence and plane-wave illumination, the laser beam is coupled to a single-mode fiber and the output is collimated. The light transmitted through the sample is collected by a 40× objective lens (NA = 0.65). Lens L_1 is set at the video port of the microscope to correct the beam divergence. An iris, I, is placed right at the image plane IP$_1$ to control the size of image. The image is transferred to the CCD using a 4-f system composed of lenses L_2 ($f = 1000$ mm) and L_4 ($f = 700$ mm). At the Fourier plane FP, we place the phase-contrast filter (PCF), which is made of nematic liquid crystal 2×2 cm^2 sandwiched between layers of indium tin oxide (ITO) electrodes. At the center of the PCF, a 150-μm diameter circular portion of the ITO electrode is removed such that only the outer part of PCF can have biased-phase modulation. The driving electric field is a square wave with 5-kHz frequency, which is generated by an I/O board. The Fourier lens, L_2, spatially decomposes the image field into its average component DC and a spatial varying or

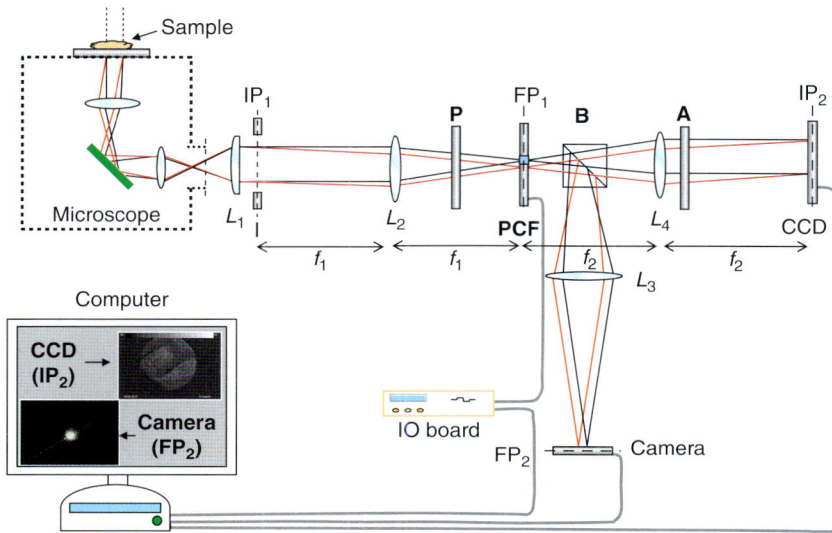

FIGURE 11.2   f-FPM experimental set_p (notations are explained in text). (*From Ref. 4.*)

scattered field AC at the Fourier plane, FP_1. The position of the DC component is adjusted to overlap with the central pinhole of the PCF, such that only the scattered field AC undergoes phase-shifting when voltage is applied. The DC and phase-shifted AC components interfere at the image plane of the 4-f system and the resulting image is captured by a CCD camera. For alignment purposes, we simultaneously image the Fourier plane, FP. Using the beam splitter, B, and lens, L, the image of PCF is captured by a video camera. Since there is a continuous transition between the AC and DC components and the size of the PCF pinhole is finite, the aperture at IP_1 was used to controllably select the DC component, i.e., the spatial low-frequency content of the image. If the DC spot size is too small relative to the nonmodulated central aperture of PCF, part of the AC component can pass through the central hole and, therefore, the reference beam itself shows strong spatial modulation.

The near-video-rate imaging capability of f-FPM was demonstrated by continuously acquiring quantitative-phase images of dissolving sugar crystals in water. The dissolving rate of sugar was adjusted by controlling a heating coil that surrounded the sample. Figure 11.3 shows a series of quantitative phase images of a dissolving sugar crystal. As can be seen, the dynamic change in crystal thickness and volume can be quantitatively monitored.

Further, we used f-FPM to image live cells in culture. Live human epithelial (HeLa) cells were grown in the Dulbecco's modified Eagle medium (DMEM) containing 10% fetal calf

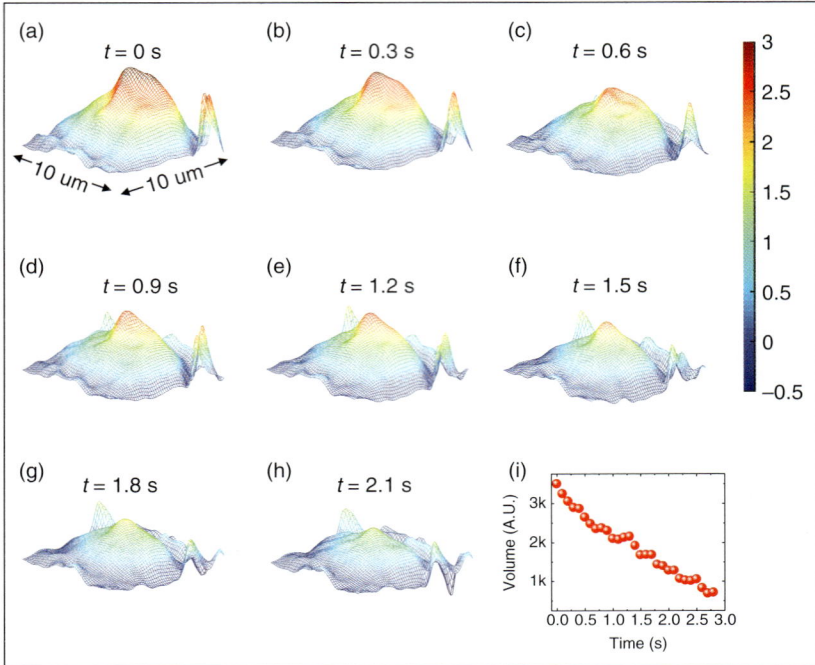

FIGURE 11.3   Sequential surface f-FPM images of a sugar crystal melting in the water (time between images is 300 ms). (*From Ref. 4.*)

serum (FCS). These cells were imaged directly in culture conditions, without additional preparation. Figure 11.4*a* shows a f-FPM image of a live HeLa cell. Since the optical path length is integrated along the beam propagation direction, the sharpness of subcellular structures tends to decrease. Differential interference contrast (DIC) microscopy, on the other hand, produces high-contrast images of cells and organelles because it is sensitive to phase gradients, rather than the phase itself. Using this idea, we numerically processed the quantitative-phase image to implement what we refer to as digital DIC. This operation is possible because of the knowledge of the quantitative-phase map provided by f-FPM. Before computing its spatial gradient, the quantitative-phase image is filtered numerically to reduce the contribution of the medium spatial-frequency range. We convert the phase image to an intensity map by computing the cosine of the phase, which is what would be measured if the sample field were interfered with a plane-wave reference. To provide flexibility in adjusting contrast, we apply a variable-phase shift to the argument of this cosine function. Finally, we take the gradient of this intensity

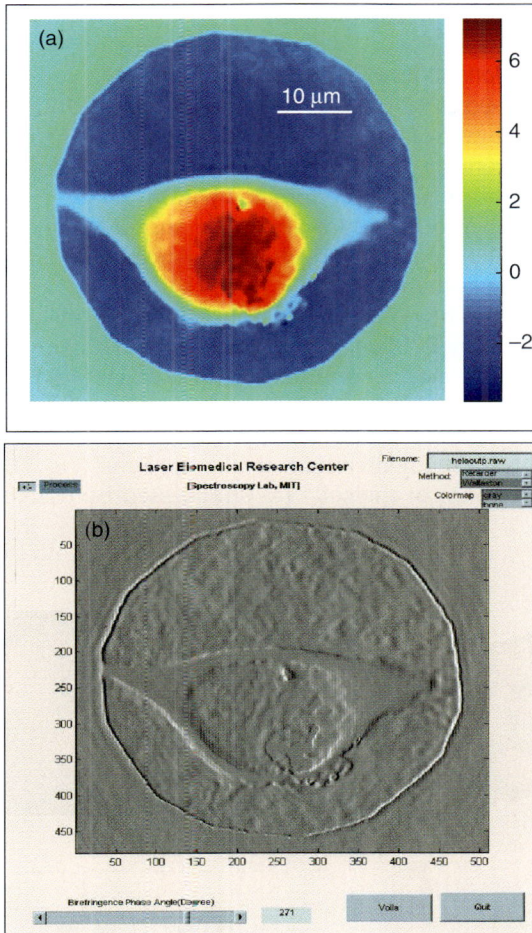

FIGURE 11.4 (a) Phase images of HeLa cell using f-FPM. (b) Graphic user interface for processing of digital DIC of a HeLa cell. (*From Ref. 4.*)

image. Figure 11.4b displays the digital DIC image of a HeLa cell with an addition phase shift by 89°. As can be seen, the cell and nucleus boundary are clearly visible with higher contrast than in the quantitative-phase image. The graphic interface provides a slide-bar control-phase shift, which can be viewed as the digital analogous to the Wollaston prism in typical DIC microscopes. With this numerical processing, a single f-FPM measurement can provide complementary views of the same sample, which is particularly appealing as the contrast is intrinsically generated, without sample preparation.

11.1.3   Biological Applications

Slow Fluctuations in Red Blood Cell Membranes

In 2006, we used FPM to measure nanoscale fluctuations associated with the membrane of live erythrocytes during time intervals ranging from a few seconds to hours.[5] The experimental results demonstrated the existence of dynamic subdomains within the cell, which exhibit oscillations at various frequencies. The nature of these fluctuations suggests that they are driven by deterministic phenomena associated with this living system.

Fresh human blood sandwiched between two coverglasses was directly imaged by FPM using a 40× objective for 45 minutes at a repetition rate of 2 frames per minute. A typical wide-field FPM image of the blood smear showing the discoid appearance of individual cells is presented in Fig. 11.5a. For analysis of cell dynamics, individual cells were segmented from the background (Fig. 11.5b). Translational movement of the cell was suppressed by tracking the cell

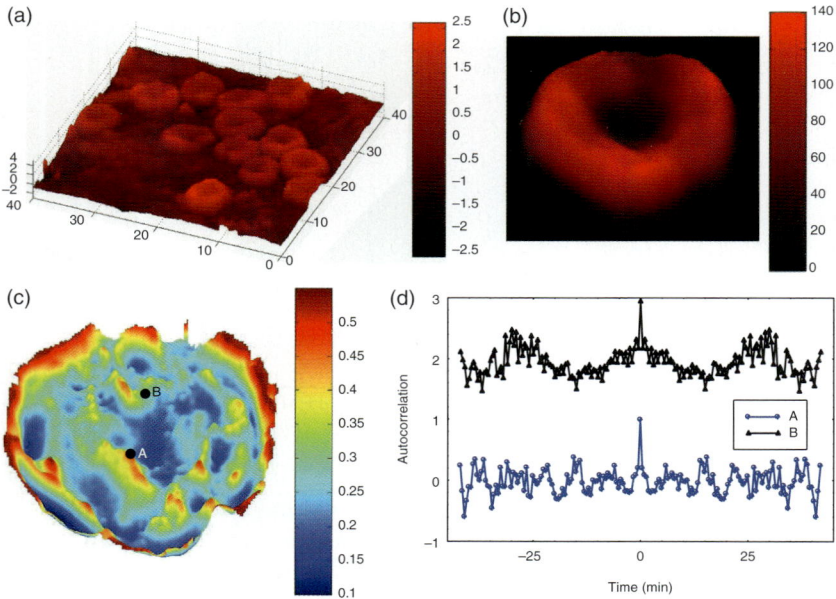

FIGURE 11.5   (a) FPM image of blood smear; the colorbar indicates thickness in microns. (b) Surface image of a single red blood cell; the colorbar shows the phase shift in nanometer. (c) Average frequency map of the cell, calculated from the FPM time-lapse data set; the colorbar has units of min^{-1}. (d) Normalized autocorrelations associated with temporal fluctuations of the points shown in (c). The top curve was shifted for better visibility. Adapted from Figs. 1-2, Ref. 5. SPIE used with permission, J. Biomed. Opt. Lett, G. Popescu, K. Badizadegan, R.R. Dasari, & M.S. Feld, "Observation of dynamic subdomains in red blood cells," 11, 040503 (2006).

centroid, such that time-series of individual cell fluctuations were obtained over the period of observation. These datasets contain unique $h(x, y; t)$ information about the membrane dynamics. The power spectra associated with these fluctuations were measured across the cell, and the entire cell was mapped out in terms of the average frequency of fluctuations. Figure 11.5c shows this map and suggests that the cell can be divided into areas of independent dynamics with different average oscillation frequencies. In addition, the autocorrelation analysis of fluctuations within each domain shows clear periodic patterns (Fig. 11.5d). The presence of sinusoidal patterns in the fluctuation signals indicates that the cell dynamics is nonrandom and possibly associated with deterministic phenomena within the cell.[5]

These measurements were too slow at the time to study further the rheology and ATP-dependent activity in RBCs. These phenomena were investigated later using *diffraction-phase microscopy*,[6–10] a faster technique described in Sec 11.2. Nevertheless, these initial measurements indicated the potential of QPI methods for studying such interesting nanoscale dynamics in live cells.

Cell Growth

In 2008, we reported FPM measurements of cell growth.[11] Several decades ago, it has been shown that the optical phase-shift through the cells is a measure of the cellular *dry mass* content.[12,13] Optical interferometry provides access to the phase information of a given transparent sample; the main challenge is to suppress the environmental noise, which hinders the ability to measure optical path-length shifts quantitatively.

It was shown earlier that FPM provides quantitative phase images of live cells with high-transverse resolution and low noise over extended periods of time. The general expression for the spatially resolved quantitative-phase images obtained from a cell sample is given by

$$\varphi(x,y) = \frac{2\pi}{\lambda} \int_0^{h(x,y)} [n_c^z(x,y,z) - n_0] dz \qquad (11.3)$$

In Eq. (11.3), λ is the wavelength of light, h is the local thickness of the cell, and n_0 is the refractive index of the surrounding liquid. The quantity, n_c^z, is the refractive index of cellular material, which is generally an inhomogeneous function in all three dimensions. Without loss of generality, Eq. (11.3) can be rewritten in terms of an axially averaged refractive index, n_c, as

$$\varphi(x,y) = \frac{2\pi}{\lambda} [n_c(x,y) - n_0] h(x,y) \qquad (11.4)$$

However, it has been shown that the refractive properties of a cell composed mainly of protein has, to a good approximation, a simple dependence on protein concentration,[12,13]

$$n_c(x,y) = n_0 + \alpha C(x,y) \tag{11.5}$$

In Eq. (11.5), α is referred to as the refraction increment (units of mL/g), and C is the concentration of dry protein in the solution (in g/mL). Using this relationship, the dry mass surface density, σ, of the cellular matter is obtained from the measured phase map as

$$\sigma(x,y) = \frac{\lambda}{2\pi\alpha} \varphi(x,y) \tag{11.6}$$

In order to illustrate the potential of FPM for measuring the dry mass distribution of live cells, we used the FPM instrument for imaging confluent monolayers of HeLa cells. Figure 11.6 shows an example

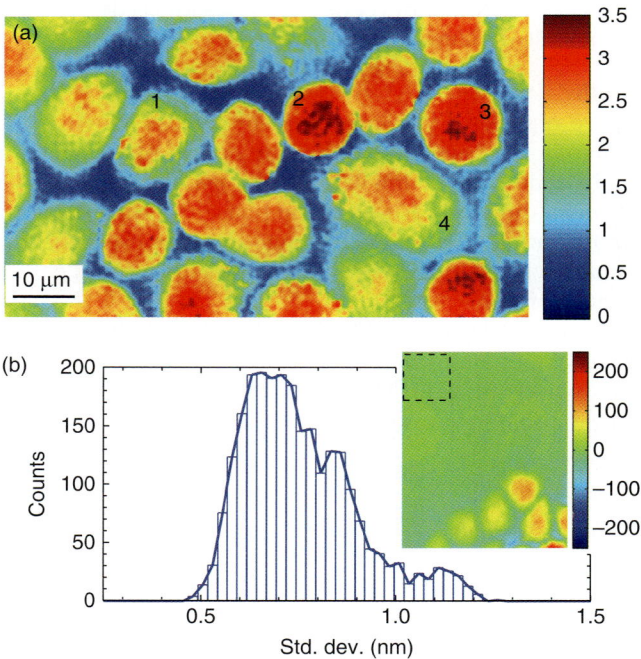

FIGURE 11.6 (a) Dry mass density distribution $\sigma(x, y)$ obtained using FPM. The colorbar has units of picograms/μm^2. (b) Histogram of the path-length standard deviation corresponding to the pixels within the 15- $\times$ 15-μm^2 selected area shown in inset. The colorbar of the inset indicates optical path-length in units of nanometers, which sets the ultimate sensitivity to dry mass changes to 4 fg/μm^2. (From Fig. 4, Ref. 11.) The American Physiological Society, used with permission, Am. J. Physiol. Cell Physiol., G. Popescu, Y. Park, N. Lue, C. Best-Popescu, L. Deflores, R.R. Dasari, M.S. Feld & K. Badizadegan, "Optical imaging of cell mass & growth dynamics," 295, C538-44 (2008).

of the dry mass density distribution, σ (units of pg/μm²), obtained from nearly confluent HeLa cells. In applying Eq. (11.6), we used α = 0.2 mL/g for the refraction increment, which corresponds to an average of reported values.[14] Quantitative information about the dry mass of cells thus allows investigation of cell movement, growth, or shape change in a totally noninvasive manner. In order to quantify the phase stability of the instrument, we recorded 240 phase images over 2 hours from a cell sample that contained regions with no cells. We measured the path-length standard deviation of each pixel within a 15- × 15-μm² region. The average of these standard deviations had a value of 0.75 nm, which indicates that the sensitivity to changes in dry mass surface density has a value of 3.75 fg/μm².

Quantitative knowledge about cell growth can provide information about cell cycling, functioning, and disease.[15,16] We employed FPM to quantify growth of HeLa cells in culture. Data containing time-resolved FPM images were acquired at a rate of four frames a minute over periods of up to 12 hours and the dry mass surface density information was extracted as presented above. Each cell fully contained in the field of view was segmented using a MATLAB program based on iterative image thresholding and binary dilation, which was developed in our laboratory. Figure 11.7a shows the

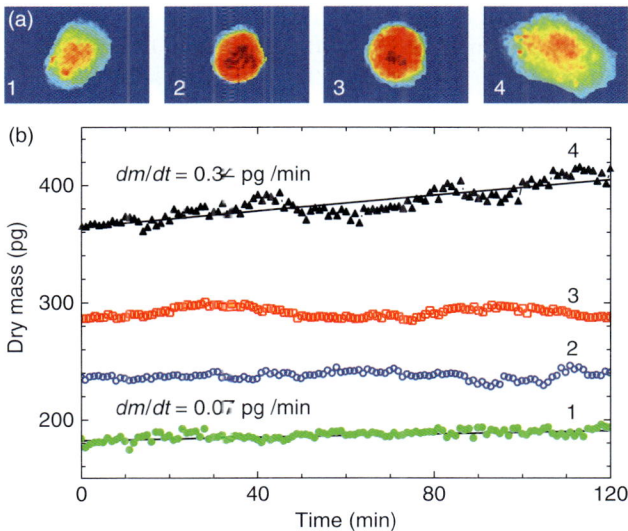

FIGURE 11.7 (a) Images of the cells segmented from Fig. 11.6, as indicated. (b) Temporal evolution of the total dry mass content for each cell. The solid lines for cells 1 and 4 indicate fits with a linear function. The values of the slope obtained from the fit are indicated. (From Fig. 4, Ref. 11). The American Physiological Society, used with permission, Am. J. Physiol. Cell Physiol., G. Popescu, Y. Park, N. Lue, C. Best-Popescu, L. Deflores, R.R. Dasari, M.S. Feld & K. Badizadegan, "Optical imaging of cell mass & growth dynamics," 295, C538-44 (2008).

segmented images of the four cells shown in Fig. 11.6. We monitored the total dry mass of each cell over a period of 2 hours. The results are summarized in Fig. 11.7b. Cell 4 exhibits linear growth, as does cell 1, although it is reduced by a factor of almost 5. In contrast, cell 2 shows virtually no change in mass, whereas cell 3 appears to exhibit slight oscillatory behavior, the origin of which is not clearly understood. These results demonstrate the capability of FPM to quantify small changes in cell mass and, therefore, monitor in detail the evolution of cell cycle and its relationship with function.

Understanding cell growth, a very basic and, yet, insufficiently known phenomenon, represents perhaps one of the most exiting applications of QPI, as reiterated in Sec. 15.2.

Cell Motility

The necessity for quantitative, model-based studies of cell motility and migration was recognized decades ago[17] and continues to evolve with new experimental and analytical tools.[18,19] Understanding the motion of live cells requires tracking their mass-weighted centroid displacements and performing averages over time and ensembles of cell populations. However, historically, optical techniques such as bright field microscopy,[20] phase-contrast,[21,22] and Nomarski/DIC[23] have been used for tracking the centroid of live cells. For a review and timeline, see Ref. [24]. More recently, confocal microscopy has been used to investigate motility of hydra cells,[25] and two-photon fluorescence microscopy to study the motion of lymphocytes *in vivo*.[26]

FPM provides a precise localization of the center of mass, which is instrumental for studying cell motility with high accuracy. Figure 11.8a shows the mean-squared displacement dependence for cells that, during the investigation, appeared to be in a resting state, i.e., their total dry mass value did not vary considerably over the period of observation. We will refer to these cells as being in the G0/G1 phase of the cell cycle. Remarkably, the long-time behavior ($\tau > 10$ min.) shows a power-law trend with an exponent $\alpha = 5/4$, which indicates *super-diffusion*. Similar power-law dependence was obtained on hydra-cell aggregates by Upadhyaya et al., and the underlying thermodynamics of the motion was studied using Tsallis statistics.[25] This *generalized thermodynamic* theory relies on the non-extensive entropy defined as[27]

$$S_q = \frac{1 - \int [p(x)]^q \, dx}{q - 1} \qquad (11.7)$$

with p a probability distribution function and q the non-extensiveness parameter, which for $q \to 1$ yields the Gibbs formalism, i.e., $S = -\int p(x) \ln[p(x)] dx$. We calculated the cell velocity distribution for

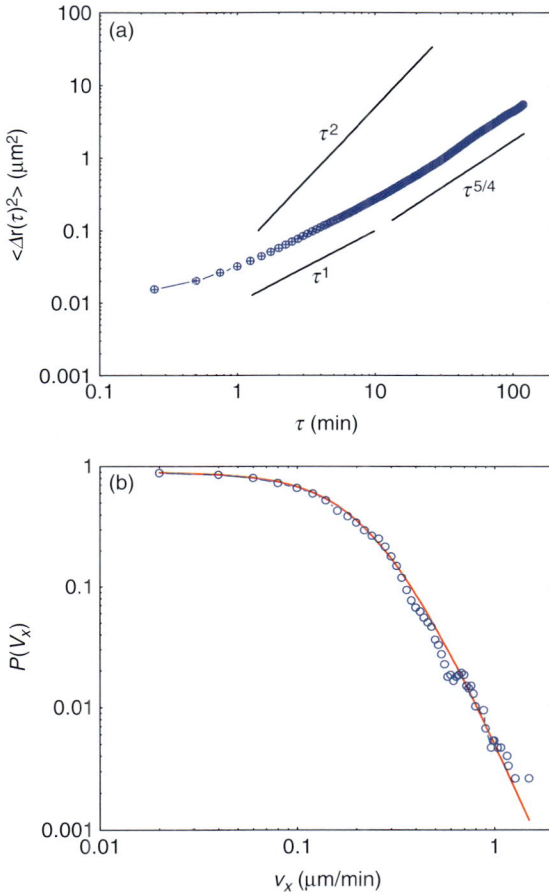

FIGURE 11.8 Mean-squared displacement of resting cells. The solid line indicates power laws with various exponents.

periods of 1 minute and found a clear non-Gaussian behavior. Figure 11.8*b* shows the probability density of the velocity projected onto the x-axis, which was fitted with the function[25]

$$P(v_x) = C_q \left[1 - (1-q) \frac{\gamma v_x^2}{2} \right]^{\frac{1}{1-q}} \tag{11.8}$$

The parameter γ in Eq. (11.8) contains information about both the apparent mass and equivalent Boltzmann coefficient. However, the cell motion is not thermally driven, and the Boltzmann factor relates to the energetics of cytoskeletal activity. As can be seen in Fig. 11.8*b*, Eq. (11.8) provides a very good description of the data, which demonstrates that this Tsallis formalism can be successfully applied to study this super-diffusive motion. The coefficient q obtained from the fit has an average value of $q = 1.56 \pm 0.04$, which agrees very well with the value of 1.5 previously reported on Hydra cells.[25]

However, at times $\tau < 10$ min, the mean-squared displacement shown in Fig. 11.8a exhibits a more complicated behavior, which cannot be modeled by a simple power-law function. Specifically, the center of mass motion appears subdiffusive at $0 < \tau < 2$ min and transitions towards a Brownian regime for $2 < \tau < 10$ min. We hypothesized that the internal mass dynamics, such as organelle transport and cytoskeletal remodeling, are responsible for the short-time behavior of the center of mass. We validated this hypothesis by performing further experiments on cells in the G2/M phase. These cells have a homogeneous distribution of internal structure and, thus, do not exhibit the short time subdiffusive motion. These findings may open the door for a new class of applications, in which cell organelles can potentially be used as intrinsic reporters on the viscoelastic properties of intracellular matter.

Although initially implemented with low-coherence fields, FPM did not reveal the type subcellular details specific to white light methods (e.g., phase contrast and differential interference contrast). This is simply because a coherence length of >10 μm is of the order of a cell thickness, which brings no benefit in terms of sectioning capability. For this reason, LED/SLD sources provide comparable results to laser illumination. Still, the most appealing feature of FPM is its high-phase stability, which is due to the common-path interferometry.

11.2 Diffraction-Phase Microscopy (DPM)

Diffraction-phase microscopy (DPM) is a quantitative-phase imaging technique that combines the single-shot feature of HPM (Sec. 9.2) with the common-path geometry associated with FPM (Sec. 11.1). The principle of DPM and main applications of DPM are described below.

11.2.1 Principle

In 2006, our group at MIT reported: "Diffraction phase microscopy for quantifying cell structure and dynamics."[28] The main advancement in DPM is the combination between the significant phase stability associated with common-path interferometers and high-acquisition speeds of off-axis methods. This combination of features allowed us to study in depth interesting biological phenomena, especially red blood cell dynamics,[6–10,29–32] which will be discussed below.

The DPM experimental setup is shown in Fig. 11.9. The second harmonic radiation of a Nd:YAG laser ($\lambda = 532$ nm) is used as illumination for an inverted microscope, which produces the magnified

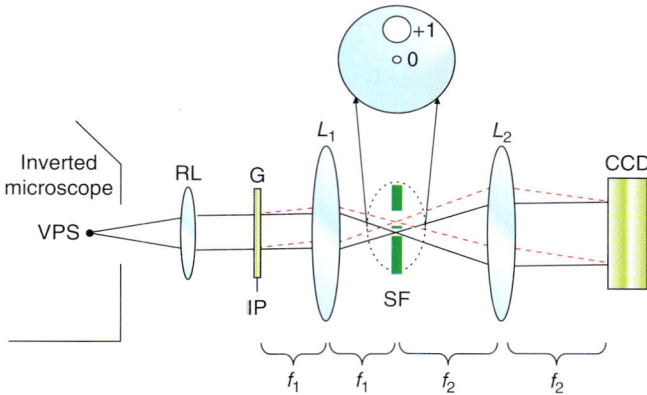

FIGURE 11.9 Experimental setup. VPS, virtual source point; G, grating; IP, image plane; L_1, L_2 lenses ($f_{1,2}$, respective focal distances); SF, spatial filter (expanded in the inset).

image of the sample at the output port. The microscope image appears to be illuminated by a virtual source point VPS. A relay lens, RL, collimates the light originating at VPS and replicates the microscope image at the plane, IP. A diffraction phase grating G (hence "diffraction-phase microscopy") is placed at this image plane and generates multiple diffraction orders containing full spatial information about the image. The goal is to select two diffraction orders (0th and 1st) that can be further used as reference and sample fields in a very compact Mach-Zender interferometer. In order to accomplish this, a standard spatial-filtering lens system, L_1–L_2, is used to select the two diffraction orders and generate the final interferogram at the CCD plane. The 0th-order beam is low-pass filtered using the spatial filter SF positioned in the Fourier plane of L_1, such that at the CCD plane it approaches a uniform field. The spatial filter allows passing the entire frequency content of the 1st diffraction order beam and blocks all the other orders. The 1st order is thus the imaging field and the 0th order plays the role of the reference field. The two beams traverse the same optical components, i.e., they propagate along a common optical path, thus significantly reducing the longitudinal phase noise. The direction of the spatial modulation was chosen at an angle of 45° with respect to the x and y axes of the CCD, such that the total field at the CCD plane has the form

$$U(x,y) = |U_0| e^{i[\phi_0 + \beta(x+y)]} + |U_1(x,y)| e^{i\phi(x,y)} \qquad (11.9)$$

In Eq. (11.9), $|U_{0,1}|$ and $\phi_{0,1}$ are the amplitudes and the phase of, respectively, the orders of diffraction 0, 1, while β represents the spatial frequency shift induced by the grating to the 0th order (i.e., the spatial frequency of the grating itself). Note that, as a consequence of the central ordinate theorem, the reference field is proportional to the spatial average of the microscope image field,

$$|U_0|e^{i\phi_0} \propto \frac{1}{A} \int |U(x,y)| e^{i\phi(x,y)} dxdy \qquad (11.10)$$

where A is the total image area. As we discussed earlier (Sec. 11.1), in FPM, the *reference* for interferometry is also the spatial average of the image field.[1]

The CCD has an acquisition rate of 291 frames/s at the full resolution of 640×480 pixels. To preserve the transverse resolution of the microscope, the spatial frequency β is chosen to match or exceed the maximum frequency allowed by the numerical aperture of the instrument. Throughout our experiments, the microscope was equipped with a $40\times$ (0.65 NA) objective, which is characterized by a diffraction-limited resolution of 0.4 μm. The microscope-relay lens combination produces a magnification of about 100, thus the diffraction spot at the grating plane has a size of approximately 40 μm. The grating pitch is 20 μm, which allows for preserving the full resolution given by the microscope objective. The L_1–L_2 lens system has an additional magnification of $f_2/f_1 = 3$, such that the sinusoidal modulation of the image is sampled by 6 CCD pixels per period. Thus, as in other off-axis methods (DHM, HPM, discussed in Chap. 9), the spatially-resolved quantitative phase image associated with the sample is retrieved from a single CCD recording via a spatial Hilbert transform.[33]

To quantify the stability of the DPM instrument and thus the sensitivity of cell topography to dynamical changes, we recorded sets of 1000 *no-sample* images, at 10.3 ms/frame, and performed noise analysis on both single points and entire field of view. The *spatial* standard deviation of the path-length associated with the full field of view had a *temporal* average of 0.7 nm. Thus, DPM provides quantitative phase images which are inherently stable to the level of subnanometer optical path-length and at an acquisition speed limited only by the detector.

The ability to study live cells was demonstrated by imaging droplets of whole blood sandwiched between cover slips, with no additional preparation. Figure 11.10 shows a quantitative phase image of live blood cell, where the normal, discocyte shape can be observed. The cell profile is shown in Fig. 11.10*b*. The traces in Fig. 11.10*c* indicate that the fluctuations in the rim of the cell are higher in amplitude than in the center. The images shown here are of lower quality than subsequent data provided by DPM, but they are important as being the first reported DPM results. DPM was further combined with

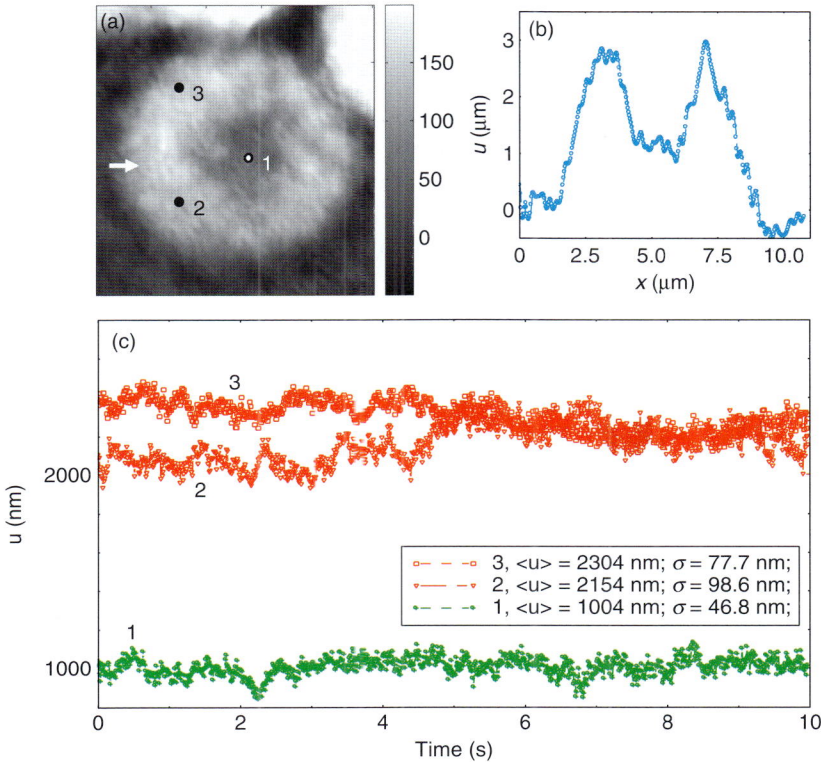

FIGURE 11.10 DPM assessment of the single cell shape and dynamics. (a) DPM image of a single RBC (the colorbar represents the path length in nanometers). (b) Thickness profile of the cell in *a*. (c) Local thickness fluctuations at the three points on the cell indicated in *a*. The respective average thickness, < *u* >, and standard deviations, σ, are indicated (*From. Ref. 28.*)

fluorescence microscopy. which provides new opportunities for in-depth biological studies. This development is discussed in the following section.

11.2.2   Further Developments

Diffraction-Phase and Fluorescence Microscopy (DPF)

Fluorescence microscopy provides images of both intrinsic and exogenous fluorophores, capable of uncovering subtle processes within the cell.[34] Significantly, recent advances of the green fluorescent protein (GFP) technology allows an intrinsically fluorescent molecule to be genetically fused to a protein of interest in live cell populations.[35–37] This type of imaging has revolutionized cell imaging

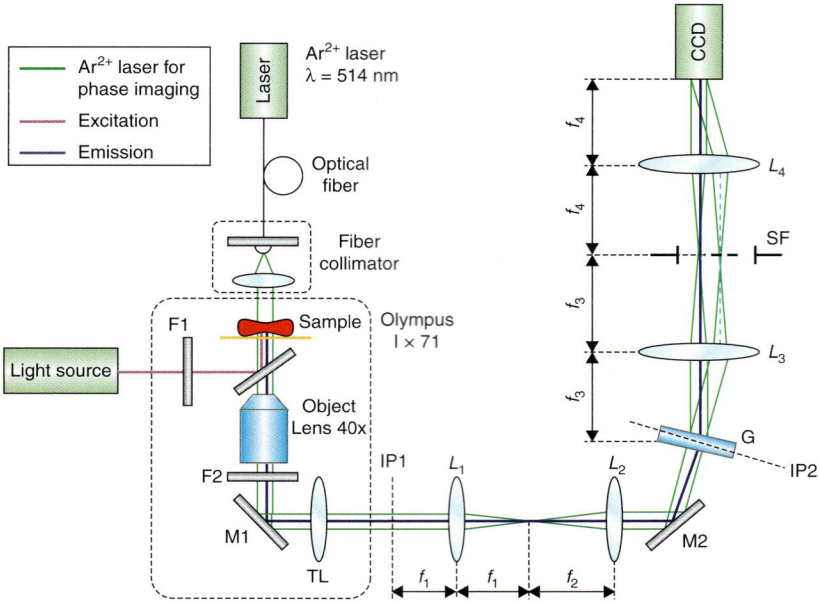

FIGURE 11.11   DPF setup. F1, 2, filters; M1, 2, mirrors; L_{1-4} lenses (f_{1-4}, respective focal lengths); G, grating; SF, spatial filter; $IP_{1,2}$, image planes; SF, spatial filter. Ref. [38]

by offering a new window into phenomena related to cell physiology. In 2006, our group at MIT demonstrated a practical combination of epi-fluorescence and DPM, referred to as the diffraction-phase and fluorescence (DPF) microscopy.[38]

The DPF experimental setup is depicted in Fig. 11.11. An inverted microscope is equipped for standard epi-fluorescence, using a UV lamp and an excitation-emission filter pair, F1, 2. An Ar^{2+} laser ($\lambda = 514$ µm) is used as an illumination source for transmission phase imaging. Through its video output port, the microscope produces the image of the sample at the image plane IP1 with magnification $M = 40$. The lens system L_1–L_2 is used to collimate the unscattered field (spatial DC component) and further magnify the image by a factor $f_2/f_1 = 3$, at the plane IP_2. An amplitude diffraction grating G is placed at IP_2 and generates multiple diffraction orders containing full spatial information about the sample image. Like in DPM, the goal is to isolate the 0th and 1st orders and to create a common-path Mach-Zender interferometer, with the 0th order as the reference beam and the 1st order as the sample beam. To accomplish this, a standard 4-f spatial filtering lens system L_3–L_4 is used. This system selects only the 0th and 1st order and generates the final interferogram at the CCD plane. The 0th-order beam is low-pass filtered using a pinhole placed at the Fourier plane

L_3 so that it becomes a plane wave after passing through lens L_4. The spatial filter allows passing the entire frequency information of the 1st-order beam and blocks the high-frequency information of the 0th beam. Compared to conventional Mach-Zender interferometers, the two beams propagate through the same optical component, which significantly reduces the longitudinal phase noise, without the need for active stabilization. The fluorescence light also passes through the grating that generates two diffraction spots in the Fourier plane of lens L_3. However, due to its spatial incoherence, the fluorescence diffraction orders are much broader than the pinhole, which blocks the 0th order almost entirely. Therefore, the fluorescence image information is carried to the CCD by the 1st order of diffraction. A commercial digital camera was used to capture both the interferogram and the fluorescence image. The CCD has a resolution of 3000×2000 pixels and each pixel is 7.8×7.8 μm in size. From the interferogram recorded, the quantitative-phase image is extracted via a spatial Hilbert transform, as described earlier in Sec. 5.2.[33] The grating period is smaller than the diffraction spot of the microscope at the grating plane so the optical resolution of the microscope is preserved.

In order to illustrate the combined phase-fluorescence imaging capability, we performed experiments of kidney (mesangial) cells in culture. The cells were imaged directly in culture dishes, surrounded by culture medium. Prior to imaging, the cells were treated with Hoest solution for 60 minutes at 38°C and 5% CO_2. This fluorescent dye binds to the DNA molecules and is commonly used to reveal the cell nuclei. Figure 11.12 shows an example of our composite investigation. The quantitative-phase image of a single cell is shown in Fig. 11.12a. Fig. 11.12b shows the fluorescence image of the same cell, where it becomes apparent that the cell contains two nuclei. Figure 11.12c shows the composite image.

Thus, by taking advantage of the difference in the spatial coherence of the two fields, the fluorescence and phase-imaging light pass through the same optics, without the need for separation by using, for instance, dichroic mirrors. While the diffraction grating provides a stable geometry for interferometry, it introduces light losses, which may affect fluorescence imaging especially of weak fluorophores. This aspect can be ameliorated by using a sinusoidal amplitude grating that maximizes the diffraction in the +1 and −1 orders.

Confocal Diffraction-Phase Microscope

In 2008, we presented a QPI method, *confocal diffraction-phase microscopy* (cDPM), which provides quantitative-phase measurements from localized sites on a sample with high sensitivity.[39] The technique combines common-path interferometry with confocal microscopy in a *transmission* geometry. The capability of the technique for static imaging was demonstrated by imaging polystyrene microspheres

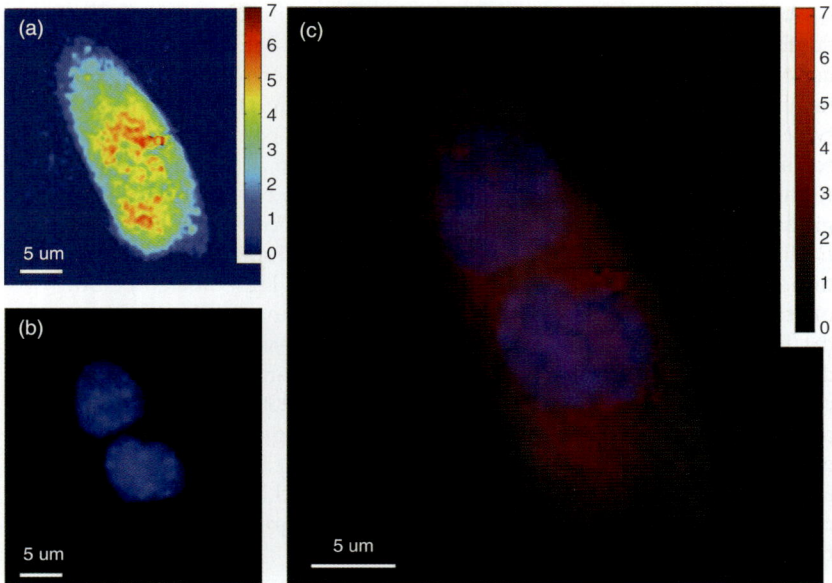

FIGURE 11.12   (a) Quantitative-phase image of a kidney cell. The colorbar indicates optical phase shift in radians. (b) Fluorescence image of the same DNA-stained cell. (c) Overlaid images from (a) and (b): red-DPM image and blue-epi-fluorescence image. Colorbar indicates phase in radians. (*From Ref. 38.*)

and live HT29 cells, while dynamic imaging is demonstrated by quantifying the nanometer scale fluctuations of red blood cell membranes.

11.2.3   Biological Applications

Fresnel Particle-Tracking Using DPM

In 2007, our group at MIT developed a novel experimental technique based on DPM that allows for tracking small particles in three dimensions with nanometer accuracy.[40] The longitudinal positioning of a micron-sized particle is determined by using the Fresnel approximation to describe the transverse distribution of the field scattered from the particle (hence the name of the technique, *Fresnel particle-tracking,* or *FTP*).

Biological cells are complex structures with multiple characteristic time and length scales.[41] The ability to measure locally how a cellular component responds to an applied shear stress at short time and small displacements has a variety of applications. The mechanical properties of the cells and intracellular material determine the proper functioning of the living system. A number of techniques have been developed for probing the rheological properties of complex fluids at the microscopic scale.[42–44] By measuring the low-coherence light scattering from embedded microparticles, volumes of 0.1 picoliters can be investigated.[45,46] The three-dimensional trajectories of embedded colloidal particles can be interpreted in terms of the viscoelastic properties of the surrounding medium.[47] This approach has revealed new and unexpected behavior of cellular matter.[48,49] However, particle-tracking techniques typically rely on imaging the particle of interest in two dimensions and assuming that the material under investigation is isotropic, which is an idealization. Tracking microparticles in the full three-dimensional space enables studying the physical properties of anisotropic media (e.g., cell mitotic spindles) and investigating surface effects at boundaries.[50,51]

Our method, referred to as Fresnel particle-tracking (FPT), relies on quantifying the wavefront of the light scattered by the particle. In order to image the phase distribution of the scattered field, we employ DPM[28] in an inverted microscope imaging geometry.[38] A similar approach for finding the plane of focus was proposed by Dubois et al. using digital holography.[52]

We imaged 2-µm-diameter polystyrene beads suspended in water undergoing Brownian motion. The goal is to infer the longitudinal position (z-axis) of the bead with respect to the plane of focus from the quantitative-phase image. To retrieve the z position we used a correlation algorithm, as follows. We modeled the particle as a phase object: the in-focus field associated with the particle is U_p, with its phase, ϕ_p, taking into account the spherical profile of the particle. When the particle is out of focus by a distance z, the scattered field, U_S, at the image plane can be obtained from the Fresnel equation, which is essentially a convolution operation between the in-focus field, U_p, and the Fresnel wavelet, U_F

$$U_S(x,y;z) \propto \int\int U_p(x',y')U_F(x-x',y-y';z)dx'dy' \qquad (11.11)$$

We numerically simulate a set of fields, U_S, for various values of z and obtain the respective two-dimensional phase distributions, ϕ_S. Experimentally, DPM provides the phase image associated with the particle at the unknown position, z_0. In order to find z_0, we cross-correlate the measured and simulated phase distributions, and identify the value of z for which the cross-correlation attains a maximum value. This process is depicted in Fig. 11.13. Note that the principle of finding the unknown depth is somewhat similar to the character recognition procedure via matched filters discussed in the holography chapter (Chap. 6).

In order to demonstrate this procedure, we first measured the z-positions of a particle fixed on a microscope slide, as the focus of the microscope objective is adjusted manually (Fig. 11.14a.). It can be seen

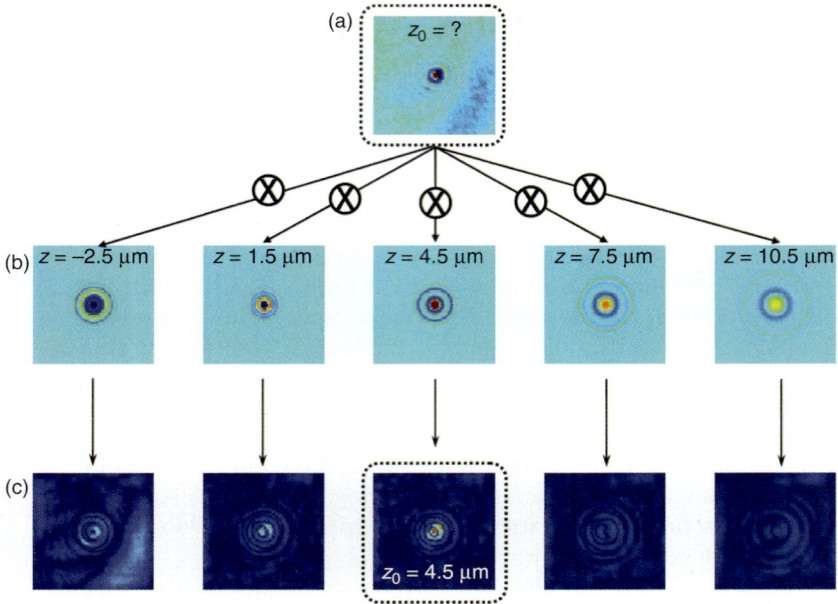

FIGURE 11.13 (*a*) Measured phase distribution of the light scattered from the bead. (*b*) Simulated wavefront distributions for various values of the particle position, *z*. (*c*) Result of the 2D cross correlations between the phase in *a* and each of the simulated phases in *b*. The cross correlation of highest value provides the value of z_0. (*From Ref. 38.*)

that our algorithm correctly retrieves the position of the bead with respect to the in-focus plane. The existing errors are largely due to the imprecision in the manual focus adjustment. Figure 11.14*b* shows the measured 3D trajectory of 660 positions for a bead undergoing Brownian motion in water. The bead positions were retrieved from the DPM images measured at rate of 33 Hz for 20 s. In the image plane, a CCD pixel corresponds to 80 nm in the sample plane. However, using pixel interpolation, we extracted the *x-y* position of the particle with precision better than this value. The overall precision of the particle *z*-axis position is defined by the stability of the DPM phase imaging. In order to assess the stability of our measuring system and, therefore, the overall robustness of our algorithm, we recorded the trajectory of a bead that is fixed on a cover slip. Remarkably, as shown in Fig. 11.14*c*, the retrieved position is confined to a $20 \times 20 \times 20$ nm^3 volume (approximately 10^{-5} femtoliters). Figure 11.14*d* shows the histogram of the rms *z*-positions that have a mean value of approximately 11 nm.

From the measured trajectories of beads suspended in water, we obtained the mean-squared displacement, defined as $MSD = \overline{[\mathbf{r}(t+\tau) - \mathbf{r}(t)]^2}$, with **r** the position vector and the horizontal bar

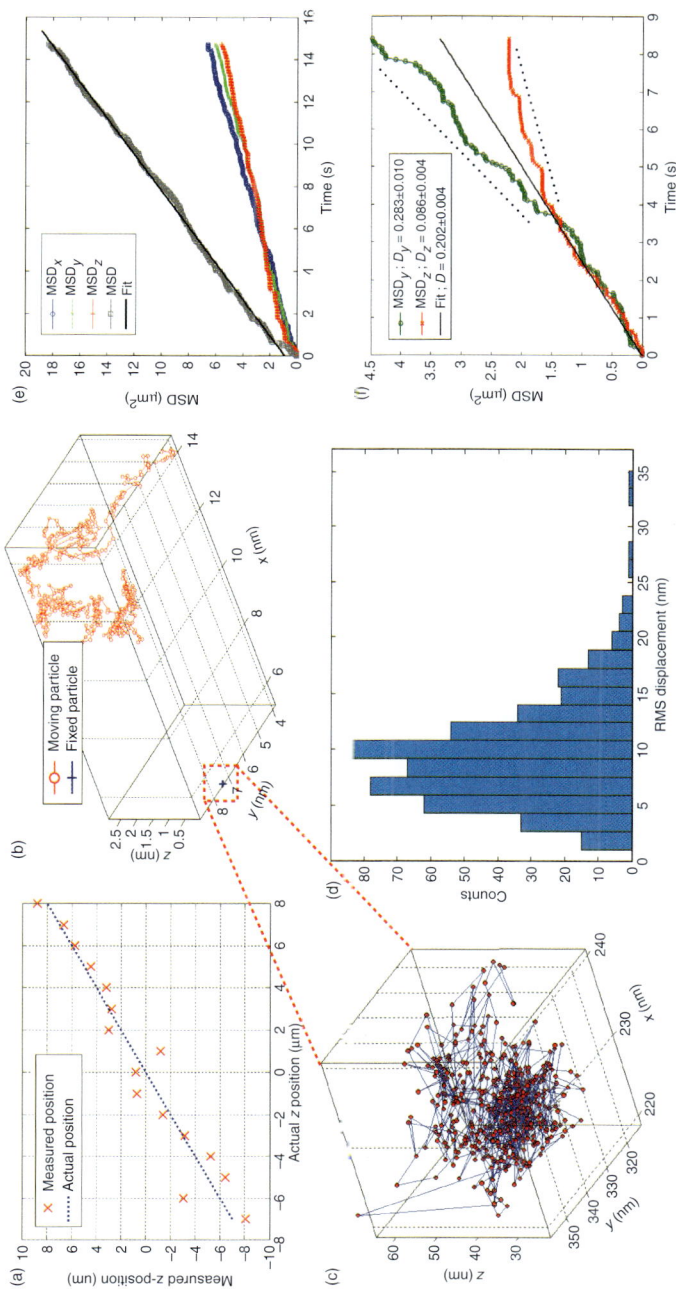

FIGURE 11.14 (a) Retrieved and expected z positions for a bead affixed to a microscope slide as the focus is adjusted. (b) 3D trajectories of a 2-m diameter bead under Brownian motion in water (red circles) and affixed to a cover glass (blue cross). (c) Retrieved trajectory for the fixed bead in b. (d) Histogram of z-axis rms displacement for the fixed bead. (e) Total and individual axis MSD for Brownian particle in water, as indicated. (f) The y- and z-axis MSD for a particle embedded in water, in the vicinity of the cover glass. (From Ref. 40.)

indicating time average. Figure 11.14*e* depicts the *MSD* dependence on the time delay for each coordinate *x, y, z,* and for the total displacement. The individual component *MSD* is defined as $MSD_\xi = \overline{[\xi(t+\tau)-\xi(t)]^2}$, where $\xi = x, y, z$, and $MSD = \sum_\xi MSD_\xi$. The solid line indicates the fit to the equation $MSD = D\tau$, where the diffusion coefficient, *D,* is given by the Stokes-Einstein equation,[53]

$$D = \frac{k_B T}{3\pi\eta d} \tag{11.12}$$

In Eq. (11.12), k_B is the Boltzmann constant, *T* is the absolute temperature (295 °K throughout our experiments), η is the water viscosity, and *d* the particle diameter. Using the viscosity as the fit parameter, we obtain $\eta = 1.051 \pm 0.005$ cp, which agrees very well with the expected value of 1.002 cp. This demonstrates the ability of FPT to measure physical properties of fluids at microscopic scales. In the proximity of physical boundaries, particles under Brownian motion experience anisotropic behavior, i.e. the displacements along different directions are statistically different. We used FPT to analyze the motion of a bead moving in the immediate vicinity of the microscope slide. Figure 11.14*f* shows the MSD associated with the *y-* and *z*-axis of this particle.

In summary, Fresnel particle-tracking is a novel technique for particle tracking in the three-dimensional space and with nanoscale precision. We anticipate that FPT will help gain a better understanding of the viscoelastic properties of cellular matter at the microscopic scale. Probing beads can be both engulfed within cells and conjugated to cell membranes, which will enable FPT to study various phenomena at the cellular and subcellular level.

Red Blood Cell Mechanics

The red blood cell (RBC) deformability in microvasculature governs the cell's ability to survive the physical demands of circulation, as well as the cells ability to transport oxygen in the body.[54] Interestingly, RBCs must periodically pass a deformability test by being forced to squeeze through narrow passages (*sinuses*) in the spleen; upon failing this mechanical assessment, the cell is destroyed and removed from circulation by *macrophages.*[55] Quantifying and understanding the mechanics of live RBCs requires sensitive probes of their structure at the nanoscale.[56–64] Such investigation promises new insights into the etiology of a number of human diseases.[65] In the healthy individual, these cells withstand repeated, large-amplitude mechanical deformations as they circulate through the microvasculature. Throughout their 120-day lifespan, the RBC's mechanical integrity degrades and ultimately they are replaced by new circulating RBCs. Certain pathological conditions such as spherocytosis, malaria, and sickle cell disease cause changes in both the equilibrium shape and mechanics of RBCs, which impact their transport function. Thus, understanding the microrheology of RBCs is highly interesting both from a basic science and a clinical point of view.

Lacking a conventional 3D cytoskeleton structure, RBCs maintain their shape and mechanical integrity through a spectrin-dominated, triangular 2D network associated with the cytosolic side of their plasma membrane.[66] This semiflexible filament network confers shear and bulk moduli to the composite membrane structure.[67] The fluid lipid bilayer is thought to be the principal contributor to the bending or curvature modulus of the composite membrane. Little is known about the molecular and structural transformations that take place in the membrane and spectrin network during the cell's response to pathophysiological conditions, which are accompanied by changes in RBC mechanics.

A number of techniques have been used recently to study the rheology of live cells.[65] Micropipette aspiration,[58] electric field deformation,[56] and optical tweezers[65] provide quantitative information about the shear and bending moduli of RBC membranes in *static* conditions. However, *dynamic,* frequency-dependent knowledge of RBC mechanics is currently very limited.[68] RBC thermal fluctuations ("flickering") have been studied for more than a century[69] to better understand the interaction between the lipid bilayer and the cytoskeleton.[28,70–72] Membrane fluctuation dynamics of RBCs can be influenced by physiological conditions and disease states. Fluctuations in phospholipid bilayer and attached spectrin network are known to be influenced by cytoskeletal defects, stress, and actin–spectrin dissociations arising from metabolic activity linked to adenosine-5'-triphosphate (ATP) concentration.[58,73–76] Nevertheless, quantifying these motions is experimentally challenging, and reliable spatial and temporal data are desirable.[71,77,78]

Existing optical methods include phase-contrast microscopy (PCM),[79] reflection interference contrast microscopy (RICM),[80] and fluorescence interference contrast (FLIC).[78] RBCs lack nuclei and organelles and can be assumed optically homogeneous, i.e. characterized by a constant refractive index. Therefore, measurement of the cell optical path-length via interferometric techniques can provide information about the physical topography of the membrane with sub-wavelength accuracy and without contact.

In 2010, Park et al. from MIT reported: "Measurement of red blood cell mechanics during morphological changes."[8] Blood samples were collected and centrifuged for 10 minutes at an acceleration of 2000 g and temperature of 5°C to separate RBCs from plasma. The cells were washed three times with a saline solution and were ultimately resuspended in phosphate buffered saline (pH 7.4), as described in Ref. [81]. The RBCs were then placed between cover slips and imaged without additional preparation.

Our samples were primarily composed of RBCs with the typical discocytic shape (DCs), but also contained cells with abnormal morphology which formed spontaneously in the suspension, such as echinocytes (ECs) (cells with a spiculated shape), and spherocytes (SCs) (cells that had maintained a roughly spherical shape). By taking into account the free-energy contributions of both the bilayer and cytoskeleton, these morphological changes have been successfully modeled.[82]

As shown above, the standard description of a RBC treats the membrane as a flat surface; in reality the membrane of a RBC is curved and has the compact topology of a sphere. The Levine group at University of California, Los Angeles, developed model that takes into account this geometrical effect and incorporates the curvature and topology of a RBC within the fluctuation analysis (for details on this theory, see Ref. [83]). Because of the membrane curvature, the bending and compression modes of a spherical membrane are coupled in the linear order in deformations. The geometric coupling of bending and compression generates undulatory dynamics consistent with the experiments, and does not require the postulate of a surface-tension-like term in the Hamiltonian of the composite membrane.[84] The coupling of the fluid to the membrane is done using the stick boundary conditions and the stress balance condition at the surface of the membrane.

To quantitatively investigate the material properties of RBCs, we analyze the spatial and temporal correlations of the out-of-plane fluctuations of the RBC membrane and interpret them using the viscoelastic continuum model of the composite spectrin-network/lipid membrane. This model, due to Kuriabova and colleagues from the Levine group,[83] accounts for the linear coupling between the bending and compression modes of a curved membrane, and thus provides a better description of the dynamics of the RBC than theories based on a flat membrane. The undulatory dynamics of a RBC is probed experimentally by measuring the spatial and temporal correlations of the out-of-plane fluctuations of the membrane. Theoretically, these correlations can be calculated using the response matrix, χ, defined above and the fluctuation-dissipation theorem.[85] The correlation of height fluctuations at two points on the membrane separated by the projected distance d and time t is defined by

$$\Lambda(\rho,\tau) = \left\langle \Delta h(\rho,\tau)\Delta h(0,0)\right\rangle_{r,t} \tag{11.13}$$

where the angular brackets denote both spatial and temporal averaging. Interestingly, the spatio-temporal correlation function in Eq. (11.13) has the same form as the one used to describe optical-field correlations (Chap. 3, Eq. 3.2.). From the results of our previous work,[86] and the fluctuation-dissipation theorem, we find that the

Fourier transform of this function in the frequency domain, ω, is given by[83]

$$\Lambda(\rho,\omega) = \left(1 - \frac{d^2}{4R^2}\right) \frac{2k_BT}{\omega} \sum_l \text{Im}[\chi_{ww}(l,\omega)]P_l\left(1 - \frac{d^2}{4R^2}\right) \qquad (11.14)$$

where $P_l(x)$ is the Legendre polynomial of lth order, R is the radius of RBC, k_b is the Boltzmann constant, and T is temperature. These height-height correlation functions are plotted in Fig. 11.15a-c. We show the spatial decay of the height-height correlations at two fixed frequencies: one corresponding to the elastic plateau (low frequency) and one to the viscously dominated regime (high frequency). At lower frequencies, we see a pronounced oscillatory behavior in the correlation function. The negative correlations are due to the dominance of the

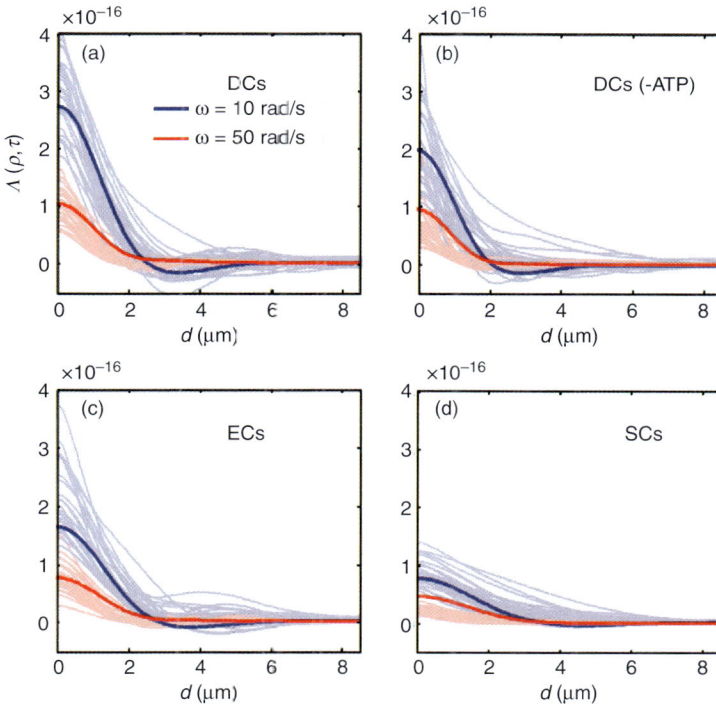

Figure 11.15   Height-height correlation function from experiments (faint lines, N = 30 per each group) and calculation (thick lines). (a-d) Height-height correlation function as a projected distance for DCs (a), ECs (b), and SCs (d) for ω = 6 rad/s (blue) and ω = 50 rad/s (red). (From Fig. 3, Ref. 8) Proc. Nat. Acad. Sci., used with permission, Y.K. Park, C.A. Best, K. Badizadegan, R.R. Dasari, M.S. Feld, T. Kuriabova, M.L. Henle, A.J. Levine, and G. Popescu, "Measurement of red blood cell mechanics during morphological changes," 107, 6731 (2010).

small-l contributions to the response function at low frequencies. At higher frequencies, we see that there is a shorter-ranged and nearly monotonic decay of the height-height correlations. Remarkably, this mode coupling predicts a decay rate $\omega \propto q$, which is identical to the behavior by a model of the membrane with an effective surface tension proposed by Gov et al.[84,87]

Further, we used the new model of RBC mechanics over the commonly occurring discocyte-echinocyte-spherocyte (DC-EC-SC) shape transition (Fig. 11.16a-c). From measurements of dynamic fluctuations on RBC membranes, we extracted the mechanical properties of the composite membrane structure. Subtracting the instant-thickness map by the averaged-thickness map provides the instantaneous displacement maps of the RBC membranes (Fig. 11.16, right column). Over the morphological transition from DC to SC, the root-mean-squared amplitude of equilibrium membrane height fluctuations $\sqrt{\langle \Delta h^2 \rangle}$ decreases progressively from 134 nm (DCs) to 92 nm (ECs) and 35 nm (SCs), indicating an increase in cell stiffness.

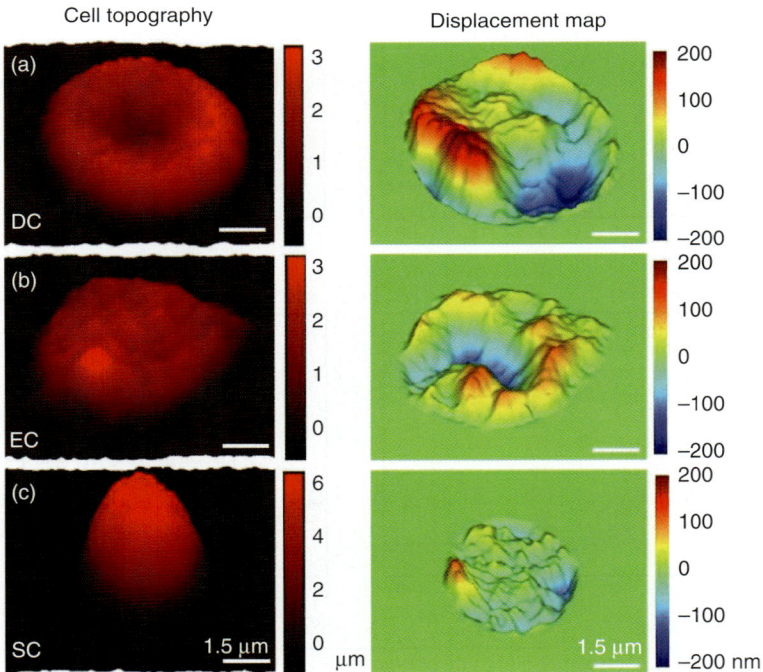

FIGURE 11.16   RBC topography (left column) and instantaneous displacement maps (right column) for a discocyte (a), echinocyte (b), and spherocyte (c), as indicated by DC, EC, and SC, respectively. (From Fig. 1, Ref. 8.) Proc. Nat. Acad. Sci., used with permission, Y.K. Park, C.A. Best, K. Badizadegan, R.R. Dasari, M.S. Feld, T. Kuriabova, M.L. Henle, A.J. Levine, and G. Popescu, "Measurement of red blood cell mechanics during morphological changes," 107, 6731 (2010).

In order to extract the material properties of the RBC, we fit our model to the measured correlation function $\Lambda(\rho, \omega)$ by adjusting the following parameters: the shear μ and bulk K moduli of the spectrin network, the bending modulus κ of the lipid bilayer, the viscosities of the cytosol η_c and the surrounding solvent η_s, and the radius of the sphere R. We constrain our fits by setting R to the average radius of curvature of the RBC obtained directly from the data and fixing the viscosities for all data sets to be $\eta_s = 1.2$ mPa·s, $\eta_s = 5.5$ mPa·s.[77,78] Finally, for a triangular elastic network we expect $\mu = \lambda$, so we set $K = 2\mu$.[79]

The parameters extracted from the fit are shown in Fig. 11.17a-c. The extracted bending modulus increases significantly during the DC-EC-SC transition ($p < 10^{-7}$). Their mean values are 6.3 ± 1.0 (DC), 11.9 ± 2.5 (EC), and 23.8 ± 4.1 (SC) in units of $k_B T$ (Fig. 11.17b). These values are in general agreement with those expected for a phospholipid bilayer $(5 - 20) \times k_B T$.[80] The increase in bending modulus suggests changes in the composition of the lipid membrane. We measured directly the change in surface area of RBCs during the transition from DC to SC morphologies and found a 31% decrease in surface area (not accounting for surface area stored in fluctuations). This surface area decrease must be accompanied by a loss of lipids, via membrane blebbing and microvesiculation. Thus, there is evidence that a significant change in lipid composition of the RBC bilayer accompanies the morphological changes from DC to EC and SC. It is thought that these changes in lipid composition generate the observed changes in the bending modulus.

The shear modulus results are shown in Fig. 11.17b and d. The fitted shear moduli are 6.4 ± 1.4 (DC), 10.7 ± 3.5 (EC), and 12.2 ± 3.0 (SC) in μNm^{-1}. These values are consistent with earlier work based on micropipette aspiration[88] and optical tweezers.[89] The magnitude of the measured shear modulus also agrees well with simple elastic models of the spectrin network. The shear modulus of SCs and ECs increased by roughly a factor of two compared to the DCs ($p < 10^{-5}$). There is, however, significant cell-to-cell variation in the shear modulus. While the histogram of shear moduli of DCs can be fit by a single Gaussian distribution centered at 6.7 μNm^{-1}, the respective shear moduli distributions for ECs and SCs are bimodal, with peaks at, respectively, 6.4 μNm^{-1} and 13.0 μNm^{-1} (ECs), and 6.8 μNm^{-1} and 12.9 μNm^{-1} (SCs) (Fig. 11.17d). These data suggest that there are essentially two independent conformations of the spectrin network: a soft configuration ($\mu \cong 7$ μNm^{-1}) and a stiff one ($\mu \cong 13$ μNm^{-1}). Essentially all DCs have the soft configuration, but the morphological transition to EC and then SC promotes the transition to the stiff network configuration. We propose that the observed morphological changes must be accompanied by modifications of the spectrin elasticity, the connectivity of the network, or its attachment to the lipid bilayer.

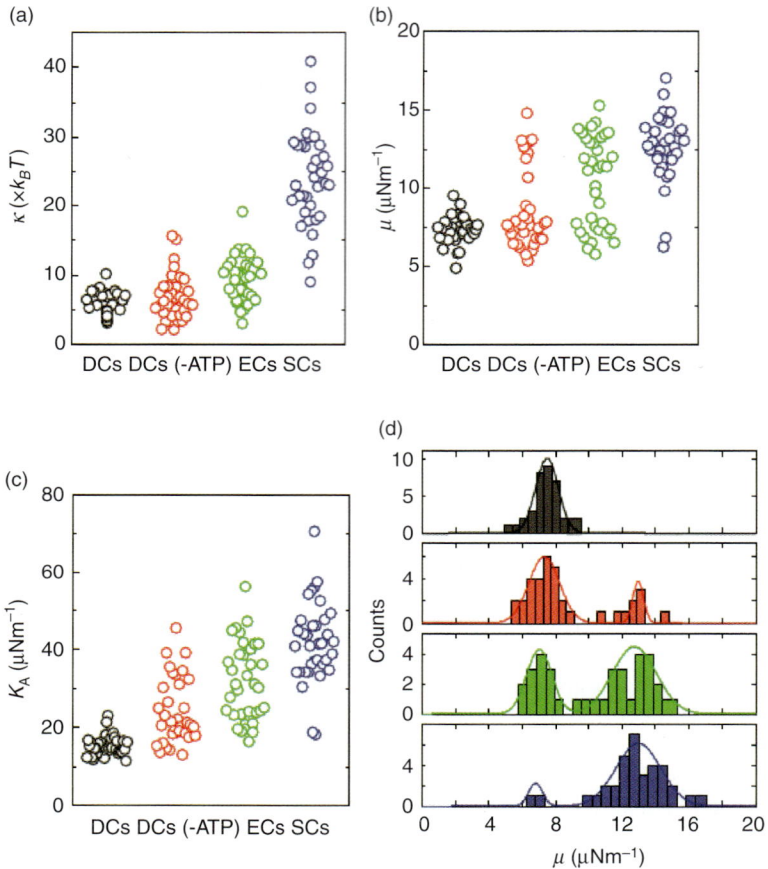

FIGURE 11.17   Shear modulus and bending modulus. (a) Shear moduli of DCs, ECs, and SCs with their mean values represented by the horizontal lines. p values verify that the differences in the shear moduli between morphological groups are statistically significant: $p < 10^{-5}$ between DCs and ECs, and between DCs and SC; $p < 10^{-4}$ between ECs and SCs. (b) bending modulus of three groups. (c) Distributions of shear moduli. The Gaussian fits are overlapped. The centers of the Gaussian fits are 7.34 (DCs), 6.97 and 12.93 (ECs), and 7.40 and 13.80 (SCs). (From Fig. 4, Ref. 8). Proc. Nat. Acad. Sci., used with permission, Y.K. Park, C.A. Best, K. Badizadegan, R.R. Dasari, M.S. Feld, T. Kuriabova, M.L. Henle, A.J. Levine, and G. Popescu, "Measurement of red blood cell mechanics during morphological changes," 107, 6731 (2010).

Imaging of Malaria-Infected RBCs

Malaria is an infectious disease caused by a eukaryotic protist of the genus *Plasmodium*.[66,90] Malaria is naturally transmitted by the bite of a female *Anopheles* mosquito. This disease is widespread in tropical and subtropical regions, including parts of the Americas (22 countries),

Asia, and Africa. There are approximately 350 to 500 million cases of malaria per year, killing between one and three million people, the majority of whom are young children in sub-Saharan Africa.[90] When a mosquito bites an infected person, a small amount of blood is taken, which contains malaria parasites. These develop within the mosquito, and about 1 week later, when the mosquito takes its next blood food, the parasites are injected with the mosquito's saliva into the person being bitten. After a period of between 2 weeks and several months (occasionally years) spent in the liver, the malaria parasites start to multiply within red blood cells, causing symptoms that include fever and headache. In severe cases, the disease worsens, leading to hallucinations, coma, and death. Five species of the *Plasmodium* parasite can infect humans; the most serious forms of the disease are caused by *Plasmodium falciparum*. Malaria caused by *Plasmodium vivax*, *Plasmodium ovale*, and *Plasmodium malariae* result in milder disease in humans that is not generally fatal.

During the intra-erythrocytic (RBC) development, the malaria parasite *Plasmodium falciparum* causes structural, biochemical, and mechanical changes to host RBCs.[91] Major structural changes include the growing of vacuole of parasites in cytoplasm of host RBCs, loss of cell volume, and the appearance of small, nanoscale protrusions, called "knobs" or "blebs," on the membrane surface.[92] From the bio-chemical standpoint, a considerable amount of hemoglobin (Hb) is digested by parasites during intra-erythrocytic development and converted into insoluble polymerized forms of heme, known as hemozoin.[93,94] Hemozoin appears as brown crystals in the vacuole of parasite in later maturation stages of *P. falciparum*-invaded human RBCs (*Pf*-RBCs).[66] Two major mechanical modifications are loss of RBC deformability[95–97] and increased cytoadherence of the invaded RBC membrane to vascular endothelium and other RBCs.[98] These changes lead to sequestration of RBCs in microvasculature in the later stages of parasite development, which is linked to vital organ dys-function in severe malaria. In the earlier stage, where deformability occurs, *Pf*-RBCs continue to circulate in the bloodstream despite infection.

In 2008, the MIT groups lead by Subra Suresh and Michael Feld collaborated on using DPM to study the modifications in RBC dynamics produced by the onset and development of malaria.[6] To investigate morphological changes of *Pf*-RBCs, we measured the instantaneous thickness profile, $h(x,y,t_0)$, of cells.[99] Figure 11.18a-d shows topographic images of healthy and *Pf*-RBCs at all stages of development. The effective stiffness map of the cell, $k_e(x,y)$ is obtained at each point on the cell, assuming an elastic restoring force associated with the membrane,

$$k_e(x,y) = k_B T / \left\langle \Delta h(x,y)^2 \right\rangle \tag{11.15}$$

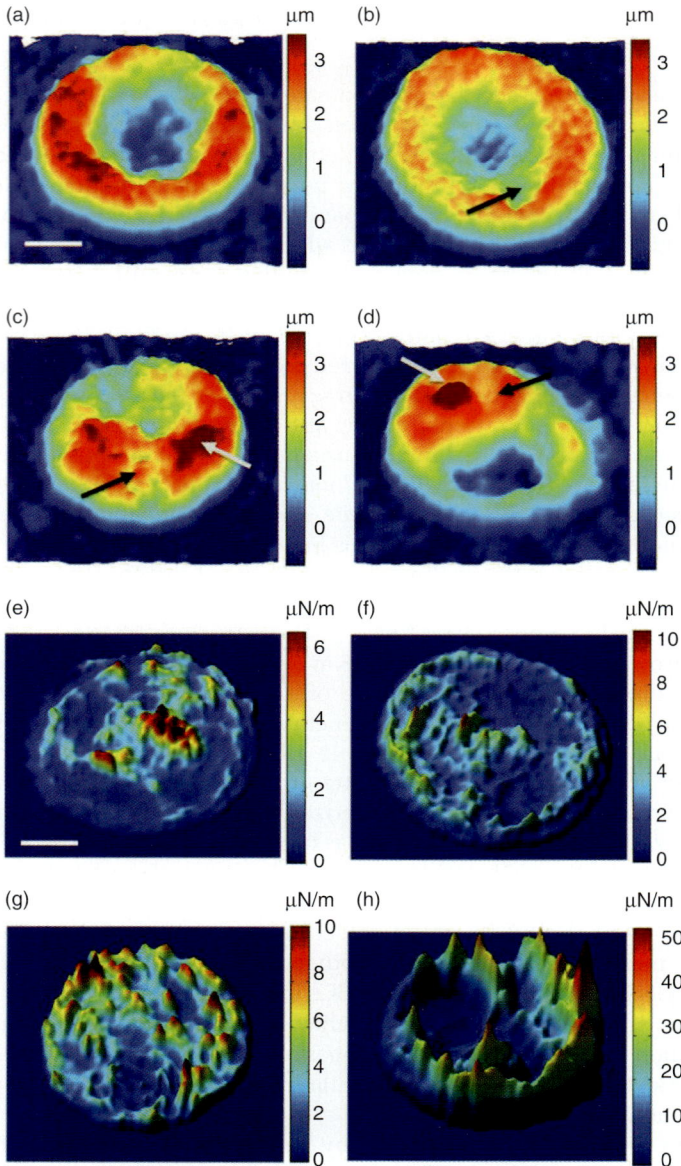

FIGURE 11.18 Topographic images and effective elastic constant maps of *Pf*-RBCs. (*a* and *e*) Healthy RBC. (*b* and *f*) Ring stage. (*c* and *g*) Trophozoite stage. (*d* and *h*) Schizont stage. The topographic images in *a–d* are the instant thickness map of *Pf*-RBCs. The effective elastic constant maps were calculated from the root-mean-squared displacement of the thermal membrane fluctuations in the *Pf*-RBC membranes. Black arrows indicate the location of *P. falciparum*, and the gray arrows the location of hemozoin. (Scalebar, 1.5 µm). (*Adapted from Fig. 3, Ref. 6*) *Proc. Nat. Acad. Sci.,* used with permission, Y.K. Park, M. Diez-Silva, G. Popescu, G. Lykotrafitis, W. Choi, M.S. Feld & S. Suresh, "Refractive index maps and membrane dynamics of human red blood cells parasitized by Plasmodium falciparun," 105, 13730 (2008).

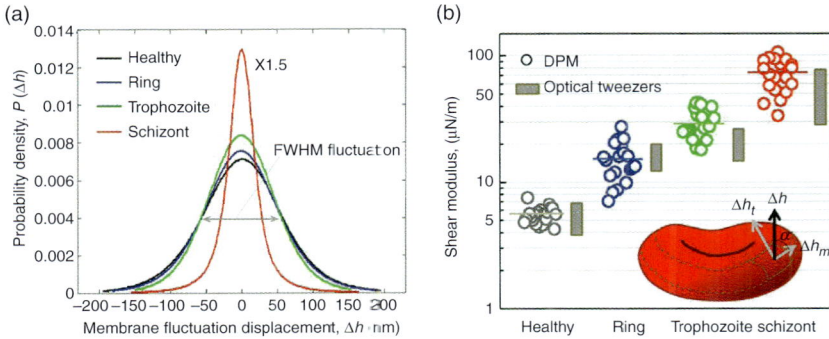

FIGURE 11.19 Membrane fluctuations and in-plane shear modulus at different intra-erythrocytic stages of *Pf*-RBCs. (*a*) Histograms of cell-thickness fluctuation of *Pf*-RBCs. (Histogram of the schizont stage is scaled down by a factor of 1.5.) (*b*) In-plane shear modulus of the RBC membrane versus developmental stage of *Pf*-RBCs. The in-plane shear modulus was calculated from the in-plane membrane displacement, Δh_n over the rims of RBCs. Also shown for comparison are the estimated from optical tweezers (22). (*Inset*) Illustration of RBC and membrane fluctuations: Δh, thickness fluctuations measured by DPM; Δh_t, in-plane membrane displacement; Δh_n, out-of-plane membrane displacement, and α, the angle between Δh and Δh_t. The measurements were performed at the room temperature (23°C) and for each group 20 samples were tested. (*From Fig. 4, Ref. 6.*) *Proc. Nat. Acad. Sci.*, used with permission, Y.K. Park, M. Diez-Silva, G. Popescu, G. Lykotrafitis, W. Choi, M.S. Feld & S. Suresh, "Refractive index maps and membrane dynamics of human red blood cells parasitized by Plasmodium falciparun," 105, 13730 (2008).

where k_B is the Boltzmann constant, T the absolute temperature, and $\langle \Delta h(x,y)^2 \rangle$ the mean-squared displacement. Representative k_e maps of RBCs at the indicated stages of parasite development are shown in Fig. 11.18*e-g*. The map of instantaneous displacement of cell-membrane fluctuation, $\Delta h(x,y,t)$, was obtained by subtracting time-averaged cell shape from each thickness map in the series. A histogram showing membrane displacements for all parasite stages is shown in Fig. 11.19*a*.

Our DPM experiments provide quantitative information from which in-plane shear modulus of RBC membranes with attached spectrin cytoskeletons could be determined. The in-plane shear modulus, G, can be obtained using the Fourier-transformed Hamiltonian (strain energy) and equipartition theorem:[100]

$$G \cong \frac{k_B T \ln(A/a)}{3\pi \langle \Delta h_t^2 \rangle} \tag{11.16}$$

where A is the projected diameter of RBC (~ 8 μm), and a is the minimum spatial wavelength measured by DPM (~ 0.5 μm). The tangential component of displacement in membrane fluctuations, Δh_t^2, was decoupled from axial membrane fluctuation, Δh^2, using the angle, α, between the direction of Δh_t and the normal direction of membrane as illustrated in Fig. 11.19*b* (*inset*). The angle, α, is extracted from cell topographical information measured by DPM. Δh_t^2 was calculated and averaged along the circumference of cell, from which in-plane shear modulus, G, was calculated. We determine that $G = 5.5 \pm 0.8 \, \mu\text{N}/\text{m}$ for healthy RBCs (Fig. 11.19*b*), which compares well with independent modulus measurements, extracted for healthy RBCs from micropipette aspiration and optical tweezers.[88,101,102] The modulus for ring ($G = 15.3 \pm 5.4 \, \mu\text{N}/\text{m}$), trophozoite ($G = 28.9 \pm 8.2 \, \mu\text{N}/\text{m}$), and schizont ($G = 71.0 \pm 20.2 \, \mu\text{N}/\text{m}$) stages is in good quantitative agreement with that inferred from large-deformation stretching with optical tweezers of *Pf*-RBCs over all stages of parasite maturation.[101]

These results demonstrate a potentially useful avenue for exploiting the *nanoscale* and *quantitative* features of QPI for measuring cell membrane fluctuations, which, in turn, can report on the onset and progression of pathological states. We will revisit this concept in Chap. 15.

References

1. G. Popescu, L. P. Deflores, J. C. Vaughan, K. Badizadegan, H. Iwai, R. R. Dasari and M. S. Feld, Fourier phase microscopy for investigation of biological structures and dynamics, *Opt. Lett.*, 29, 2503–2505 (2004).
2. H. Kadono, M. Ogusu and S. Toyooka, Phase-Shifting Common-Path Interferometer Using A Liquid-Crystal Phase Modulator, *Optics Communications*, 110, 391–400 (1994).
3. A. Y. M. Ng, C. W. See and M. G. Somekh, Quantitative optical microscope with enhanced resolution using a pixelated liquid crystal spatial light modulator, *J. Microscopy*, 214, 334 (2004).
4. N. Lue, W. Choi, G. Popescu, R. R. Dasari, K. Badizadegan and M. S. Feld, Quantitative phase imaging of live cells using fast Fourier phase microscopy, *Appl. Opt.*, 46, 1836 (2007).
5. G. Popescu, K. Badizadegan, R. R. Dasari and M. S. Feld, Observation of dynamic subdomains in red blood cells, *J. Biomed. Opt. Lett.*, 11, 040503 (2006).
6. Y. K. Park, M. Diez-Silva, G. Popescu, G. Lykotrafitis, W. Choi, M. S. Feld and S. Suresh, Refractive index maps and membrane dynamics of human red blood cells parasitized by Plasmodium falciparum, *Proc Natl Acad Sci U S A*, 105, 13730 (2008).
7. Y. K. Park, C. A. Best, T. Auth, N. Gov, S. A. Safran, G. Popescu, S. Suresh and M. S. Feld, Metabolic remodeling of the human red blood cell membrane, *Proc. Nat. Acad. Sci.*, 107, 1289 (2010).
8. Y. K. Park, C. A. Best, K. Badizadegan, R. R. Dasari, M. S. Feld, T. Kuriabova, M. L. Henle, A. J. Levine and G. Popescu, Measurement of red blood cell mechanics during morphological changes, *Proc. Nat. Acad. Sci.*, 107, 6731 (2010).
9. G. Popescu, Y. Park, W. Choi, R. R. Dasari, M. S. Feld and K. Badizadegan, Imaging red blood cell dynamics by quantitative phase microscopy, *Blood Cells Molecules and Diseases*, 41, 10–16 (2008).
10. Y. Park, M. Diez-Silva, D. Fu, G. Popescu, W. Choi, I. Barman, S. Suresh and M. S. Feld, Static and dynamic light scattering of healthy and malaria-parasite invaded red blood cells, *Journal of Biomedical Optics*, 15, 020506 (2010).
11. G. Popescu, Y. Park, N. Lue, C. Best-Popescu, L. Deflores, R. R. Dasari, M. S. Feld and K. Badizadegan, Optical imaging of cell mass and growth dynamics, *Am J Physiol Cell Physiol*, 295, C538-44 (2008).

12. R. Baber, Interference microscopy and mass determination, *Nature*, 169, 366–367 (1952).
13. H. G. Davies and M. H. F. Wilkins, Interference microscopy and mass determination, *Nature*, 161, 541 (1952).
14. R. Barer, Interference microscopy and mass determination, *Nature*, 169, 366–367 (1952).
15. I. J. Conlon, G. A. Dunn, A. W. Mudge and M. C. Raff, Extracellular control of cell size, *Nat. Cell Biol.*, 3, 918–921 (2001).
16. D. E. Ingber, Fibronectin controls capillary endothelial cell growth by modulating cell shape, *Proc Natl Acad Sci U S A*, 87, 3579–3583 (1990).
17. P. Weiss and B. Garber, Shape and Movement of Mesenchyme Cells as Functions of the Physical Structure of the Medium. Contributions to a Quantitative Morphology, *PNAS*, 38, 264–280 (1952).
18. P. K. Maini, D. L. McElwain and D. I. Leavesley, Traveling wave model to interpret a wound-healing cell migration assay for human peritoneal mesothelial cells, *Tissue Eng*, 10, 475–482 (2004).
19. M. H. Zaman, R. D. Kamm, P. Matsudaira and D. A. Lauffenburger, Computational model for cell migration in three-dimensional matrices, *Biophys J*, 89, 1389–1397 (2005).
20. J. C. M. Mombach and J. A. Glazier, Single cell motion in aggregates of embryonic cells, *Phys. Rev. Lett.*, 76, 3032–3035 (1996).
21. M. H. Gail and C. W. Boone, Locomotion of mouse fibroblasts in tissue culture, *Biophys. J.*, 10, 981–993 (1970).
22. A. Czirok, K. Schlett, E. Madarasz and T. Vicsek, Exponential distribution of locomotion activity in cell cultures, *Phys. Rev. Lett.*, 81, 3038–3041 (1998).
23. H. Grueler, Analysis of cell movement, *Blood cells*, 10, 61–77 (1984).
24. G. A. Dunn and G. E. Jones, Timeline—Cell motility under the microscope: Vorsprung durch Technik, *Nat. Rev. Mol. Cell Biol.*, 5, 667–672 (2004).
25. A. Upadhyaya, J.-P. Rieu, J. A. Glazier and Y. Sawada, Anomalous diffusion and non-Gaussian velocity distribution of Hydra cells in cellular aggregates, *Physica A*, 293, 549–558 (2001).
26. M. J. Miller, S. H. Wei, I. Parker and M. D. Cahalan, Two-photon imaging of lymphocyte motility and antigen response in intact lymph node, *Science*, 296, 1869–1873 (2002).
27. C. Tsallis, Possible Generalization of Boltzmann-Gibbs Statistics, *Journal of Statistical Phys.*, 52, 479–487 (1988).
28. G. Popescu, T. Ikeda, R. R. Dasari and M. S. Feld, Diffraction phase microscopy for quantifying cell structure and dynamics, *Opt Lett*, 31, 775–777 (2006).
29. M. Mir, M. Ding, Z. Wang, J. Reedy, K. Tangella and G. Popescu, Blood screening using diffraction phase cytometry, *J. Biomed. Opt.*, 027016-1-4 (2010).
30. M. Mir, Z. Wang, K. Tangella and G. Popescu, Diffraction Phase Cytometry: blood on a CD-ROM, *Opt Exp.*, 17, 2579 (2009).
31. G. Popescu, "Quantitative Phase Imaging of Nanoscale Cell Structure and Dynamics," in Methods in Cell Biology B. P. Jena, ed., pp. 87–115 (Academic Press, San Diego, 2008).
32. G. Popescu, Y. K. Park, R. R. Dasari, K. Badizadegan and M. S. Feld, Coherence properties of red blood cell membrane motions, *Phys. Rev. E.*, 76, 031902 (2007).
33. T. Ikeda, G. Popescu, R. R. Dasari and M. S. Feld, Hilbert phase microscopy for investigating fast dynamics in transparent systems, *Opt. Lett.*, 30, 1165–1168 (2005).
34. J. R. Lakowicz and SpringerLink (Online service), *Principles of fluorescence spectroscopy*, xxvi, 954 p. (Springer, New York, 2006).
35. M. Chalfie, Y. Tu, G. Euskirchen, W. W. Ward and D. C. Prasher, Green Fluorescent Protein as a Marker for Gene-Expression, *Science*, 263, 802–805 (1994).
36. R. M. Dickson, A. B. Cubitt, R. Y. Tsien and W. E. Moerner, On/off blinking and switching behaviour of single molecules of green fluorescent protein, *Nature*, 388, 355–358 (1997).
37. A. Miyawaki, J. Llopis, R. Heim, J. M. McCaffery, J. A. Adams, M. Ikura and R. Y. Tsien, Fluorescent indicators for Ca2+ based on green fluorescent proteins and calmodulin, *Nature*, 388, 882–887 (1997).

38. Y. K. Park, G. Popescu, K. Badizadegan, R. R. Dasari and M. S. Feld, Diffraction phase and fluorescence microscopy, *Opt. Exp.*, 14, 8263 (2006).

39. N. Lue, W. Choi, K. Badizadegan, R. R. Dasari, M. S. Feld and G. Popescu, Confocal diffraction phase microscopy of live cells, *Opt. Lett.*, 33, 2074 (2008).

40. Y. K. Park, G. Popescu, R. R. Dasari, K. Badizadegan and M. S. Feld, Fresnel particle tracking in three dimensions using diffraction phase microscopy, *Opt. Lett.*, 32, 811 (2007).

41. L. H. Deng, X. Trepat, J. P. Butler, E. Millet, K. G. Morgan, D. A. Weitz and J. J. Fredberg, Fast and slow dynamics of the cytoskeleton, *Nature Materials*, 5, 636–640 (2006).

42. F. C. MacKintosh and C. F. Schmidt, Microrheology, *Curr. Opin. Colloid Interface Sci.*, 4, 300–307 (1999).

43. T. A. Waigh, Microrheology of complex fluids, *Rep. Prog. Phys.*, 68, 685–742 (2005).

44. G. Popescu and A. Dogariu, Scattering of low coherence radiation and applications, invited review paper, *E. Phys. J.*, 32, 73–93 (2005).

45. G. Popescu and A. Dogariu, Dynamic light scattering in localized coherence volumes, *Opt. Lett.*, 26, 551–553 (2001).

46. G. Popescu, A. Dogariu and R. Rajagopalan, Spatially resolved microrheology using localized coherence volumes, *Phys. Rev. E*, 65, 041504 (2002).

47. T. Gisler and D. A. Weitz, Tracer microrheology in complex fluids, *Curr. Opin. Colloid Interface Sci.*, 3, 586–592 (1998).

48. S. H. Tsen, A. Taflove, J. T. Walsh, V. Backman and D. Maitland, Pseudo-spectral time-domain Maxwell's equations solution of optical scattering by tissue-like structures, *Lasers in Surgery and Medicine*, RS4009, 10–10 (2004).

49. A. Caspi, R. Granek and M. Elbaum, Diffusion and directed motion in cellular transport, *Phys. Rev. E*, 66, (2002).

50. M. Speidel, A. Jonas and E. L. Florin, Three-dimensional tracking of fluorescent nanoparticles with subnanometer precision by use of off-focus imaging, *Optics Letters*, 28, 69–71 (2003).

51. F. Chasles, B. Dubertret and A. C. Boccara, Full-field optical sectioning and three-dimensional localization of fluorescent particles using focal plane modulation, *Optics Letters*, 31, 1274–1276 (2006).

52. F. Dubois, C. Schockaert, N. Callens and C. Yourassowsky, Focus plane detection criteria in digital holography microscopy by amplitude analysis, *Opt. Exp.*, 14, 5895–5908 (2006).

53. A. Einstein, Investigations on the theory of the Brownian movement, *Ann. Physik.*, 17, 549 (1905).

54. N. Mohandas and P. G. Gallagher, Red cell membrane: past, present, and future, *Blood*, 112, 3939–3948 (2008).

55. S. L. Robbins, V. Kumar and R. S. Cotran Robbins and Cotran pathologic basis of disease (Saunders/Elsevier, Philadelphia, PA, 2010).

56. H. Engelhardt, H. Gaub and E. Sackmann, Viscoelastic properties of erythrocyte membranes in high-frequency electric fields, *Nature*, 307, 378–380 (1984).

57. M. P. Sheetz, M. Schindler and D. E. Koppel, Lateral mobility of integral membrane proteins is increased in spherocytic erythrocytes, *Nature*, 285, 510-511 (1980).

58. D. E. Discher, N. Mohandas and E. A. Evans, Molecular maps of red cell deformation: hidden elasticity and in situ connectivity, *Science*, 266, 1032–035 (1994).

59. C. P. Johnson, H. Y. Tang, C. Carag, D. W. Speicher and D. E. Discher, Forced unfolding of proteins within cells, *Science*, 317, 663–666 (2007).

60. C. F. Schmidt, K. Svoboda, N. Lei, I. B. Petsche, L. E. Berman, C. R. Safinya and G. S. Grest, Existence of a flat phase in red cell membrane skeletons, *Science*, 259, 952–955 (1993).

61. J. D. Wan, W. D. Ristenpart and H. A. Stone, Dynamics of shear-induced ATP release from red blood cells, *Proceedings of the National Academy of Sciences of the United States of America*, 105, 16432–16437 (2008).

62. Y. Cui and C. Bustamante, Pulling a single chromatin fiber reveals the forces that maintain its higher-order structure, *Proceedings of the National Academy of Sciences of the United States of America*, 97, 127–132 (2000).

63. T. R. Hinds and F. F. Vincenzi, Evidence for a calmodulin-activated Ca2+ pump ATPase in dog erythrocytes, *Proc Soc Exp Biol Med*, 181, 542–549 (1986).

64. M. Schindler, D. E. Koppel and M. P. Sheetz, Modulation of membrane protein lateral mobility by polyphosphates and polyamines, *Proc Natl Acad Sci U S A*, 77, 1457–1461 (1980).

65. G. Bao and S. Suresh, Cell and molecular mechanics of biological materials, *Nature Mat.*, 2, 715–725 (2003).

66. B. Bain, A beginner's guide to blood cells, *Blackwell Pub, Malden, MA* (2004).

67. J. B. Fournier, D. Lacoste and E. Rapha, Fluctuation Spectrum of Fluid Membranes Coupled to an Elastic Meshwork: Jump of the Effective Surface Tension at the Mesh Size, *Phys. Rev. Lett.*, 92, 18102 (2004).

68. M. Puig-de-Morales-Marinkovic, K. T. Turner, J. P. Butler, J. J. Fredberg and S. Suresh, Viscoelasticity of the human red blood cell, *Am. J. Physiol., Cell Physiol.*, 293, 597–605 (2007).

69. T. Browicz, Further observation of motion phenomena on red blood cells in pathological states, *Zentralbl. Med. Wiss.*, 28, 625–627 (1890).

70. N. Gov, A. G. Zilman and S. Safran, Cytoskeleton confinement and tension of red blood cell membranes, *Phys. Rev. Lett.*, 90, 228101 (2003).

71. A. Zilker, M. Ziegler and E. Sackmann, Spectral-Analysis Of Erythrocyte Flickering In The 0.3-4-Mu-M-1 Regime By Microinterferometry Combined With Fast Image-Processing, *Phys. Rev. A*, 46, 7998–8002 (1992).

72. S. Tuvia, S. Levin, A. Bitler and R. Korenstein, Mechanical Fluctuations of the Membrane-Skeleton Are Dependent on F-Actin ATPase in Human Erythrocytes, *Proc. Natl. Acad. Sci. U. S. A.*, 141, 1551–1561 (1998).

73. N. S. Gov and S. A. Safran, Red Blood Cell Membrane Fluctuations and Shape Controlled by ATP-Induced Cytoskeletal Defects, *Biophys. J.*, 88, 1859 (2005).

74. C. L. L. Lawrence, N. Gov and F. L. H. Brown, Nonequilibrium membrane fluctuations driven by active proteins, *J. Chem. Phys.*, 124, 074903 (2006).

75. S. Tuvia, S. Levin, A. Bitler and R. Korenstein, Mechanical fluctuations of the membrane-skeleton are dependent on F-actin ATPase in human erythrocytes, *J. Cell Biol.*, 141, 1551–1561 (1998).

76. J. Li, M. Dao, C. T. Lim and S. Suresh, Spectrin-Level Modeling of the Cytoskeleton and Optical Tweezers Stretching of the Erythrocyte, *Biophys. J.*, 88, 3707–3719 (2005).

77. F. Brochard and J. F. Lenron, Frequency spectrum of the flicker phenomenon in erythrocytes, *J. Physique*, 36, 1035–1047 (1975).

78. Y. Kaizuka and J. T. Groves, Hydrodynamic damping of membrane thermal fluctuations near surfaces imaged by fluorescence interference microscopy, *Phys. Rev. Lett.*, 96, 118101 (2006).

79. F. Brochard and J. F. Lennon, Frequency spectrum of the flicker phenomenon in erythrocytes, *J. Physique*, 36, 1035–1047 (1975).

80. A. Zilker, H. Engelhardt and E. Sackmann, Dynamic Reflection Interference Contrast (Ric-) Microscopy—A New Method To Study Surface Excitations Of Cells And To Measure Membrane Bending Elastic-Moduli, *J. Physique*, 48, 2139–2151 (1987).

81. C. A. Best, J. E. Cluette-Brown, M. Teruya, A. Teruya and M. Laposata, Red blood cell fatty acid ethyl esters: a significant component of fatty acid ethyl esters in the blood, *J Lipid Res*, 44, 612–620 (2003).

82. H. W. G. Lim, M. Wortis and R. Mukhopadhyay, Stomatocyte-discocyte-echinocyte sequence of the human red blood cell: evidence for the bilayer-couple hypothesis from membrane mechanics, *Proc Natl Acad Sci U S A*, 99, 16766–16769 (2002).

83. T. Kuriabova and A. J. Levine, Nanorheology of viscoelastic shells: Applications to viral capsids, *Physical Review E*, 77, 031921 (2008).

84. N. Gov, A. G. Zilman and S. Safran, Cytoskeleton Confinement and Tension of Red Blood Cell Membranes, *Phys. Rev. Lett.*, 90, 228101 (2003).

85. P. M. Chaikin and T. C. Lubensky, Principles of Condensed Matter Physics (Cambridge University Press, New, York 1995).

86. T. Kuriabova and A. Levine, Nanorheology of viscoelastic shells: Applications to viral capsids, *Phys. Rev. E*, 77, 31921 (2008).

87. G. Popescu, T. Ikeda, K. Goda, C. A. Best-Popescu, M. Laposata, S. Manley, R. R. Dasari, K. Badizadegan and M. S. Feld, Optical Measurement of Cell Membrane Tension, *Phys. Rev. Lett.*, 97, 218101 (2006).

88. R. Waugh and E. A. Evans, Thermoelasticity of red blood cell membrane, *Biophys. J.*, 26, 115–131 (1979).

89. M. Dao, C. T. Lim and S. Suresh, Mechanics of the human red blood cell deformed by optical tweezers, *J. Mech. Phys. Solids*, 51, 2259–2280 (2003).

90. Wikipedia, *Malaria*, http://en.wikipedia.org/wiki/Main_Page).

91. S. Suresh, J. Spatz, J. P. Mills, A. Micoulet, M. Dao, C. T. Lim, M. Beil and T. Seufferlein, Connections between single-cell biomechanics and human disease states: gastrointestinal cancer and malaria, *Acta Biomaterialia*, 1, 15–30 (2005).

92. A. Kilejian, Characterization of a Protein Correlated with the Production of Knob-Like Protrusions on Membranes of Erythrocytes Infected with Plasmodium falciparum, *Proc. Natl. Acad. Sci. U.S.A.*, 76, 4650–4653 (1979).

93. I. W. Sherman, Biochemistry of Plasmodium (malarial parasites), *Microbiological Reviews*, 43, 453 (1979).

94. D. E. Goldberg, A. F. G. Slater, A. Cerami and G. B. Henderson, Hemoglobin Degradation in the Malaria Parasite Plasmodium falciparum: An Ordered Process in a Unique Organelle, *Proc. Natl. Acad. Sci. U.S.A.*, 87, 2931–2935 (1990).

95. G. B. Nash, E. O'Brien, E. C. Gordon-Smith and J. A. Dormandy, Abnormalities in the mechanical properties of red blood cells caused by Plasmodium falciparum, *Blood*, 74, 855–861 (1989).

96. H. A. Cranston, C. W. Boylan, G. L. Carroll, S. P. Sutera, I. Y. Gluzman and D. J. Krogstad, Plasmodium falciparum maturation abolishes physiologic red cell deformability, *Science*, 223, 400–403 (1984).

97. M. Paulitschke and G. B. Nash, Membrane rigidity of red blood cells parasitized by different strains of Plasmodium falciparum, *J Lab Clin Med*, 122, 581–589 (1993).

98. L. H. Miller, D. I. Baruch, K. Marsh and O. K. Doumbo, The pathogenic basis of malaria, *Nature*, 415, 673–679 (2002).

99. Y. K. Park, M. Diez-Silva, G. Popescu, G. Lykotrafitis, W. Choi, M. S. Feld and S. Suresh, Refractive index maps and membrane dynamics of human red blood cells parasitized by Plasmodium falciparum, *Proc. Natl. Acad. Sci. U. S. A.*, 105, 13730 (2008).

100. J. C. M. Lee and D. E. Discher, Deformation-Enhanced Fluctuations in the Red Cell Skeleton with Theoretical Relations to Elasticity, Connectivity, and Spectrin Unfolding, *Biophys. J.*, 81, 3178–3192 (2001).

101. S. Suresh, J. Spatz, J. P. Mills, A. Micoulet, M. Dao, C. T. Lim, M. Beil and T. Seufferlein, Connections between single-cell biomechanics and human disease states: gastrointestinal cancer and malaria, *Acta Biomater.*, 1, 15–30 (2005).

102. J. P. Mills, M. Diez-Silva, D. J. Quinn, M. Dao, M. J. Lang, K. S. W. Tan, C. T. Lim, G. Milon, P. H. David, O. Mercereau-Puijalon, S. Bonnefoy and S. Suresh, Effect of plasmodial RESA protein on deformability of human red blood cells harboring Plasmodium falciparum, *Proc. Natl. Acad. Sci. U.S.A.*, 104, 9213–9217 (2007).

CHAPTER 12

White Light Methods

As already established in Chapter 8, *spatial-phase sensitivity* is an important figure of merit of QPI methods. Much of the spatial nonuniformities that have the effect of smoothing out the details in a quantitative-phase image are caused by *speckle*. This random interference phenomenon occurs due to coherent superposition of various fields both from the specimen itself and spurious reflections from interfaces in the optical setup, or unwanted scattering from dirt and imperfection associated with the optics (Ref. 1, for example, provides a detailed description of speckle). Using broadband illumination fields or, equivalently, short coherence light, limits the occurrence of this undesirable random interference patterns (speckle).[2] White light provides a coherence length of the order of 1 μm. Therefore, the superposition between various field components is *coherent* only if the pathlength differences are of this order or shorter. Overall, white light has the effect of removing the speckle structure almost entirely. Not surprisingly, speckle was observed in imaging experiments and became a subject of intense research only after the invention of lasers. This is clearly illustrated by a simple comparison between a white light phase-contrast image and another obtained by laser light. We will revisit this comparison in the context of QPI with white light and laser in Section 12.2. There are many interferometric methods that have been implemented with white light. In the following, we will discuss two white light QPI methods that have been successfully employed in biological investigations: QPI using the transport of intensity equation and spatial light interference microscopy.

12.1   QPI Using the Transport of Intensity Equation

In 1998, Paganin and Nugent reported "Noninterferometric phase imaging with partially coherent light."[3] This is a QPI method that does not involve a typical interferometric geometry, but instead uses the fact that the image field itself is an interferogram.[4] The technique is based on the theoretical formalism developed by Teague[5] and operates with a regular bright field microscope. The essence of this approach is that shifting a given specimen through focus generates images in which the intensity and phase distributions are mathematically coupled. Thus, measuring several intensity images around the focus provides the quantitative phase image of the in-focus field.

12.1.1   Principle

Consider a scalar field at a certain plane (say, the image plane), which can be written in terms of an amplitude and phase as

$$U(\mathbf{r}) = \sqrt{I(\mathbf{r})} \cdot e^{i\phi(\mathbf{r})} \tag{12.1}$$

where $\mathbf{r} = (x, y)$, I is the intensity, and $\phi(\mathbf{r})$ the spatial phase distribution of interest. Under the approximation of a slowly varying field along the optical axis (i.e., paraxial approximation), the propagation obeys the *transport of intensity equation*[5]

$$k_0 \frac{\partial I(\mathbf{r})}{\partial z} = -\nabla[I(\mathbf{r})\nabla\phi(\mathbf{r})] \tag{12.2}$$

where k_0 is the wavenumber, $k_0 = \frac{2\pi}{\lambda}$ and $\nabla = (\partial/\partial x, \partial/\partial y)$. Equation (12.2) indicates that knowledge of the intensity distribution and its axial derivative (i.e., derivative along z) yields information about the transverse phase distribution. The intensity distribution can be directly measured at the image plane, and its z-derivative is obtained by defocusing the image slightly in both the positive and negative z-direction.

Assuming weakly scattering objects, the intensity on the right-hand side can be approximated as uniform and, thus, pulled out of the divergence operator. With this, Eq. (12.2) becomes

$$\frac{\partial I(\mathbf{r})}{\partial z} = -\frac{I_0}{k_0}\nabla^2\phi(\mathbf{r}) \tag{12.3}$$

where I_0 is the (uniform) intensity distribution at the plane of focus. Note that now Eq. (12.3) indicates direct experimental access to the Laplacian of ϕ via measurements of longitudinal gradient of I. Experimentally, the gradient along z is measured by shifting the specimen over small distances around the plane of focus (see Fig. 12.1),

$$\frac{\partial I(\mathbf{r},0)}{\partial z} \approx \frac{1}{2\Delta z}[I(\mathbf{r},\Delta z) - I(\mathbf{r},-\Delta z)]$$
$$= g(\mathbf{r}) \tag{12.4}$$

where Δz is the shift amount, of the order of the fraction of wavelength, and $g(\mathbf{r})$ denotes the measured derivative.

Combining Eqs. (12.3) and (12.4), the inverse Laplace operation in Eq. (12.4) can be performed in the frequency domain. Thus, taking

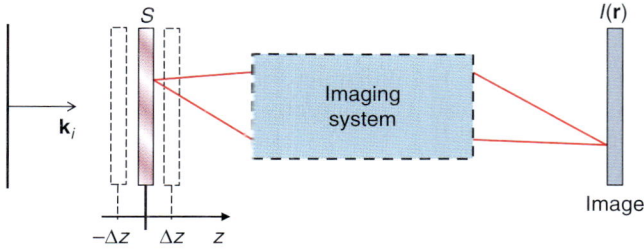

FIGURE 12.1 QPI via transport of intensity equation. Intensity images are recorded for each of the specimen positions: $z = -\Delta z$, $z = 0$, $z = \Delta z$.

the Fourier transform of Eq. (12.3) with respect to $\mathbf{r} = (x, y)$ and using $g(\mathbf{r})$ as the measured data, we obtain (via the differentiation theorem of the Fourier transform, see Appendix B)

$$\tilde{\phi}(k_{\perp}) = \frac{k_0}{I_0} \cdot \frac{\tilde{g}(k_{\perp})}{k_{\perp}^2} \tag{12.5}$$

In Eq. (12.5), $k_{\perp} = (k_x, k_y)$ is the conjugate variable to $\mathbf{r}$. Finally, the quantitative phase image, $\phi(\mathbf{r})$, is obtained by Fourier transforming Eq. (12.5) back to the spatial domain. Note that this phase retrieval works under the assumption that there are no zeros in the measured intensity, that is, in the absence of phase discontinuities. The problem of phase retrieval by this method in the presence of vortices is discussed in Ref. 6.

Of course, the method requires a certain degree of spatial coherence, over which ϕ can be properly defined. The problem of phase retrieval using partially coherent fields is further discussed in Refs. 7, 8. Since the method assumes field propagation in the paraxial approximation (the spatial analog to *quasi-monochromatic* fields), existence of spatial coherence is somewhat implicit. Note that, although not using an additional (reference) field for interference, the phase is retrieved interferometricaly, simply because the image itself is an interferogram (as first described by Abbe in 1873[4]). This method benefits from operating with a commercial microscope and is inherently common path, which confers stability. Most importantly, the white light illumination provides spatial uniformity.

12.1.2 Further Developments

Interestingly, this method has been implemented with hard X-rays before light.[9] In 2000, Allman et al. demonstrated phase measurements based on the same principle using neutron radiation,[10] while Bajt et al. demonstrated quantitative-phase imaging in a transmission electron microscope.[11]

Below, we discuss the main applications to biology using optical domain QPI using transport of intensity equation.

12.1.3   Biological Applications

In 1998, Barty et al. demonstrated the potential of this QPI method for imaging cells.[12] Figure 12.2 illustrates this capability with a measurement on a live cheek cell. For comparison, a DIC image of the same specimen is presented (Fig. 12.2*a*). Clearly, the phase image renders structural details in the cell, which perhaps are very difficult to reveal by laser illumination.

Later, the same approach was used to study red blood cell volumetry under various conditions of osmotic stress.[13] Figure 12.3 shows how the phase measurements are performed using three intensity recordings. Figure 12.4 summarizes the results of the osmotic study.

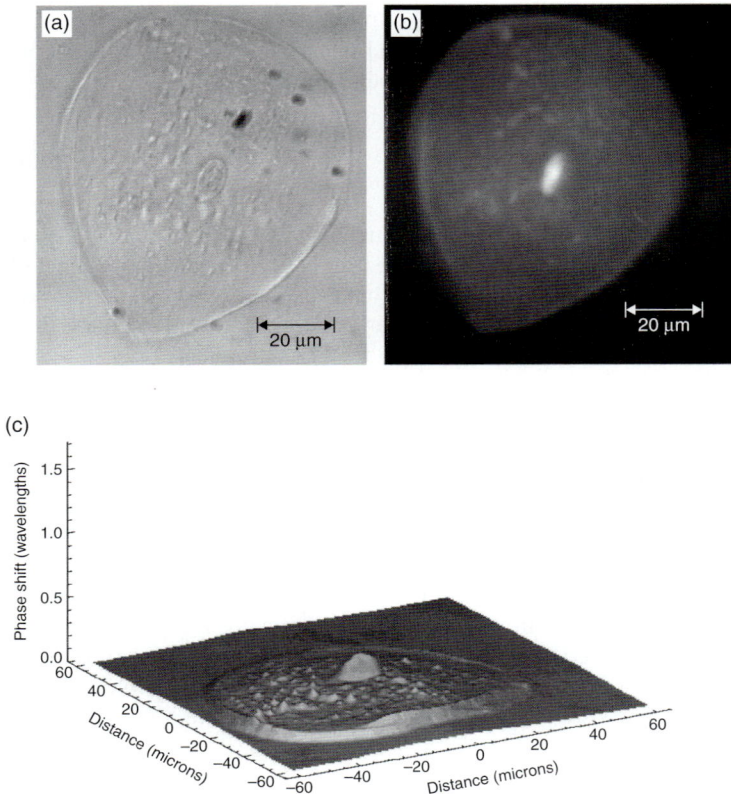

FIGURE 12.2   Comparison of the recovered phase-amplitude image of an unstained cheek cell recovered from images taken at +/−(2+/−0.5) μm on either side of best focus. (*a*) Nomarski DIC image of the cell, (*b*) recovered phase image. The surface plot in (*c*) demonstrates that the artifact level outside the cell is low and that both the nucleus and the mitochondria within the cell membrane are clearly resolved. The Optical Society, used with permission, *Opt. Lett:* A. Barty, K.A. Nugent, D. Paganin & A. Roberts, "Quantitative optical phase microscopy," 23, 817–819 (1998).

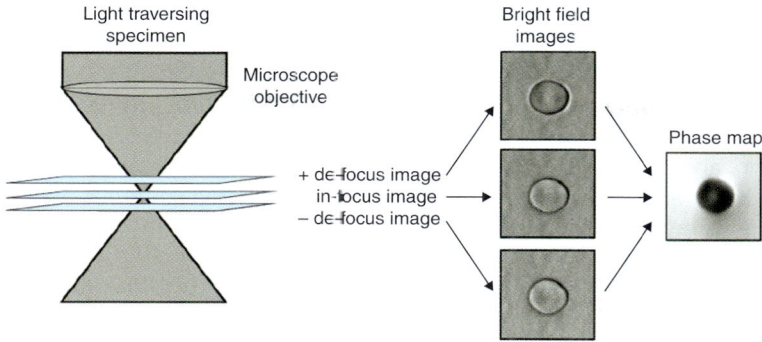

FIGURE 12.3 Schematic representation of the QPM phase map generation. The phase map is calculated from an in-focus and a pair of equidistant de-focus (+/−0.5 µm) bright field images of rat erythrocytes. While there is little visually discernible difference in the appearance of the bright field images, the phase map generated shows clear evidence of centrally modulated optical thickness. Cellular Physiology & Biochemistry used with permission, C. L. Curl, et al., "Single cell volume measurement by quantitative phase microscopy (QPM): A case study of erythrocyte morphology," 17, 193–200 (2006).

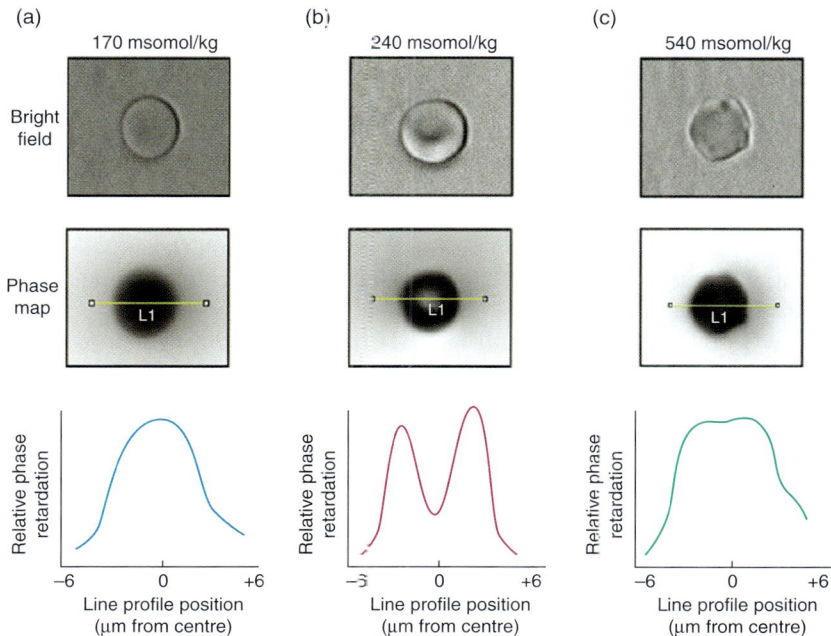

FIGURE 12.4 Rat erythrocyte morphology in buffer solutions of differing osmolality, as indicated. Cellular Physiology & Biochemistry used with permission, C.L. Curl, et al., "Single cell volume measurement by quantitative phase microscopy (QPM): A case study of erythrocyte morphology," 17, 193–200 (2006).

For each panel (A, B, and C), the upper image is the bright field in-focus view, the center image is the calculated phase map, and the lower plot shows the line profile from the phase map obtained at the line indicated. In Panel A (hypotonic solution), the line profile plot through the cell diameter shows phase modulation indicative of spherical morphology. In Panel B (near-isotonic solution), the expected biconcave shape is recovered. In Panel C (hypertonic solution), the line profile plot through the cell diameter shows phase modulation indicative of crenated morphology.

These examples show the potential of this white light QPI method to perform noninvasive cell imaging. While the theoretical algorithm is accurate within certain limits (paraxial approximation, no phase discontinuities), the technique benefits from spatial uniformity and operates with existing microscopes, without other hardware modifications besides shifting the specimen through focus.

12.2   Spatial Light Interference Microscopy (SLIM)

Our group at the University of Illinois at Urbana-Champaign developed SLIM as a QPI method that combines the spatial uniformity associated with white light illumination and the stability of common path interferometry.[14] SLIM is implemented as an *add-on* module to existing phase-contrast microscopes and, thus, inherently overlays the quantitative-phase image with the fluorescence channels of the microscope. SLIM provides speckle-free images, which confers *spatially sensitive* optical path-length measurements (0.3 nm). The common path geometry enables *temporally sensitive* optical path-length measurement (0.03 nm). Due to the short coherence length of the white light field, SLIM can render 3D tomographic images of transparent structures just by scanning the specimen through focus. The tomographic capability, referred to as spatial light interference tomography (SLIT) is described in more detail in Section 14.1.

The principle and main applications of SLIM are presented below.

12.2.1   Principle

A schematic of the instrument setup is depicted in Fig. 12.5a. SLIM was developed by producing additional spatial modulation to the image field outputted by a commercial phase contrast microscope (PCM) (for a review of PCM, see Section 5.3). Specifically, in addition to the $\pi/2$ shift introduced in PC between the scattered and unscattered light from the sample,[16] we generated further phase shifts, by increments of $\pi/2$ and recorded additional images for each phase map, as in traditional phase-shifting interferometry (see Section 8.2.). Thus, the objective exit pupil, which contains the phase-shifting ring, is imaged via lens L_1 onto the surface of a reflective liquid crystal phase modulator (LCPM). The active pattern on the LCPM is calculated to precisely match the size and position of the phase ring image,

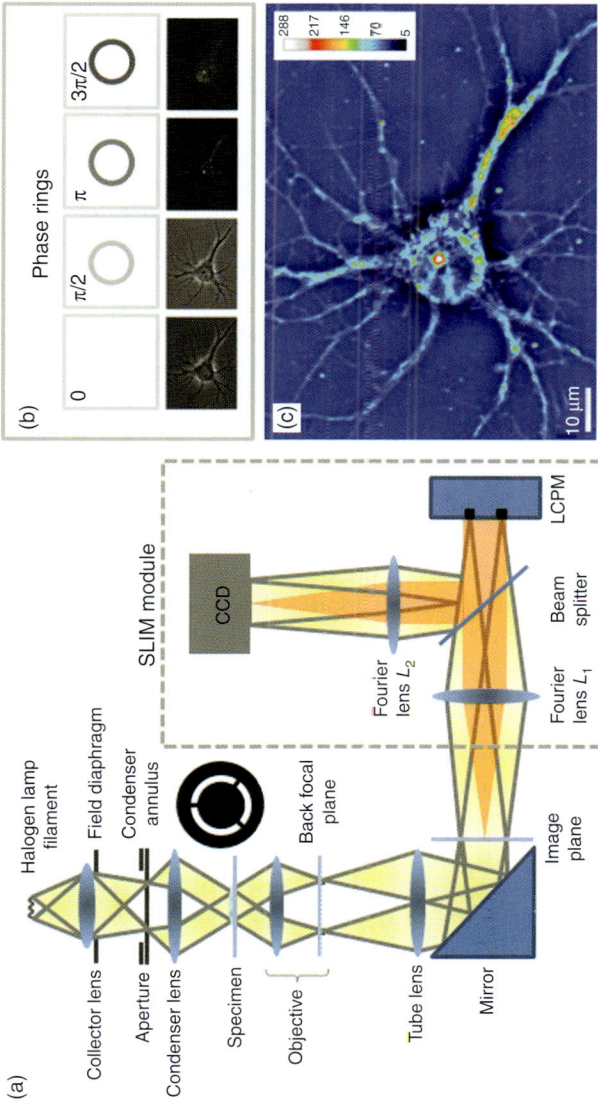

FIGURE 12.5 SLIM principle. (*a*) Schematic setup for SLIM. The SLIM module is attached to a commercial phase contrast microscope (Axio Observer Z1, Zeiss, in this case). The lamp filament is projected onto the condenser annulus. The annulus is located at the focal plane of the condenser, which collimates the light towards the sample. For conventional phase contrast microscopy, the phase objective contains a phase ring, which delays the unscattered light by a quarter wavelength and also attenuates it by a factor of 5. The image is delivered via the tube lens to the image plane, where the SLIM module processes it further. The Fourier lens L1 relays the back focal plane of the objective onto the surface of the liquid crystal phase modulator (LCPM, Boulder Nonlinear). By displaying different masks on the LCPM, the phase delay between the scattered and unscattered components is modulated accurately. Fourier lens L2 reconstructs the final image at the CCD plane, which is conjugated with the image plane. (*b*) The phase rings and their corresponding images recorded by the CCD. (*c*) SLIM quantitative-phase image of a hippocampal neuron. The color bar indicates optical path-length in nanometers. (*From Ref. 14.*)

such that additional phase delay can be controlled between the scattered and unscattered components of the image field. Thus, four images corresponding to each phase shift are recorded (Fig. 12.5b), to produce a *quantitative-phase image* that is uniquely determined. Fig. 12.5c depicts the quantitative-phase image associated with a cultured hippocampal neuron, which is proportional to

$$\phi(x,y) = k_0 \int_0^{h(x,y)} [n(x,y,z) - n_0] dz$$

$$= k_0 \Delta \bar{n}(x,y) h(x,y)$$

(12.6)

In Eq. (12.6), $k_0 = 2\pi/\lambda$, $n(x,y,z) - n_0$ is the local refractive index contrast between the cell and the surrounding culture medium, $\Delta \bar{n}(x,y) = \dfrac{1}{h(x,y)} \int_0^{h(x,y)} [n(x,y,z) - n_0] dz$ is the axially-averaged refractive index contrast, $h(x,y)$ is the local thickness of the cell, and λ the mean wavelength of the illumination light. The typical irradiance at the sample plane is ~1 nW/μm^2.

The exposure time was 10 to 50 ms, for all the images presented here. This level of exposure is 6 to 7 orders of magnitude below that of typical confocal microscopy and is, therefore, much safer for live cell imaging.[17]

Typically in phase-shifting interferometry, the illuminating field is assumed to be monochromatic. For broadband fields, the *phase shift* in question must be properly defined.[18] Thus, if a non-monochromatic field is *spatially coherent*, the phase information is that of an effective monochromatic field oscillating at the average frequency (see Ref. 19 for a recent reiteration of this concept in the context of X-rays). This concept is key in SLIM.

The SLIM principle relies on the spatial decomposition of a statistically homogeneous field, U, into its *spatial average* (i.e., unscattered) and a *spatially varying* (scattered) component, as described in the section on phase-contrast microscopy (PCM) (Section 5.3.)

$$U(\mathbf{r};\omega) = U_0(\omega) + U_1(\mathbf{r};\omega)$$

$$= |U_0(\omega)| e^{i\phi_0(\omega)} + |U_1(\mathbf{r};\omega)| e^{i\phi_1(\mathbf{r};\omega)}$$

(12.7)

where $\mathbf{r} = (x,y)$. Using the *spatial* Fourier representation $\tilde{U}(\mathbf{q};\omega)$ of U, it becomes apparent that the average field U_0 is proportional to the DC component $\tilde{U}(0;\omega)$, whereas U_1 describes the non-zero frequency content of U. Thus, the image field U can be regarded as the interference between its spatial average and its spatially varying component. As mentioned already, the description of an arbitrary image as an interference phenomenon was recognized more than a century ago by Abbe in the context of microscopy (see Section 5.1. for a review): "The microscope image is the interference effect of a diffraction phenomenon."[20] Further, describing an image as an interferogram set the basis for Gabor's in-line holography (Section 6.1).[21] The spatially-resolved cross-spectral density can be written as

$$W_{01}(\mathbf{r};\omega) = \langle U_0(\omega) \cdot U_1^*(\mathbf{r};\omega) \rangle \tag{12.8}$$

where * stands for complex conjugation and angular brackets indicate ensemble average. If ω_c is the mean frequency of the power spectrum, $S(\omega) = \langle |U_0(\omega)|^2 \rangle$, W_{01} can be written as

$$W_{01}(\mathbf{r};\omega - \omega_0) = |W_{01}(\mathbf{r};\omega - \omega_0)| e^{i[\Delta\phi(\mathbf{r};\omega-\omega_0)]} \tag{12.9}$$

It follows that the temporal cross-correlation function is obtained by Fourier transforming Eq. (12.9) (see Ref. 22),

$$\Gamma_{01}(\mathbf{r};\tau) = |\Gamma_{01}(\mathbf{r};\tau)| \cdot e^{i[\omega_0\tau + \Delta\phi(\mathbf{r};\tau)]} \tag{12.10}$$

with $\Delta\phi(\mathbf{r}) = \phi_0 - \phi_1(\mathbf{r})$ representing the spatially varying phase difference of the cross-correlation function. Equation (12.10) indicates that, for spatially coherent illumination, the spatially varying phase of the cross-correlation function can be retrieved through measurements at various time delays, τ. This phase information is equivalent to that of a purely monochromatic light at frequency ω_0. For the white light field used in our experiments, the optical spectrum was measured separately and the autocorrelation function was inferred, as shown in Fig. 12.6. It can be seen that, indeed, the white light field behaves as a monochromatic field oscillating at the mean frequency ω_0. Using Eq. (12.10), one obtains the following irradiance distribution in the plane of interest as a function of the time delay, τ

$$I(\mathbf{r};\tau) = I_0 + I_1(\mathbf{r}) + 2|\Gamma_{01}(\mathbf{r};\tau)| \cos[\omega_0\tau + \Delta\phi(\mathbf{r})] \tag{12.11}$$

When the delay, τ, between U_0 and U_1 is varied, interference is obtained simultaneously at each point $r = (x, y)$ of the image. The average U_0 is constant over the entire plane and can be regarded as the common reference field of an array of interferometers. In addition, U_0 and U_1 traverse similar optical paths, i.e., they form a common-path interferometer. Thus, the influence of inherent phase noise due to vibration or air fluctuations is inherently minimized, allowing for a precise retrieval of $\Delta\phi$.

For modifications of the time delay around $\tau = 0$ that are comparable to the optical period, $|\Gamma|$ can be assumed to vary slowly at each point. This is clearly illustrated by the oscillating component of the autocorrelation function shown in Fig. 12.6c. Thus, using Eq. (12.11), the spatially varying phase of Γ can be reconstructed as

$$\Delta\phi(\mathbf{r}) = \arg\left[\frac{I(\mathbf{r};\tau_3) - I(\mathbf{r};\tau_1)}{I(\mathbf{r};\tau_0) - I(\mathbf{r};\tau_2)}\right] \tag{12.12}$$

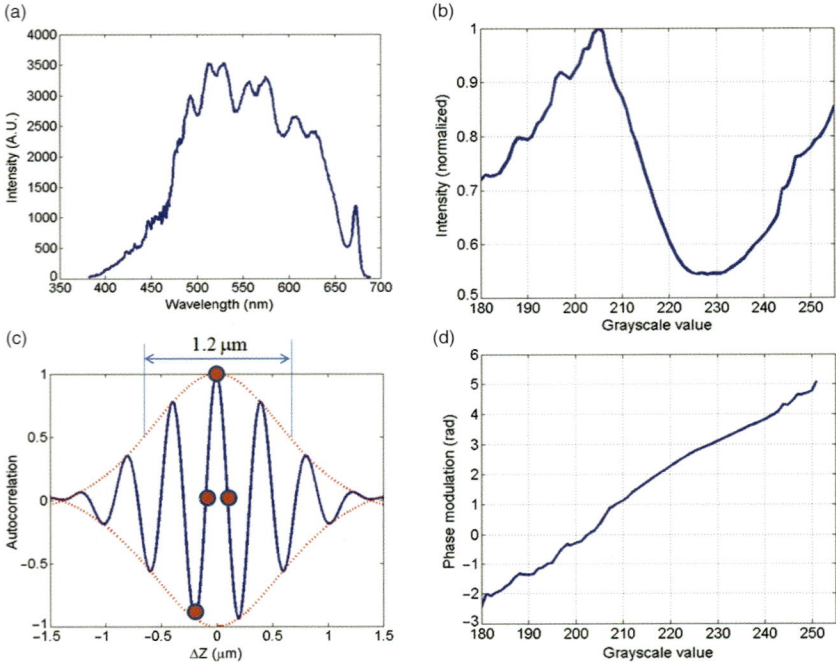

FIGURE 12.6 (a) Spectrum of the white light emitted by the halogen lamp. The center wavelength is 552.3 nm. (b) The autocorrelation function (blue solid line) and its envelope (red dotted line). The 4 circles indicate the phase shifts produced by LCPM. The refractive index of the medium is 1.33. (c) Intensity modulation obtained by displaying different grayscale values on the LCPM. (d) Phase vs. gray scale calibration curve obtained by Hilbert transform of the signal in (c). (*From Ref. 14.*)

where τ_k is given by $\omega_0 \tau_k = k \cdot \pi/2$, $k = 0,1,2,3$. If we define $\beta(\mathbf{r}) = |U_1(\mathbf{r})|/|U_0|$, then the phase associated with the image field $U = U_0 + U_1$ can be determined as

$$\phi(\mathbf{r}) = \arg\left[\frac{\beta(\mathbf{r})\sin(\Delta\phi(\mathbf{r}))}{1+\beta(\mathbf{r})\cos(\Delta\phi(\mathbf{r}))}\right] \qquad (12.13)$$

Equation (12.13) shows how the quantitative-phase image is retrieved via four successive intensity images measured for each phase shift.

In order to assess the overall *phase accuracy* of SLIM, we imaged an amorphous carbon film and compared the profilometry results against atomic-force microscopy (AFM). The topography measurements by SLIM and AFM, respectively, are summarized in Fig. 12.7a and b. As illustrated by the topography histograms presented in Fig. 12.7c, the two types of measurement agree within a fraction of a nanometer. Note that both SLIM and AFM are characterized by much smaller errors than suggested by the widths of the histogram modes, as these

(a)

(b)

(c)

(d)

(e)

(f)

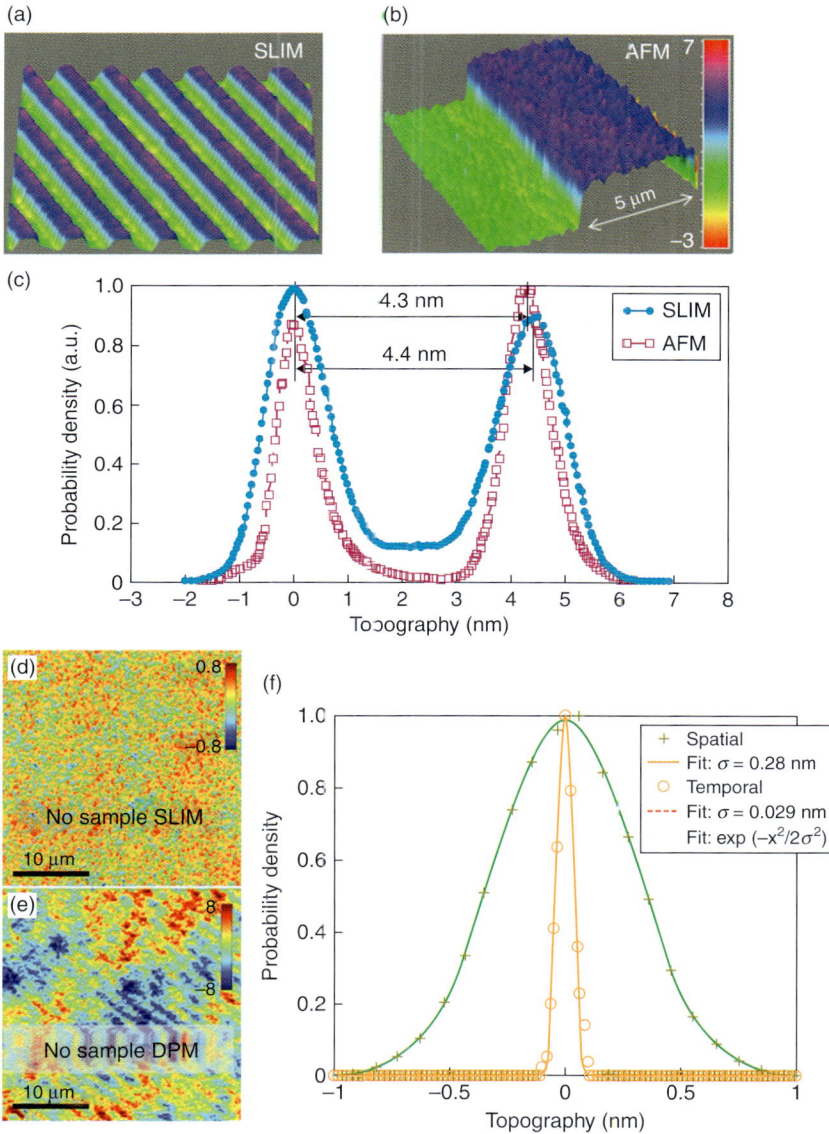

FIGURE 12.7   Comparison between SLIM and AFM. (a) SLIM image of an amorphous carbon film (40×/0.75 NA objective). (b) AFM image of the same sample. The colorbar indicates thickness in nanometer. (c) Topographical histogram for AFM and SLIM, as indicated. (d) Topography noise in SLIM; colorbar in nanometers. (e) Topography noise associated with diffraction-phase microscopy, a laser-based technique; colorbar in nanometers. (f) Optical path-length noise level measured spatially and temporally, as explained in text. The solid lines indicate Gaussian fits, with the standard deviations as indicated. (From Ref. 14.)

widths also reflect irregularities in the surface topography due to the fabrication process itself. Compared to AFM, SLIM is non-contact, parallel, and faster by more than three orders of magnitude. Thus, SLIM can optically measure an area of $75 \times 100 \ \mu m^2$ in 0.5 second (Fig. 12.7a) compared to a 10- $\times$ 10-μm^2 field of view measured by AFM in 21 minutes (Fig. 12.7b). Of course, unlike AFM, SLIM provides *nanoscale accuracy* in topography within the diffraction-limited *transverse resolution* associated with the optical microscope.

In Fig. 12.7d-e, we show a comparison between the SLIM speckle-free image and a similar background image obtained by diffraction phase microscopy (DPM)[23] (see Sec. 11.2), a laser-based technique that was interfaced with the same microscope (Fig. 12.7e). We note that, as already described in Sec. 11.2., DPM is a robust and high-performance laser technique, allowing sub-nanometer path-length temporal stability.[23-30] Still, due to the lack of speckle effects granted by its broad spectral illumination, SLIM's spatial uniformity and accuracy for structural measurements is at least an order of magnitude better than DPM's. In order to quantify this spatiotemporal phase sensitivity, we imaged the SLIM background (i.e., no sample) repeatedly to obtain a 256-frame stack. Figure 12.7f shows the spatial and temporal histograms associated with the optical path-length shifts across a 10- $\times$ 10-μm field of view and, respectively, over the entire stack. These noise levels, 0.3 nm and 0.03 nm, represent the ultimate limit in optical path-length sensitivity across the frame and from frame to frame, respectively. There are several error sources that can potentially be diminished further: residual mechanical vibrations in the system that are not "common path," minute fluctuations in the intensity and spectrum of the thermal light source, digitization noise from the CCD camera (12 bit in our case), and the stability (repeatability) of the liquid crystal modulator (8 bit).

The LCPM maximum refresh rate is 60 Hz, which ultimately allows for 15 SLIM images per second (using the *grouped* processing described in Sec. 8.5.1. and Fig. 8.7.). However, here we show quantitative-phase images obtained at 2.6 frames/s, as the camera has a maximum acquisition rate of 11 frames/s. Nevertheless, the acquisition speed is not a limitation of principle and can be easily pushed to video rate by employing faster phase modulation and a CCD camera.

Remarkably, SLIM is able to reveal topography of single atomic layers.[31] We performed measurements on graphene flakes. Graphene is a two-dimensional lattice of hexagonally arranged and sp^2-bonded carbon atoms, i.e., a monolayer of the bulk material graphite. The graphene sample was obtained here by mechanically exfoliating a natural graphite crystal using adhesive tape.[32] The exfoliated layers were deposited on a glass slide, which was then cleaned using isopropanol and acetone to remove excess adhesive

(a)

(b)

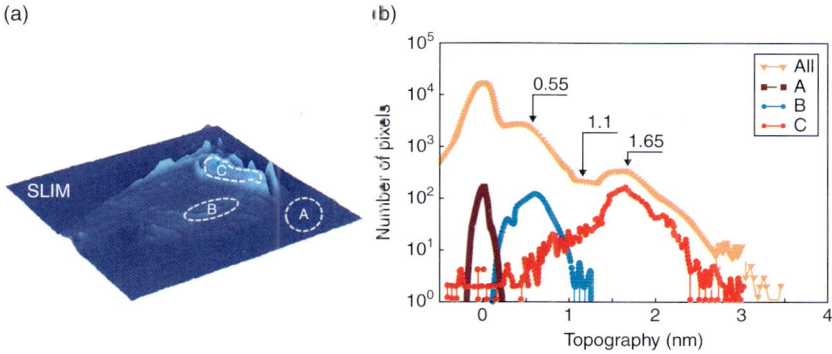

FIGURE 12.8 SLIM topography of graphene. (a) Quantitative-phase image of a graphene flake, objective 40×/0.75. (b) Topography histogram for the various regions indicated in (a). (From Ref. 31.)

residue. Single-layer (graphene) and few-layer graphite flakes are routinely obtained in this process, with lateral dimensions up to several tens of microns. Figure 12.8a shows the SLIM image of such a graphene flake. Qualitatively, it demonstrates that the background noise is below the level of the sample itself. To quantify the nanoscale profile of this structure we transformed the phase distribution ϕ into thickness h, via $h = \phi\lambda/2\pi(n-1)$, with $n = 2.6$ the refractive index of graphite.[33] Thus, we generated the topography histogram of the entire sample and individual regions, as shown in Fig. 12.8b. The overall histogram exhibits local maxima at topography values of 0 nm (background), 0.55 nm, 1.1 nm, 1.65 nm. These results indicate that the topography of the graphene sample has a *staircase* profile, in increments of 0.55 nm. This compares with reported values in the literature for the thickness of *individual atomic layers* of graphene via AFM in air (~1 nm step size) or scanning tunneling microscopy (STM) in ultra-high vacuum (~0.4-nm step size).[34,35] The difference between air and vacuum measurements indicate the presence of ambient species (nitrogen, oxygen, water, organic molecules) on the graphene sheet in air. Thus, SLIM provides topographical accuracy that compares with AFM, but its acquisition time is much faster and, of course, it operates in non-contact mode.

The extreme *spatial-phase sensitivity* of SLIM is essential in biological studies of structure and dynamics. It might be possible in the near future to sense motion due to a small number of molecules, say, diffusing in a live cell. An important feature of SLIM is that it provides phase images that are inherently overlaid with the other channels of the microscope, most importantly, fluorescence. This aspect is detailed in the next section.

12.2.2   Further Developments

SLIM-Fluorescence Multimodal Capability

Simultaneous fluorescence imaging complements SLIM's unique structural information with the ability to study cellular constituents with molecular specificity. Figure 12.9a and b shows SLIM imaging of axons and dendrites identified by fluorescent staining for somato-dendritic microtubule associated protein 2 (MAP2) of primary hippocampal neurons cultured for 19 days *in vitro* (DIV). Fine axonal processes are also distinguishable from dendrites by SLIM, in which the quantitative-phase imaging channel reveals changes in the local refractive index of structures reminiscent of actin-rich synaptic connections (Fig. 12.9b). As shown in Fig. 12.9c, these inhomogeneities are observed along dendrites where the spines develop. In order to quantify these structural differences observed by SLIM, we traced individual neurites using NeuronJ, a semi-automatic algorithm implemented in Java.[36]

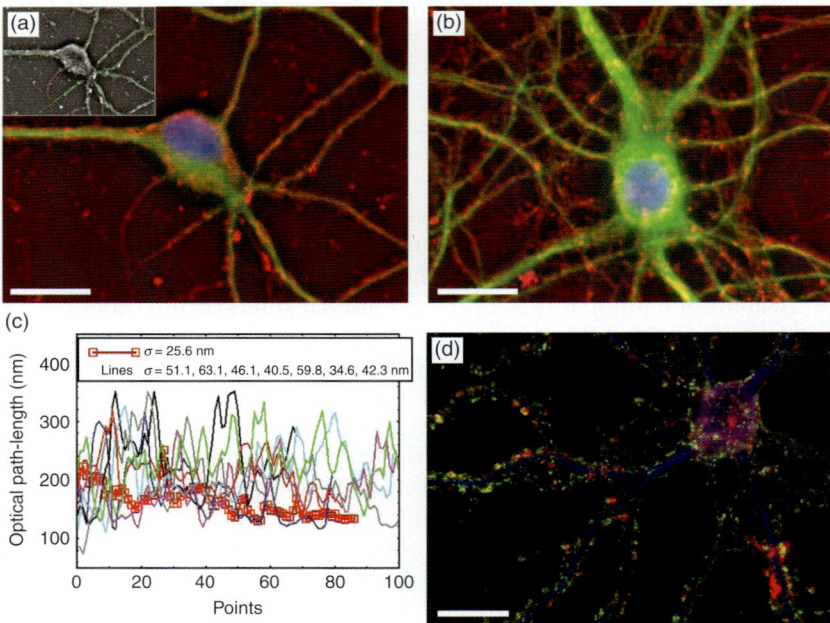

FIGURE 12.9   SLIM-fluorescence multimodal imaging. (a and b) Combined multimodal images of cultured neurons (19 DIV) acquired through SLIM (red) and fluorescence microscopy. Neurons were labeled for somatodendritic MAP2 (green), and nuclei (blue). (c) Optical path-length fluctuations along the dendrites (lines) and axon (markers) retrieved from the inset of (a). (d) Synaptic connections of a mature hippocampal neuron (33 DIV) immunochemically labeled for synapsin (green), MAP2 (blue), and f-actin using rhodamine phalloidin (red). All scale bars are 20 μm (40×/0.75NA). (*From Ref. 14.*)

The results demonstrate the optical path-length fluctuations for each trace, wherein we found that the standard deviation of the path-length fluctuations along the axons, $\sigma = 25.6$ nm, is the lowest among all neurites. This result indicates that subtle inhomogeneities are associated with the connecting synaptic structures, which can be revealed by SLIM as path-length changes. By 3 weeks in dispersed culture, the majority of dendritic spines mature to form presynaptic buttons[37] on the dendritic shafts of hippocampal neurons.[38] These are comparable to synaptic elaborations on a mature hippocampal neuron (33 DIV) with labeled f-actin, synapsin, and MAP2 (Fig. 12.9d). Thus, SLIM may offer a window into studying the dynamic processes associated with the formation and transition of collateral filopodia into spines, and the dynamics of plasticity-related changes in spine structure.[39-41] Note that SLIM can be used to image cellular dynamics over extended periods of time without loss in performance or damage to the specimen.

Computational Imaging

From the quantitative phase measured, other representations of the information can be obtained numerically. Figure 12.10a shows two cardiac myocytes captured by SLIM. Phase-contrast images are obtained naturally as one channel of SLIM as already shown earlier in Fig. 12.5b. The halo effect is clearly seen in the PCM image. The spatial gradient of the SLIM image simulates DIC microscopy, as

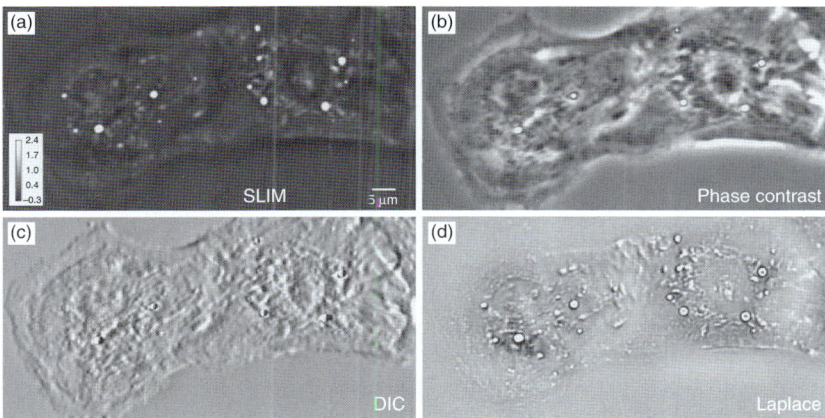

FIGURE 12.10   SLIM computational imaging. (a) SLIM image of two cardiac myocytes in culture; colorbar indicates phase shift in radians. The objective is ZEISS Plan-Neofluar 40×/0.75. (b) Phase-contrast qualitative imaging. (c) Simulated DIC based on the phase measurement of SLIM. (d) Laplacian of SLIM image. Since the quantitative-phase information is obtained, all other microscopy such as DIC, PC, and dark field can be numerically simulated through SLIM imaging. Objective: Zeiss EC Plan-Neofluar 40×/0.75. [Fig. 2, SPIE used with permission, J. Biomed. Opt., Z. Wang, L. Millet, V. Chan, H. Ding, M.U. Gillette, R. Bashir & G. Popescu, "Label-free intracellular transport measured by Spatial Light Interference Microscopy," 16(2), (2011). (From Ref. 53.)]

show in Fig. 12.10c. The shadow artifact, typical for DIC, can be seen clearly. This is the result of the one dimensional derivative that generates DIC. In common DIC microscopes, the operation is performed by overlaping two replicas of the image field that are slightly shifted in the transverse plane, within the diffraction spot. Note that, not surprisingly, this gradient image is very sensitive to edges in the image.

Further, we show that the Laplacian of the image, a second-order derivative operation, is often more powerfull than DIC in revealing fine structures within the cell, as it does not contain shadow artiacts. Note that, as described in the image-filtering section in Chap. 4 (Sec. 4.5), the derivative operations are essentially high-pass frequency operations. This is especially evident in Fig. 12.10d where the small particles in the cardiac myocytes are resoved more clearly than in the SLIM image. It is known that heart cells are very active, i.e., energy-consuming; therefore these types of cells are rich in mitochondria, which are responsible for the energy supply of the cell metabolism.[42] Mitochondria are most likely the predominant type of visible particles, especially in the area surrounding the cell nucleus. Because phase contrast mixes the intensity information with phase information, it suffers from phase ambiguity and is qualitative. Thus, some of the particles in Fig. 12.10b are bright and some are dark, while all the particles show high-phase values in the SLIM image.

12.2.3   Biological Applications

Cell Dynamics

Two-dimensional SLIM dynamic imaging of live cells has been performed in our laboratory over various time scales, from a fraction of second to several days. Figure 12.11 summarizes the dynamic measurements obtained via 397 SLIM images of a mixed glial-microglial cell culture over a period of 13 minutes. In order to illustrate microglial dynamics, we numerically suppressed the translation motion via an algorithm implemented in ImageJ.[43] Phase-contrast images, which are part of the measured data set, are also presented for comparison (Fig. 12.11b and c). These results demonstrate that PC cannot provide quantitative information about dynamic changes in optical path-length because the light intensity is not linearly dependent on phase. In addition, the cell size is significantly overestimated by PCM due to the well-known *halo artifact*, which makes the borders of the cell appear bright (Fig. 12.11b and d shows the same field of view).[16] By contrast, SLIM reveals details of intracellular dynamics, as evidenced by the time-traces (Fig. 12.11e and f). Path-length fluctuations associated with two arbitrary points on the cell reveal interesting, periodic behavior (Fig. 12.11f). At different sites on the cell, the rhythmic motions have different periods, which may indicate different rates of metabolic or phagocytic activity. This periodicity can be observed

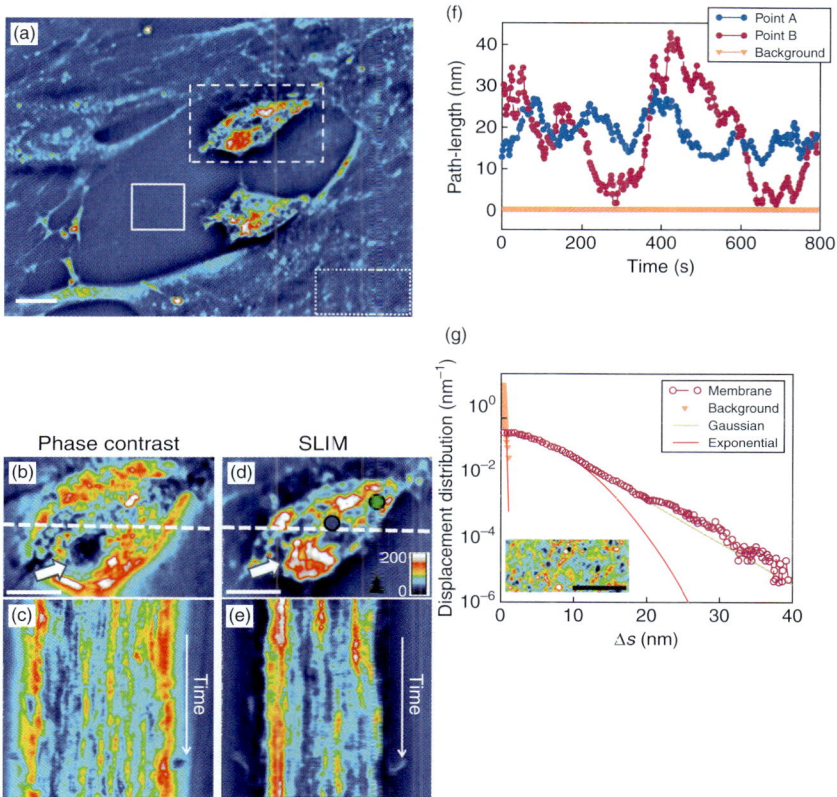

FIGURE 12.11   SLIM dynamic imaging of mixed glial-microglial cell culture. (a) Phase map of two microglia cells active in a glia cell environment. Solid-line box indicates the background used in g, dashed-line box crops a reactive microglia cell used in b-e and dotted-line box indicates glial cell membrane used in g. (b) Phase-contrast image of the cell shown in a. (c) Registered time-lapse projection of the corresponding cross-section through the cell as indicated by the dash line in b. The signal represents intensity and has no quantitative meaning. (d) SLIM image of the cell in b; the fields of view are the same. The arrows in b and d point to the nucleus which is incorrectly displayed by PC as a region of low signal. (e) Registered time-lapse projection of the corresponding cross-section through the cell as indicated by the dash line in d. The colorbar indicates path-length in nanometer. (f) Path-length fluctuations of the points on the cell (indicated in d) showing periodic intracellular motions (blue- and green-filled circles). Background fluctuations (black) are negligible compared to the active signals of the microglia. (g) Semi-logarithmic plot of the optical path-length displacement distribution associated with the glial cell membrane indicated by the dotted box in a. The solid lines show fits with a Gaussian and exponential decay, as indicated in the legend. The distribution crosses over from a Gaussian to an exponential behavior at approximately 10 nm. The background path-length distribution, measured from the solid line box, has a negligible effect on the signals from cells and is fitted very well by a Gaussian function. The inset shows an instantaneous path-length displacement map associated with the membrane. (h) Dry mass change of the cell in d. The linear mass increase is indicated. The dry mass fluctuation associated with the background is negligibly small. Objective: 40×/0.75. (From Ref. 14.)

in the coordinated cell behavior as the cell extends broad, dynamic filopodial ruffles under, and above, the neighboring glial cells.

Further, we studied glial cell membrane fluctuations. Due to the extremely low noise level of SLIM, the probability distribution of path-length displacements between two successive frames was retrieved with a dynamic range of over five orders of magnitude (Fig. 12.11g). Note that these optical path-length fluctuations, Δs, are due to both membrane displacements and local refractive index changes caused by cytoskeleton dynamics and particle transport. Remarkably, this distribution can be fitted very well with a Gaussian function up to path-length displacements $\Delta s = 10$ nm, at which point the curve crosses over to an exponential decay. The normal distribution suggests that these fluctuations are the result of numerous uncorrelated processes governed by equilibrium. On the other hand, exponential distributions are indicative of deterministic motions, mediated by metabolic activity.

Using the procedure outlined in Ref. 44, we use the quantitative-phase information rendered by SLIM to extract the non-aqueous, i.e., dry mass, of this microglia cell. The approximately linear mass increase of 4.6 fg/s evidenced in Fig. 12.11h is most likely because the cell is continuously scavenging the neighboring glia cells. In the next section, we show detailed cell growth experiments performed on *Escherichia coli*.

Cell Growth

Due to its improved sensitivity and stability, SLIM can further be used to accurately measure cell growth. The question of how single cells grow has been described as, "One of the last big unsolved problems in cell biology."[45] The reason that this debate has persisted despite decades of effort is primarily due to the following reasons: cells are small, difficult to measure and on average only double in size between divisions. Until recently, the state of the art method to assess a single cell growth curve was using Coulter counters to measure the volume of a large number of cells, in combination with careful mathematical analysis.[46] For relatively simple cells such as *E. coli*, traditional microscopy techniques have also been used to assess growth in great detail.[47] However these types of methods make the assumption that volume is a good surrogate for mass, although it is known that the volume can change disproportionately to mass as a result of osmotic response.[48] Recently, this technological barrier has been overcome due to the advent of techniques utilizing novel microfabricated devices to "weigh" single cells.[49] Although these methods are impressive in terms of accuracy and throughput, it is well recognized that the "ideal" method should be noninvasive, and have the ability to measure the mass and volume of single cells with the required accuracy.

Here we demonstrate that SLIM is capable of making such measurements with the required accuracy. As already described in the context of Fourier-phase microscopy (Sec. 11.1.), the principle behind using interferometric techniques such as FPM and SLIM to measure cell dry mass was established in the early 1950s when it was recognized that the optical-phase shift accumulated through a live cell is linearly proportional to the dry mass (non-aqueous content) of the cell.[50-52] Recently, it has been shown theoretically and experimentally that the surface integral of the cell phase map is invariant to small osmotic changes,[48] which establishes that quantitative-phase imaging methods can be used for dry mass measurements. The dry mass density at each pixel is calculated as $\rho(x,y) = \dfrac{\lambda}{2\pi\gamma}\phi(x,y)$, where λ is the centre wavelength, γ is the average refractive increment of protein $(0.2\ \mathrm{mL/g})$[44], and $\phi(x,y)$ is the measured phase. The total dry mass is then calculated by integrating over the region of interest in the dry mass density map (see Sec. 11.1.3.2 for details on measuring cell dry mass). Thus, converting phase sensitivity into dry mass, SLIM offers exquisite sensitivity to dry mass change, i.e., 1.5 fg/μm^2 spatially and 0.15 fg/μm^2 temporally.

The question of greatest interest here is whether single cells are growing exponentially, linearly, or in some other manner. In order to answer this question, we imaged *Escherichia coli* (*E. coli*) cells growing on an agar substrate at 37°C using SLIM. The evolution of single cells was tracked using the Schnitzcell semiautomatic software developed in Michael Elowitz's laboratory at Caltech. In addition to mass, this analysis enables us to assess multiple parameters, including length and volume. Figure 12.12*a* shows the dry mass growth curves for a family of *E. coli* cells. By measuring all the cells in our field of view in the same manner we can assess the growth characteristic of a population of cells simultaneously. Thus, Fig. 12.12*b* shows the growth rate of 22 single cells as a function of mass, $dM(t)/dt$. The average of these data (black circles) indicates that the growth rate is proportional to the mass, $dM(t)/dt = \alpha M(t)$, which indicates that the cells follow an exponential growth pattern. These results are in excellent agreement with recent measurements by Godin et al.[49]

Our results demonstrate that rich information can be captured from biological structures using SLIM. Because of its implementation with existing phase-contrast microscopes, SLIM has the potential to elevate phase-based imaging from observing to quantifying over a broad range of spatiotemporal scales, in a variety of biological applications. We anticipate that the studies allowed by SLIM will further our understanding of the basic phenomena related to cell division, growth, transport, and cell-cell interaction. Because of the extremely short coherence length of this illumination light, approximately 1.2 μm in water, SLIM provides *label-free* optical sectioning, allowing a three-dimensional view of live cells, which reflects the scattering potential distribution. This extension to 3D imaging, called *spatial light interference tomography (SLIT)* is discussed further in Sec. 14.1.

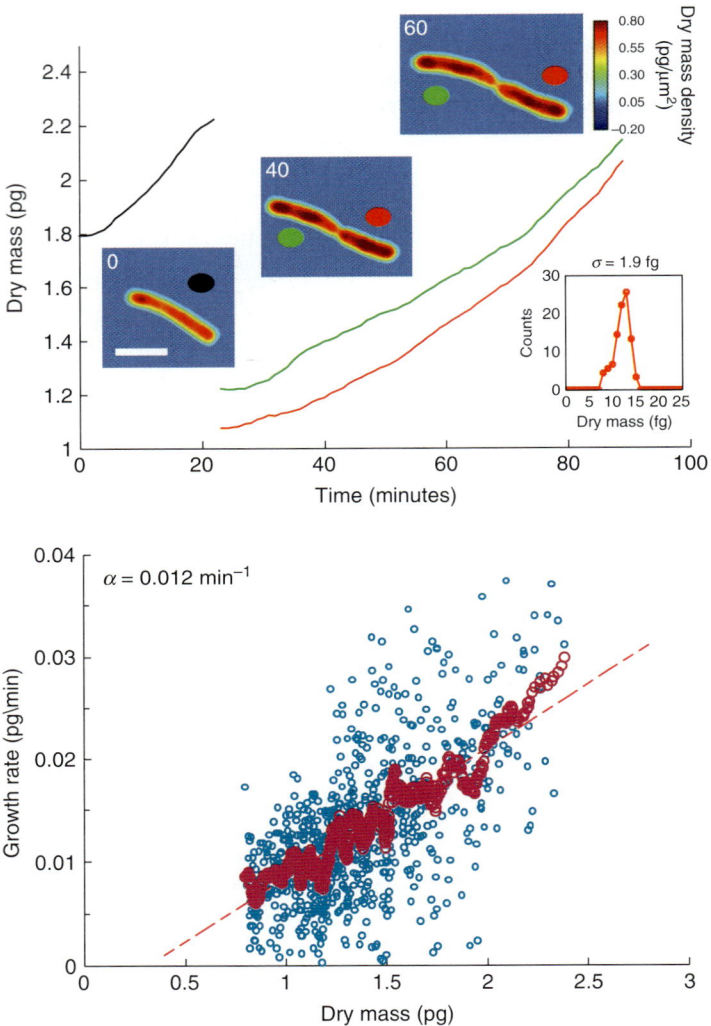

Figure 12.12 SLIM measurements of *E.coli* growth. (*a*) Dry mass vs. time for a cell family, growth curves for each cell are indicated by the colored circles on the images. The inset shows the histogram of the dry mass noise associated with the background of the same projected area as the average cell (standard deviation σ is shown). (*b*) Growth rate vs. mass of 22 cells measured in the same manner, faint circles indicate single data points from individual cell growth curves, dark circles show the average and the dashed line is a linear fit through the averaged data, the slope of this line, 0.012 min^{-1}, is a measure of the average growth constant for this population. The linear relationship between the growth rate and mass suggests that on average, *E.coli* cells exhibit a simple exponential growth pattern. Images show single cell dry mass density maps at the indicated time points, the white scale bar is 2 μm and the colorbar is in pg/μm^2. M. Mir, Z. Wang, Z. Shen, M. Bednarz, R. Bashir, I. Golding, S. G. Prasanth and G. Popescu, *Measuring Cell Cycle Dependent Mass Growth* Proc. Nat. Acad. Sci., under review). (*From Ref. 14.*)

References

1. J. W. Goodman, *Statistical optics* (Wiley, New York, 2000).
2. J. W. Goodman, *Speckle phenomena in optics: theory and applications* (Roberts & Co., Englewood, Colo., 2007).
3. D. Paganin and K. A. Nugent, Noninterferometric phase imaging with partially coherent light, *Phys. Rev. Lett.*, 80, 2586–2589 (1998).
4. E. Abbe, Beiträge zur Theorie des Mikroskops und der mikroskopischen Wahrnehmung, *Arch. Mikrosc. Anat.*, 9, 431 (1873).
5. M. R. Teague, Deterministic phase retrieval: a Green's function solution, *JOSA*, 73, 1434–1441 (1983).
6. L. J. Allen, H. M. L. Faulkner, K. A. Nugent, M. P. Oxley and D. Paganin, Phase retrieval from images in the presence of first-order vortices, *Phys. Rev. E*, 6303, art. no.-037602 (2001).
7. E. D. Barone-Nugent, A. Barty and K. A. Nugent, Quantitative phase-amplitude microscopy I: optical microscopy, *J Microsc*, 206, 194–203 (2002).
8. T. E. Gureyev, A. Roberts and K. A. Nugent, Partially Coherent Fields, The Transport-Of-Intensity Equation, And Phase Uniqueness, *J. Opt. Soc. Am. A-Opt. Image Sci. Vis.*, 12, 1942–1946 (1995).
9. K. A. Nugent, T. E. Gureyev, D. F. Cookson, D. Paganin and Z. Barnea, Quantitative phase imaging using hard x rays, *Phys. Rev. Lett.*, 77, 2961–2964 (1996).
10. B. E. Allman, P. J. McMahon, K. A. Nugent, D. Paganin, D. L. Jacobson, M. Arif and S. A. Werner, Imaging-Phase radiography with neutrons, *Nature*, 408, 158–159 (2000).
11. S. Bajt, A. Barty, K. A. Nugent, M. McCartney, M. Wall and D. Paganin, Quantitative phase-sensitive imaging in a transmission electron microscope, *Ultramicroscopy*, 83, 67–73 (2000).
12. A. Barty, K. A. Nugent, D. Paganin and A. Roberts, Quantitative optical phase microscopy, *Opt. Lett.*, 23, 817–819 (1998).
13. C. L. Curl, C. J. Bellair, P. J. Harris, B. E. Allman, A. Roberts, K. A. Nugent and L. M. D. Delbridge, Single cell volume measurement by quantitative phase microscopy (QPM): A case study of erythrocyte morphology, *Cellular Physiology and Biochemistry*, 17, 193–200 (2006).
14. Z. Wang, L. J. Millet, M. Mir, H. Ding, S. Unarunotai, J. A. Rogers, M. U. Gillette and G. Popescu, Spatial light interference microscopy (SLIM), *Optics Express*, 19, 1016 (2011).
16. F. Zernike, How I discovered phase contrast, *Science*, 121, 345 (1955).
17. J. B. Pawley Handbook of biological confocal microscopy (Springer, New York, 2006).
18. Z. Wang and G. Popescu, Quantitative phase imaging with broadband fields, *Applied Physics Letters*, 96, 051117 (2010).
19. E. Wolf, *Solution of the Phase Problem in the Theory of Structure Determination of Crystals from X-Ray Diffraction Experiments*, *Phys. Rev. Lett.*, 103, 075501 (2009).
20. F. Zernike, Phase contrast, a new method for the microscopic observation of transparent objects, Part 1, *Physica*, 9, 686–698 (1942).
21. D. Gabor, A new microscopic principle, *Nature*, 161, 777 (1948).
22. L. Mandel and E. Wolf, Optical coherence and quantum optics (Cambridge University Press, Cambridge ; New York, 1995).
23. G. Popescu, T. Ikeda, R. R. Dasari and M. S. Feld, Diffraction phase microscopy for quantifying cell structure and dynamics, *Opt Lett*, 31, 775–777 (2006).
24. Y. K. Park, C. A. Best, K. Badizadegan, R. R. Dasari, M. S. Feld, T. Kuriabova, M. L. Henle, A. J. Levine and G. Popescu, Measurement of red blood cell mechanics during morphological changes, *Proc. Nat. Acad. Sci.*, 107, 6731 (2010).
25. Y. K. Park, C. A. Best, T. Auth, N. Gov, S. A. Safran, G. Popescu, S. Suresh and M. S. Feld, Metabolic remodeling of the human red blood cell membran, *Proc. Nat. Acad. Sci.*, 107, 1289 (2010).

26. M. Mir, Z. Wang, K. Tangella and G. Popescu, Diffraction Phase Cytometry: blood on a CD-ROM, *Opt. Exp.*, 17, 2579 (2009).

27. Y. K. Park, M. Diez-Silva, G. Popescu, G. Lykotrafitis, W. Choi, M. S. Feld and S. Suresh, Refractive index maps and membrane dynamics of human red blood cells parasitized by Plasmodium falciparum, *Proc Natl Acad Sci U S A*, 105, 13730 (2008).

28. H. F. Ding, Z. Wang, F. Nguyen, S. A. Boppart and G. Popescu, Fourier Transform Light Scattering of Inhomogeneous and Dynamic Structures, *Physical Review Letters*, 101, 238102 (2008).

29. G. Popescu, Y. K. Park, R. R. Dasari, K. Badizadegan and M. S. Feld, Coherence properties of red blood cell membrane motions, *Phys. Rev. E.*, 76, 031902 (2007).

30. Y. K. Park, G. Popescu, R. R. Dasari, K. Badizadegan and M. S. Feld, Fresnel particle tracking in three dimensions using diffraction phase microscopy, *Opt. Lett.*, 32, 811 (2007).

31. Z. Wang, I. S. Chun, X. L. Li, Z. Y. Ong, E. Pop, L. Millet, M. Gillette and G. Popescu, Topography and refractometry of nanostructures using spatial light interference microscopy, *Optics Letters*, 35, 208–210 (2010).

32. K. S. Novoselov, A. K. Geim, S. V. Morozov, D. Jiang, Y. Zhang, S. V. Dubonos, I. V. Grigorieva and A. A. Firsov, Electric Field Effect in Atomically Thin Carbon Films, *Science*, 306, 666–669 (2004).

33. P. Blake, E. W. Hill, A. H. C. Neto, K. S. Novoselov, D. Jiang, R. Yang, T. J. Booth and A. K. Geim, Making graphene visible, *Applied Physics Letters*, 91, (2007).

34. M. Ishigami, J. H. Chen, W. G. Cullen, M. S. Fuhrer and E. D. Williams, Atomic Structure of Graphene on SiO_2, *Nano Letters*, 7, 1643–1648 (2007).

35. A. Shukla, R. Kumar, J. Mazher and A. Balan, Graphene made easy: High quality, large-area samples, *Solid State Communications*, 149, 718–721 (2009).

36. E. Meijering, M. Jacob, J. C. F. Sarria, P. Steiner, H. Hirling and M. Unser, Design and validation of a tool for neurite tracing and analysis in fluorescence microscopy images, *Cytometry Part A*, 58A, 167–176 (2004).

37. M. Papa, M. C. Bundman, V. Greenberger and M. Segal, Morphological Analysis of Dendritic Spine Development in Primary Cultures of Hippocampal-Neurons, *Journal of Neuroscience*, 15, 1–11 (1995).

38. N. E. Ziv and S. J. Smith, Evidence for a role of dendritic filopodia in synaptogenesis and spine formation, *Neuron*, 17, 91–102 (1996).

39. Y. Goda and G. W. Davis, Mechanisms of synapse assembly and disassembly, *Neuron*, 40, 243–264 (2003).

40. C. L. Waites, A. M. Craig and C. C. Garner, Mechanisms of vertebrate synaptogenesis, *Annual Review of Neuroscience*, 28, 251–274 (2005).

41. M. S. Kayser, M. J. Nolt and M. B. Dalva, EphB receptors couple dendritic filopodia motility to synapse formation, *Neuron*, 59, 56–69 (2008).

42. B. Alberts, *Essential cell biology : an introduction to the molecular biology of the cell* (Garland Pub., New York, 2004).

43. P. Thevenaz, U. E. Ruttimann and M. Unser, A pyramid approach to subpixel registration based on intensity, *IEEE Transactions on Image Processing*, 7, 27–41 (1998).

44. G. Popescu, Y. Park, N. Lue, C. Best-Popescu, L. Deflores, R. R. Dasari, M. S. Feld and K. Badizadegan, Optical imaging of cell mass and growth dynamics, *Am J Physiol Cell Physiol*, 295, C538–44 (2008).

45. J. B. Weitzman, Growing without a size checkpoint, *J Biol*, 2, 3 (2003).

46. A. Tzur, R. Kafri, V. S. LeBleu, G. Lahav and M. W. Kirschner, Cell growth and size homeostasis in proliferating animal cells, *Science*, 325, 167–171 (2009).

47. G. Reshes, S. Vanounou, I. Fishov and M. Feingold, Cell shape dynamics in Escherichia coli, *Biophysical Journal*, 94, 251–264 (2008).

48. G. Popescu, Y. Park, N. Lue, C. Best-Popescu, L. Deflores, R. Dasari, M. Feld and K. Badizadegan, Optical imaging of cell mass and growth dynamics, *Am. J. Physiol.: Cell Physiol.*, 295, C538 (2008).

49. M. Godin, F. F. Delgado, S. Son, W. H. Grover, A. K. Bryan, A. Tzur, P. Jorgensen, K. Payer, A. D. Grossman, M. W. Kirschner and S. R. Manalis, Using buoyant mass to measure the growth of single cells, *Nat Methods*, 7, 387–390 (2010).

50. H. G. Davies and M. H. Wilkins, Interference microscopy and mass determination, *Nature*, 169, 541 (1952).
51. R. Barer, Interference microscopy and mass determination, *Nature*, 169, 366–367 (1952).
52. M. Pluta, Advanced light microscopy (Polish Scientific Publishers, Warszawa, 1988).
53. Z. Wang, L. Millet, V. Chan, H. Ding, M. U. Gillette, R. Bashir and G. Popescu, Label-free intracellular transport measured by Spatial Light Interference Microscopy, *J. Biomed. Opt.*, 16(2), (2011).
54. M. Mir, Z. Wang, Z. Shen, M. Bednarz, R. Bashir, I. Golding, S. G. Prasanth and G. Popescu, Measuring Cell Cycle Dependent Mass Growth, *Proc. Nat. Acad. Sci.*, (under review).

CHAPTER 13

Fourier Transform Light Scattering

The basic formalism of light interaction with inhomogeneous media, or *light scattering*, has been already presented in Chap. 2. Here we review the importance of light-scattering methods for biological applications and present recent developments that connect this field with QPI. Specifically, we will introduce the principle of Fourier transform light scattering (FTLS), which allows extracting both static and dynamic light-scattering information from cells and tissues via measurements of quantitative-phase imaging.[1-8] FTLS is the *spatial analog* to Fourier transform infrared spectroscopy (FTIR), which also provides frequency domain information via measurements in the reciprocal space. Thus, while FTIR provides (*temporal* or *optical*) frequency information via measurements of temporal field autocorrelations, FTLS reveals the angular distribution (*spatial frequency* content) of the scattering field via measurements in the spatial domain (i.e., quantitative-phase and amplitude imaging). In the following, we describe the FTLS and review its applications to studying both static and dynamic light scattering by biological tissues and live cells.

13.1 Principle

13.1.1 Relevance of Light-Scattering Methods

Elastic (static) light scattering (ELS) broadly impacted our understanding of inhomogeneous matter, from atmosphere and colloidal suspensions to rough surfaces and biological tissues.[9] In ELS, by measuring the angular distribution of the scattered field, one can infer noninvasively quantitative information about the sample structure (i.e., its spatial distribution of refractive index). Dynamic (quasi-elastic) light scattering (DLS) is the extension of ELS to dynamic inhomogeneous systems.[10] The temporal fluctuations of the optical field scattered at a particular angle by an ensemble of particles under Brownian motion relates to the diffusion coefficient of the particles. *Diffusing-wave spectroscopy* integrates the principle of DLS in highly scattering media.[11] More recently, dynamic scattering from probe particles was used to study the mechanical properties of the surrounding complex fluid of interest.[12] Thus, *microrheology* retrieves viscoelastic properties of complex fluids over various temporal and length scales, and has become a subject to intense current research especially in the context of cell mechanics.[13]

Light-scattering studies have the benefit of providing information intrinsically averaged over the measurement volume. However, it is often the case that the spatial resolution achieved is insufficient. "Particle tracking" microrheology alleviates this problem by measuring the particle displacements in the imaging (rather than scattering) plane.[14,15] However, the drawback in this case is that relatively large particles are needed such that they can be tracked individually, which also limits the throughput required for significant statistical average. Recently, new research has been devoted toward developing new, CCD-based light-scattering approaches that extend the spatio-temporal scales of investigation.[16-20] In particular, it has been shown that the density correlation function can be experimentally retrieved via the two-point intensity correlation in the *Fresnel zone* of the scattered light.[21] This approach was coined *near-field scattering* and its further refinements were successfully combined with Schlierein microscopy.[22,23]

Light scattering has been widely used for studying biological samples, as it is noninvasive, requires minimum sample preparation, and extracts rich information about morphology and dynamic activity.[24-32] Applying DLS measurements to monitor the dynamic changes inside biological samples has emerged as an important approach in the field.[10,11,33-35] More recently, efforts have been devoted to study the mechanical properties of a complex fluid by detecting dynamic scattering from probing particles.[12,13,15,36-42] Thus, the viscoelastic properties of complex fluids are retrieved over various temporal scales. This method was further extended to the cell membranes with attached micron-sized beads as probes. This provides a new way to study the microrheology of live cells.[43] The information is extracted from the temporal light-scattering signals associated with the activity of individual cell component such as filamentous actin and microtubules. The polymerization and depolymerization of these cytoskeleton structures are highly dynamic processes and have important functional roles.[44]

Light-scattering measurements provide information that is intrinsically averaged over the measurement volume. Thus, the spatial resolution is compromised and the scattering contributions from individual components are averaged. Particle-tracking microrheology has been recently proposed to measure the particle displacements in the imaging (rather than scattering) plane,[14] in which the spatial resolution is preserved. However, the drawback is that relatively large particles are needed, such that they can be tracked individually, which also limits the throughput required for significant statistical average. Recently, *phase-sensitive* methods have been employed to directly extract the refractive index of cells and tissues.[45,46] From this information, the angular scattering can be achieved via the Born approximation.[47]

13.1.2 Fourier Transform Light Scattering (FTLS)

In 2008, our group at University of Illinois at Urbana-Champaign developed Fourier transform light scattering (FTLS) as an approach to study both static and dynamic light scattering.[1] FTLS combines the high spatial resolution associated with optical microscopy and intrinsic averaging of light-scattering techniques.[48] This approach relies on measuring amplitude and phase information via diffraction-phase microscopy.[49] Due to its common path interferometric geometry, DPM is stable in optical path-length, to the subnanometer level. This feature allows FTLS to perform studies on static and dynamic samples with unprecedented sensitivity, as described below. Using a different interferometer, more similar to Hilbert-phase microscopy rather than DPM, Choi et al. reported "field-based angle-resolved light scattering."[50]

Figure 13.1 depicts the experimental setup that incorporates the common path interferometer, a version of which has been already described in Sec. 11.2. Here the interferometer is interfaced with a commercial, computer-controlled microscope. The second harmonic of a Nd:YAG laser ($\lambda = 532$ nm) is used to illuminate the sample in

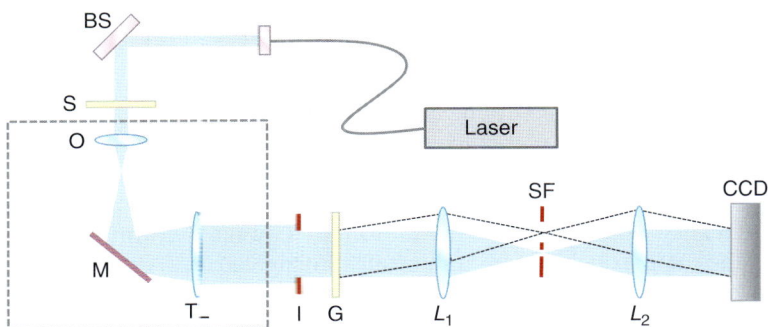

FIGURE 13.1 FTLS experimental setup. BS, beam splitter; S, sample; O, objective lens; M, mirror; TL, tube lens; I, iris; G, grating; SF, spatial filter; L_1 and L_2, lenses. (From Ref. 3)

transmission. To ensure full spatial coherence, the laser beam is coupled into a single mode fiber and further collimated by a fiber colli-
mator. The light scattered by the sample is collected by the objective
lens of the microscope (AxioObserver Z1, Zeiss) and imaged at the
side port of the microscope. A diffraction grating, G, is placed at this
image plane, thus generating multiple diffraction orders containing
full spatial information about the image. In order to establish a common-
path Mach-Zehnder interferometer, a standard spatial filtering lens
system L_1-L_2 is used to select the two diffraction orders and generate
the final interferogram at the CCD plane. The 0th-order beam is low-
pass filtered using the spatial filter SF positioned in the Fourier plane
of L_1, such that at the CCD plane it approaches a uniform field. Simultaneously, the spatial filter allows passing the entire frequency content of the 1st-diffraction-order beam and blocks all the other orders.
The 1st order is thus the imaging field and the 0th order plays the role
of the reference field. The two beams propagate along a common
optical path, thus significantly reducing the longitudinal-phase noise.
The direction of the spatial modulation is along the x-axis, such that
the total field at the CCD plane has the form

$$U(x,y) = |U_0|e^{i(\phi_0 + \beta x)} + |U_1(x,y)|e^{i\phi_1(x,y)}$$ (13.1)

In Eq. (13.1), $|U_{0,1}|$ and $\phi_{0,1}$ are the amplitudes and the phases of the
orders of diffraction 0, 1, while β represents the spatial frequency shift
induced by the grating to the 0th order. To preserve the transverse
resolution of the microscope, the spatial frequency, β, exceeds the
maximum frequency allowed by the numerical aperture of the instrument. The L_1-L_2 lens system has an additional magnification of $f_2/f_1 =$
5, such that the sinusoidal modulation of the image is sampled by
four CCD pixels per period. The interferogram is spatially high-pass
filtered to isolate the cross term,

$$R(x,y) = |U_0||U_1(x,y)|\cos[\phi_1(x,y) - \phi_0 - \beta x]$$ (13.2)

which can be regarded as the real part, $R(x,y)$, of a complex analytic signal. The imaginary component, proportional to $\sin[\phi_1(x,y) - \phi_0 - \beta x]$,
is obtained via a spatial Hilbert transform (see discussion in Chap. 8
on off-axis interferometry and Refs. 49, 51, 52). Thus, from a single
CCD exposure, we obtain the spatially-resolved phase and amplitude
associated with the image field. From this image field information
$\tilde{U}$, the complex field can be numerically propagated at arbitrary planes;
in particular, the far-field angular scattering distribution $\tilde{U}$ can be
obtained simply via a Fourier transformation,[48]

$$\tilde{U}(\mathbf{q},t) = \int U(\mathbf{r},t)e^{-i\mathbf{q}\cdot\mathbf{r}}d^2\mathbf{r}$$ (13.3)

With time-lapse image acquisition, the temporal-scattering signals are
recorded and the sampling frequency is only limited by the acquisition

speed of the camera. The power spectrum is obtained through Fourier transformation of these time-resolved scattering signals.

As system calibration, we applied FTLS to dilute microsphere water suspensions sandwiched between two coverslips, the scattering of which can be easily modeled by Mie theory as shown in Ref. 48. The measured complex field associated with such samples can be expressed as[48]

$$U(\mathbf{r};t) = \iint_A U_F(\mathbf{r}') \sum_{i=1}^{N} \delta\{[\mathbf{r} - \mathbf{r}_i(t)] - \mathbf{r}'\} d^2\mathbf{r}' \qquad (13.4)$$

In Eq. (13.4), U_F is the (time-invariant) complex field associated with each particle, δ is the 2D Dirac function describing the position (x_i, y_i) of each of the N-moving particles, and the integral is performed over the microscope field of view A.

Figure 13.2a and b shows the amplitude and phase distributions obtained by imaging 3-μm polystyrene beads at a particular point in

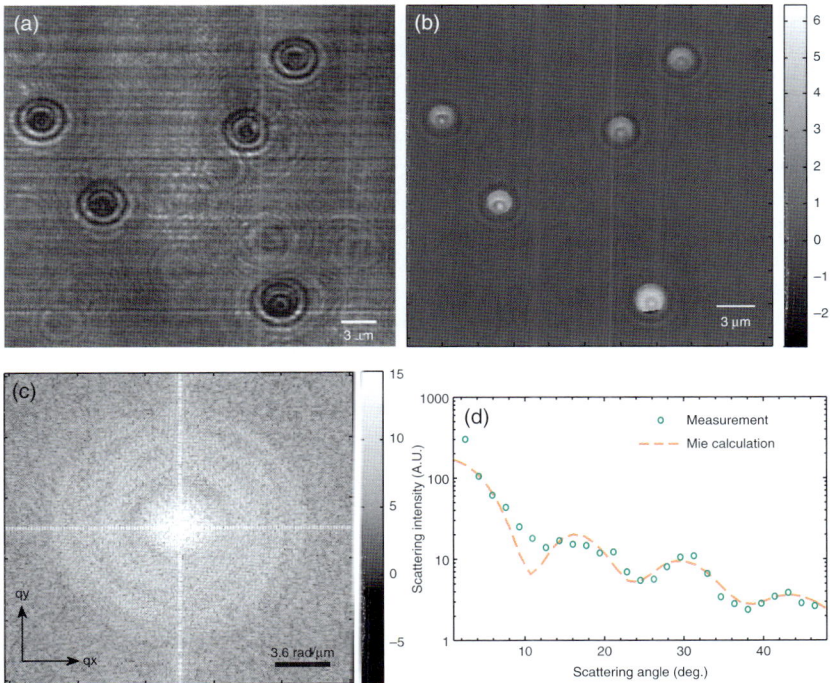

FIGURE 13.2 FTLS reconstruction procedure of angular scattering from 3-μm beads. (a) Amplitude image. (b) Reconstructed phase image. (c) Scattering wave vector map. (d) Retrieved angular scattering and comparison with Mie calculation. [APS, used with permission, *Physical Review Letters*, H. F. Ding, Z. Wang, F. Nguyen, S.A. Boppart, & G. Popescu, "Fourier transform light scattering of inhomogeneous and dynamic structures," 101, (2008). (*From Fig. 1, Ref. 1.*)]

time. For ELS studies, prior to processing the interferogram, we subtract a background image obtained as the intensity map without sample or reference beam. The scattered far-field is obtained by Fourier transforming Eq. (13.4) in space. This angular field distribution factorizes into a *form* field $\tilde{U}_F$, which is determined by the angular scattering of a single particle, and a *structure* field $\tilde{U}_S$, describing the spatial correlations in particle positions,[48]

$$\tilde{U}(\mathbf{q};t) = \tilde{U}_F(\mathbf{q})\tilde{U}_S(\mathbf{q};t) \qquad (13.5)$$

where $\mathbf{q}$ is the spatial-wave vector and $\tilde{U}_S(\mathbf{q};t) = \sum_i e^{i\mathbf{q}\cdot\mathbf{r}_i(t)}$. Figure 13.2c shows the resulting intensity distribution $|\tilde{U}_F(\mathbf{q})|^2$ for the beads in Fig. 13.2a and b. As expected for such sparse distributions of particles, the form function is dominant over the entire angular range. However, by finding the phase-weighted centroid of each particle, FTLS can retrieve independently the structure function whenever it has a significant contribution to the far-field scattering, e.g., in colloidal crystals. The scattered intensity (e.g., Fig. 13.2c) is azimuthally averaged over rings of constant scattering wave numbers, $q = (4\pi/\lambda)\sin(\theta/2)$, with θ the scattering angle, as exemplified in Fig. 13.2d. In order to test the ability of FTLS to retrieve quantitatively the form function of the spherical dielectric particles, we used Mie theory for comparison.[9] The oscillations in the angular scattering establish the quantitative agreement between the FTLS measurement and Mie theory, which contrasts with the common measurements on colloidal suspensions, where the signal is averaged over a large number of scatterers.

Acquiring sets of time-lapse phase and amplitude images, we studied the dynamic light-scattering signals from micron-sized particles undergoing Brownian motion in liquids of various viscosities. Thus, the power spectrum of the scattered intensity can be expressed for each wave vector as

$$P(\mathbf{q},\omega) = \left|\int \tilde{U}(\mathbf{q},t)e^{-i\omega t}dt\right|^2 \qquad (13.6)$$

Figure 13.3b shows the power spectrum associated with 3-μm beads in water. The experimental data is fitted with a Lorentzian function, which describes the dynamics of purely viscous fluids, $P(q,\omega) \propto 1/[1+(\omega/Dq^2)^2]$, where $D = k_B T/4\pi\eta a$, the 2D diffusion coefficient, k_B is the the Boltzmann constant, T the absolute temperature, and η the viscosity and a the radius of the bead. The fits describe our data very well and allow for extracting the viscosity of the surrounding liquids as the only fitting parameter. The measured vs. expected values of the power-spectrum bandwidth are plotted in

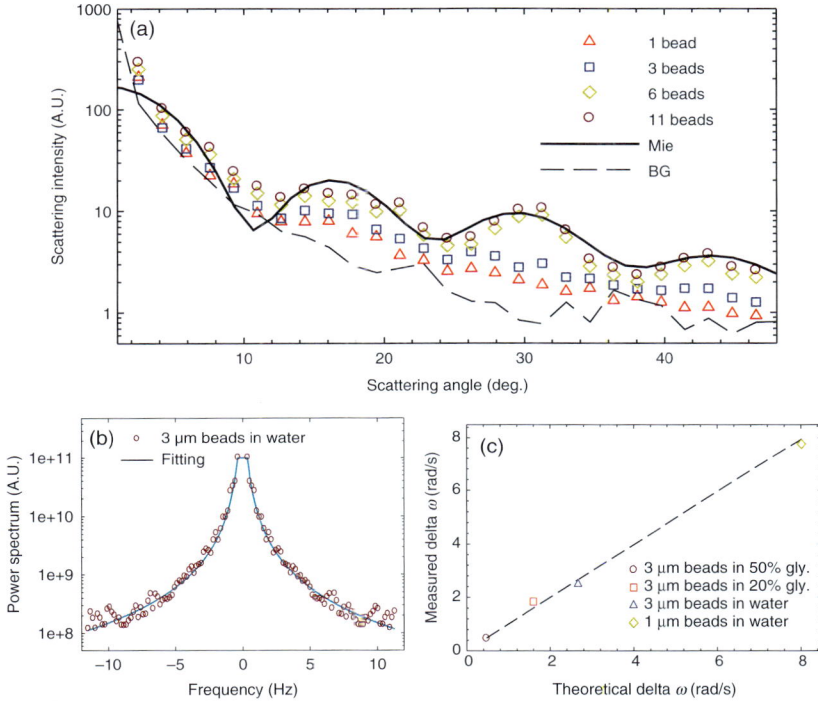

FIGURE 13.3 (*a*) Angular scattering signals associated with 1, 3, 6, and 11 particles, as indicated. The solid line indicates the Mie calculation and the dash line the background signal, measured as described in text. (*b*) Power spectrum of scattering intensity fluctuations associated with 3-μm particles in water. The solid line indicates the fit with Eq. (13.5). (*c*) Measured spectral bandwidth vs. expected bandwidth for different particles and liquids, as indicated. [APS, used with permission, *Physical Review Letters,* H. F. Ding, Z. Wang, F. Nguyen, S.A. Boppart, & G. Popescu, "Fourier transform light scattering of inhomogeneous and dynamic structures," 101, (2008). (*From Fig. 2, Ref. 1.*)]

Fig. 13.3*c*, which shows very good agreement over more than a decade in bandwidths (or, equivalently, viscosities).

In summary, Fourier transform light scattering (FTLS) is a novel experimental approach that uses QPI to perform light-scattering studies on inhomogeneous and dynamic media.[1,4,5,7,8] In FTLS, the optical phase and amplitude of a coherent image field are quantified and propagated numerically to the scattering plane. Because it detects all the scattered angles (spatial frequencies) simultaneously in each point of the image, FTLS can be regarded as the *spatial* equivalent of Fourier transform infrared spectroscopy (FTIR), where all the *temporal* frequencies are detected at each moment in time.

13.2   Further Developments

More recently, FTLS was implemented with spatial light interference microscopy (SLIM), which was discussed in Chap. 12. In this case, the spatial sensitivity to phase changes is significantly superior to laser methods and much finer details can be resolved, for example, within live cells. Of course, the phase shift in this case, and, thus, the scattering information, corresponds to the mean wavelength of this broad spectrum field (522 nm is the mean wavelength in our experiments). This interpretation assumes that dispersion effects are negligible, which is a safe assumption for optically thin objects, such as most cells and tissue slices.

Figure 13.4 shows such an example where the angular scattering associated with a putative dendrite is retrieved over an angular range of (0, 40°).[5] This example illustrates very well the advantage of measuring scattering information from image-plane measurements and the power of FTLS. Note that, while its diameter is approximately 1 μm, the pathlength shift associated with this axon is in the order of

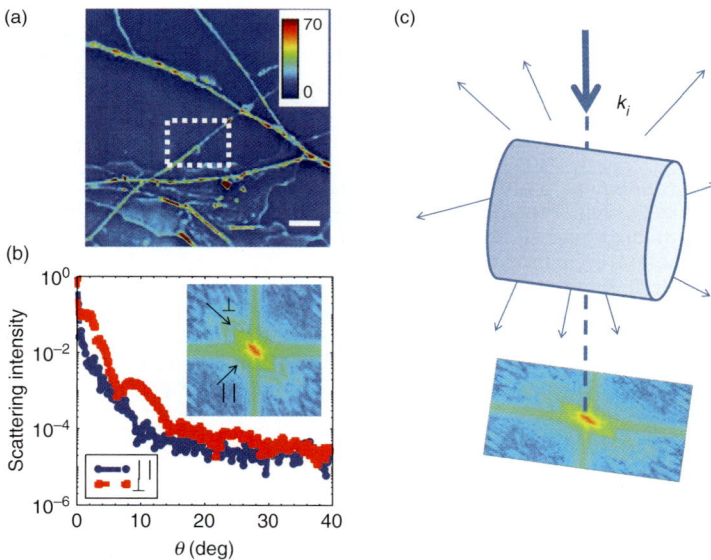

FIGURE 13.4   (a) SLIM image of a neuronal structure. The colorbar indicates optical path length in nanometer (nm) and the scale bar measures 10 μm. (b) Angular scattering associated with the putative axon selected by the dotted box in (a). The two profiles are along the perpendicular and parallel directions to the axon as indicated in the scattering map inset. (c) Illustration of traditional light-scattering measurements, where the axon is illuminated by a plane wave and the detection takes place in the far zone. [IEEE, used with permission, *IEEE Journal of Selected Topics in Quantum Electronics*, H. Ding, E. Berl, Z. Wang, L.J. Millet, M.U. Gillette, J. Liu, M. Boppart, and G. Popescu, "Fourier transform light scattering of biological structures and dynamics," (2010). (*From Fig. 4, Ref. 5.*)]

only 30 nm, which indicates a refractive index contrast roughly of $\Delta n \simeq 0.03$. Now, let us compare this with a polystyrene particle suspended in water, where the refractive index contrast is $1.59 - 1.33 = 0.26$. The scattering cross section of a 1-μm segment of the axon has a scattering cross section, $\sigma = 0.04$ μm^2, that is, essentially, a factor $(0.26/0.03)^2 \simeq 75$ times smaller than that of a 1-μm polystyrene particle in water (!). Clearly, attempting to measure the angular scattering from this structure using traditional, goniometer-based measurements would most likely be extremely challenging, perhaps requiring photon counting detectors, simply because the total scattered power is a very small quantity. A quick estimation of this power can be obtained by starting with the SLIM typical irradiance at the sample plane, which is 10^{-9} W/μm^2. If the axon is illuminated with this irradiance, the total scattered power is of the order of the irradiance times the scattering cross section, i.e., 1 pW = 10^{-12} W, which amounts to approximately 10 million visible photons per second, scattered in all directions, roughly $10^7/4\pi$ photons/srad. Further, the *angular resolution* allowed by FTLS is essentially the ratio between the wavelength of light and the field of view, $l/L \simeq 0.01$ rad. In terms of solid angle, this resolution corresponds to 10^{-4} srad. In order to match this angular resolution, a goniometer-based measurement would have to detect about 1000 photons/s at each angle. Finally, the dynamic range provided by the FTLS measurement is approximately five orders of magnitude (Fig. 13.4b), which is, for all practical purposes, out of reach for any implementation of traditional angular scattering.

This quick estimation emphasizes that measuring such weakly scattering structures with a conventional angular scattering method requires, at best, extremely sensitive detectors and a very long acquisition time to cover the same angular range with a similar resolution to FTLS. How is it possible that a QPI method such as SLIM can acquire these data in a fraction of a second with only a CCD of common sensitivity? The answer is, again, in the understanding of the image formation as an interference between the fields that scatter at various angles (as described by Abbe and discussed earlier many times[53]). First, the image-plane measurement detects the total power of *all* the scattering components simultaneously, which means higher signals. Second, the light that does not scatter is also present at the same plane, which adds further power to the total signal. Third, and perhaps most importantly, the scattered and unscattered fields interfere with very low contrast at the image plane, i.e., the image field is very uniform because of the transparency of the specimen (see Sec. 5.2. on imaging phase objects). Therefore, the finite dynamic range of the CCD is put to great use at the image plane, as it is able to capture fine details in the intensity spatial fluctuations across the field of view. By contrast, if we attempt to measure the angular scattering distribution in Fig. 13.4b directly in far field with the same CCD, clearly the dynamic range necessary is orders of magnitude beyond what is available today.

These FTLS measurements demonstrate once more that the highest sensitivity is achieved by measuring at the plane where the field of interest is the smoothest. This important experimental aspect has been discussed in the context of holography in the introductory chapter (Chap. 1). Next we discuss the main biological applications demonstrated by FTLS to date.

13.3   Biological Applications

13.3.1   Elastic Light Scattering of Tissues

FTLS was employed to determine experimentally the scattering phase functions of red blood cells (RBCs) and tissue sections, which has important implications in the optical screening of various blood constituents and tissue diagnosis.[54,55] Figure 13.5a shows a quantitative

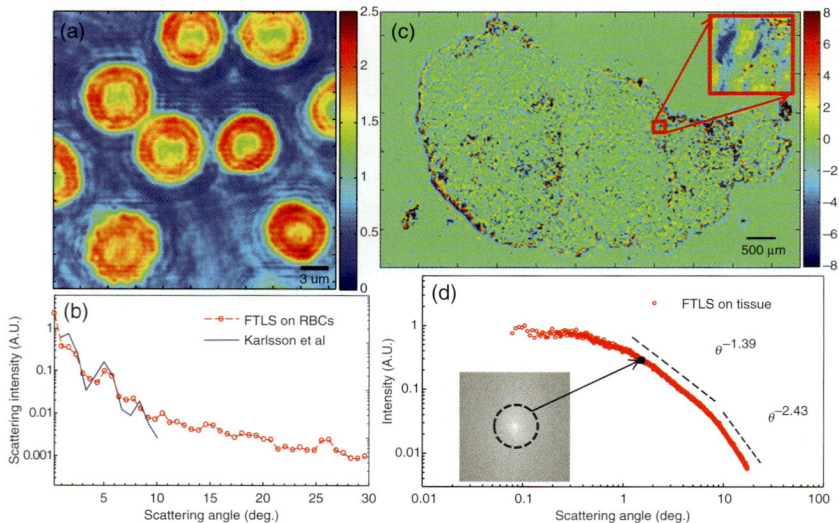

FIGURE 13.5   (a) Spatially-resolved phase distribution of red blood cells. The colorbar indicates phase shift in radians. (b) Scattering phase function associated with the cells in (a). The FDTD simulation by Karlsson et al. is shown for comparison (the x-axis of the simulation curve was multiplied by a factor of 532/633, to account for the difference in the calculation wavelength, 633 nm, and that in our experiments, 532 nm). (c) Giga-pixel quantitative-phase image of a mouse breast tissue slice. Colorbar indicates phase shift in radians. (d) Angular scattering from the tissue in c. The inset shows the 2D scattering map, where the average over each ring corresponds to a point in the angular scattering curve. The dashed lines indicate power laws of different exponents, as indicated. [APS, used with permission, *Physical Review Letters*, H. F. Ding, Z. Wang, F. Nguyen, S.A. Boppart, & G. Popescu, "Fourier transform light scattering of inhomogeneous and dynamic structures," 101, (2008). (*From Fig. 3, Ref. 1.*)]

phase image of RBCs prepared between two cover slips, with the identifiable "dimple" shape correctly recovered. The corresponding angular scattering is presented in Fig. 13.5b, where we also plot for comparison the results of a finite difference time domain (FDTD) simulation previously published by Karlsson et al.[56] Significantly, over the 10° range available from the simulation, our FTLS measurement and the simulation agree very well.

In order to extend the FTLS measurement towards extremely low-scattering angles, we scanned large fields of view by tiling numerous high-resolution microscope images. Figure 13.5c presents a quantitative-phase map of a 5-μm thick tissue slice obtained from the mammary tissue of a rat tumor model by tiling ~1000 independent images. This 0.3 gigapixel composite image is rendered by scanning the sample with a submicron precision computerized translation stage. The scattering-phase function associated with this sample is shown in Fig. 13.5d. We believe that such a broad angular range, of almost three decades, is measured here for the first time and cannot be achieved via any single measurement. Notably, the behavior of the angular scattering follows power laws with different exponents, as indicated by the two dashed lines. This type of measurements over broad spatial scales may bring new light into unanswered questions, such as tissue architectural organization and possible self-similar behavior.[57]

Upon propagation through inhomogeneous media such as tissues, optical fields suffer modifications in terms of irradiance, phase, spectrum, direction, polarization, and coherence, which can reveal information about the sample of interest. We use FTLS to extract quantitatively the scattering mean free path, l_s, and anisotropy factor, g, from tissue slices of different organs.[58] This direct measurement of tissue scattering parameters allows predicting the wave transport phenomenon within the organ of interest at a multitude of scales. The scattering mean free path, l_s, was measured by quantifying the attenuation due to scattering for each slice via the Lambert-Beer law, $l_s = -d/\ln[I(d)/I_0]$, where d is the thickness of the tissue, $I(d)$ is the irradiance of the unscattered light after transmission through the tissue, and I_0 is the total irradiance, i.e., the sum of the scattered and unscattered components. The unscattered intensity $I(d)$, i.e., the spatial DC component, is evaluated by integrating the angular scattering over the diffraction spot around the origin. The resulting l_s values for 20 samples for each organ, from the same rat are summarized in Fig. 13.6a.

The anisotropy factor, g, is defined as the average cosine of the scattering angle,

$$g = \int_{-1}^{1} \cos(\theta) p[\cos(\theta)] d[\cos(\theta)] / \int_{-1}^{1} p[\cos(\theta)] d[\cos(\theta)] \qquad (13.7)$$

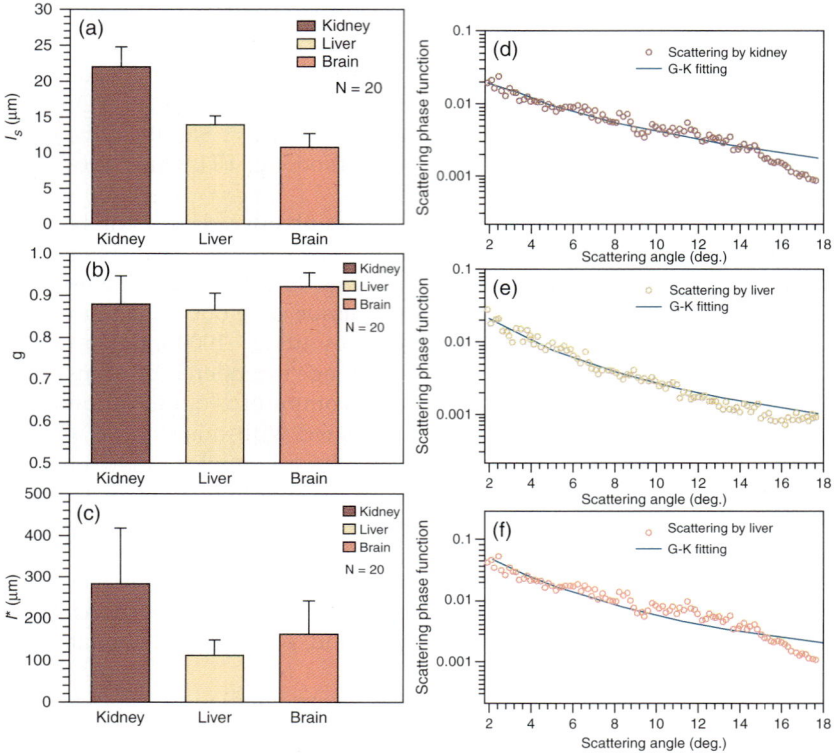

FIGURE 13.6 FTLS measurements of the scattering mean free path, l_s. (a) Anisotropy factors, (b) and transport mean free path, (c) for the three rat organs with 20 samples per group. The error bars correspond to the standard deviations (N = 20). (d-f) The angular scattering plots associated with the scattering maps in Fig. 13.2(d-f). The dash lines indicate fits with the G-K phase function. (*From Ref. 8.*)

where p is the normalized angular scattering, i.e., the phase function. Note that the description applies to tissue slices of thickness $d < l_s$, thus, it cannot be used directly in Eq. (13.7) to extract g, since g values in this case will be thickness-dependent. This is so because the calculation in Eq. (13.7) is defined over tissue of thickness $d = l_s$, which describes the average scattering properties of the tissue (i.e., independent of how the tissue is cut). Under the weakly scattering regime of interest here, this angular scattering distribution, p, is obtained by propagating the complex field numerically through $N = l_s/d$ layers of $d = 5$ μm thickness,[58]

$$p(\mathbf{q}) \propto \left| \iint [U(\mathbf{r})]^N e^{i\mathbf{q}\cdot\mathbf{r}} d^2\mathbf{r} \right|^2 \qquad (13.8)$$

Now Eq. (13.8) applies to a slice of thickness, l_s. It reflects that, by propagating through N weakly scattering layers of tissue, the total phase accumulation is the sum of the phase shifts from each layer, as is typically assumed in phase imaging of transparent structures.[59] The angular scattering distribution, or phase function, $p(\theta)$ is obtained by performing azimuthal averaging of the scattering map, $p(\mathbf{q})$, associated with each tissue sample. The maximum scattering angle was determined by the numeric aperture of the objective lens and it is about 18° for our current setup (10× objective applied for tissue study). The angular-scattering data were further fitted with Gegenbauer Kernel (GK) phase function[60]

$$P(\theta) = ag \cdot \frac{(1-g^2)^{2a}}{\pi[1+g^2-2g\cos(\theta)]^{(a+1)}[(1+g)^{2a}-(1-g)^{2a}]} \qquad (13.9)$$

Note that g can be estimated directly from the angular scattering data via its definition (Eq. 13.7). However, because of the limited angular range measured, g tends to be overestimated by this method, and, thus, the GK fit offers a more reliable alternative than the widely used Henyey-Greenstein (HG) phase function with the parameter $a = 1/2$. The representative fitting plots for each sample are shown in Fig. 13.6d-f. The final values of g are included in Fig. 13.6b and agree very well with previous reports in the literature.[61] Thus the transport mean free path, which is the renormalized scattering length to account for the anisotropic phase function, can be obtained as $l^* = l_s/(1-g)$. The l^* values for 20 samples from each organ are shown in Fig. 13.6c.

Remarkably, from these measurements of thin, singly scattering slices, the behavior of light transport in thick, strongly scattering tissue can be inferred. This is somewhat analogous to measuring the absorption length (or coefficient) of a solution from measurements on thin samples (e.g., dye solution in a thin cuvette), which is typically encountered in practice with measurements using spectrophotometers. Similarly, the samples can be much thinner than the absorption length, as long as the detection is sensitive enough, that is, as long as attenuations much lower than $1/e$ can be detected reliably. In the case of FTLS, the phase map captures fine details about the tissue optics, which translates into extreme sensitivity to attenuations due to scattering. This concept connects once more with the fact that a phase measurement in the image plane can provide scattering signals with extremely high dynamic range (see, for example, Fig. 13.4).

Still, while the concept of extracting tissue optics from QPI measurements is very appealing, the procedure outlined here, involving separate estimation of the DC component and fit with an empirical model (Eq. 13.9) is not ideal for routine applications, such as those pertaining to pathology. Thus, very recent research devoted to this subject revealed that the scattering properties of bulk tissue, i.e., l_s and g, can be *directly mapped* onto the tissue slice, using a certain mathematical model, called the *scattering-phase theorem*. This subject will be discussed further in Sec. 15.3.2.

13.3.2   Elastic Light Scattering of Cells

Light-scattering investigations can noninvasively reveal subtle details about the structural organization of cells.[30,31,62-65] FTLS was employed to measure scattering phase functions of different cell types and demonstrate its capability as a new modality for cell characterization.[5] We retrieved the scattering phase functions from three cell groups (Fig. 13.7a-c): red blood cells, myoblasts (C2C12), and neurons.

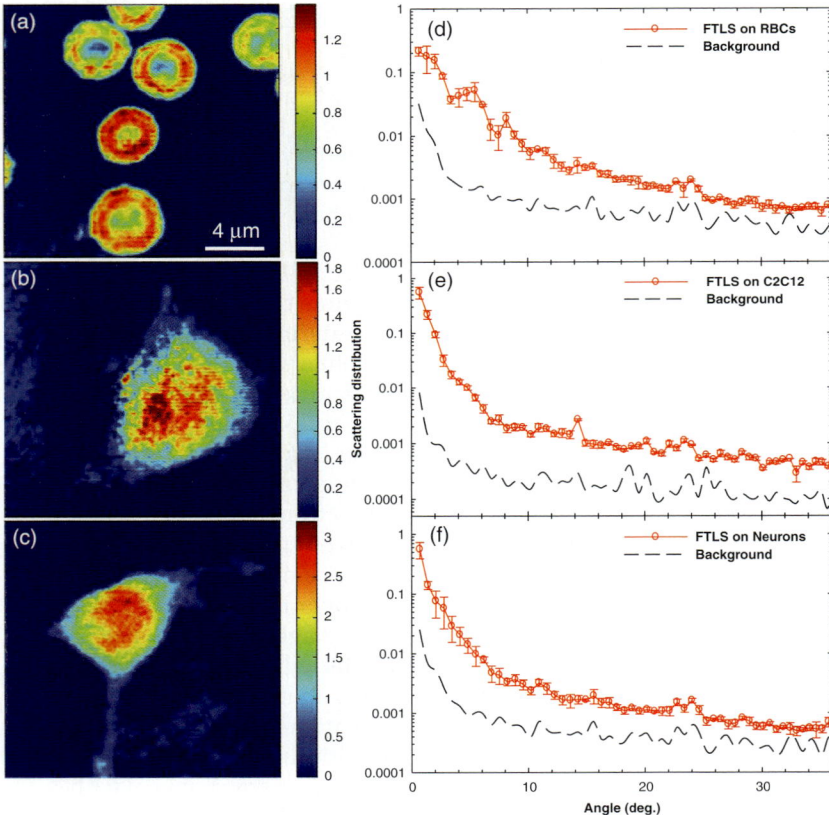

FIGURE 13.7   (*a*) Quantitative-phase images of red blood cells; (*b*) C2C12 cell, and (*c*) neuron; the scale bar is 4 μm and the colorbar indicates phase shift in radians. (*d-f*) Respective scattering phase functions measured by FTLS. [American Scientific Publishers, used with permission, *Journal of Computational and Theoretical Nanoscience*, H. Ding, Z. Wang, F. Nguyen, S.A. Boppart, L.J. Millet, M.U. Gillette, J. Liu, M. Boppart & G. Popescu, "Fourier transform light scattering (FTLS) of cells and tissues," 7, 1546 (2010). (*From Fig. 5, Ref. 5.*)]

Figure 13.7*d-f* shows the angular scattering distributions associated with these samples. For each group, we performed measurements on different fields of view. Remarkably, FTLS provides these scattering signals over approximately 35° in scattering angle (for a 40×, NA = 0.65, microscope objective) and several decades in intensity. For comparison, the scattering signature of the background (i.e., culture medium with cells thresholded out of the field of view) was also measured. These signals (Fig. 13.7*d-f*) incorporate noise contributions from the beam inhomogeneities, impurities on optics, and residues in the culture medium. The measurements demonstrate that FTLS is sensitive to scattering from single cells.

The FTLS data was further analyzed with a statistical algorithm based on the principle component analysis (PCA) aimed at maximizing the differences among the cell groups and providing an automatic means for cell sorting.[66] This statistical method mathematically transforms the data to a new coordinate system that reveals the maximum variance by multiplying the data with the chosen individual vectors. Our procedure can be summarized as follows. First, we average the n ($n = 1...45$) measurements for the three cell types (15 measurements per group), to obtain the average scattered intensity, $\overline{I(\theta_m)} = \dfrac{1}{45} \sum_{n=1...45} I_n(\theta_m)$, with $m = 1...35$ denoting the number of scattering angles. Second, we generate a matrix ΔY_{nm} of variances, where n indexes the different measurements and m the scattering angles. The covariance matrix associated with ΔY, $Cov(\Delta Y)$, is calculated and its eigenvalues and eigenvectors extracted. The three principal components are obtained by retaining three eigenvectors corresponding to the largest eigenvalues. In order to build the training set, 45 measurements (i.e., 15 per cell type) were taken and processed following the procedures described above.

Figure 13.8*a* and *b* shows a representation of the data where each point in the plot is associated with a particular FTLS measurement. In addition to the 15 measurements per group for the training sets, we performed, respectively, 15, 15, and 10 test measurements for neurons, RBCs, and C2C12 cells. The additional test measurements allowed us to evaluate the sensitivity and specificity of assigning a given cell to the correct group.[67] We obtained sensitivity values of 100%, 100%, and 70%, and specificities of 100%, 88%, and 100% for RBCs, neurons, and C2C12 cells, respectively.[5]

In sum, we showed that FTLS can be used to differentiate between various cell types. Due to the particular imaging geometry used, scattering phase functions associated with single cells can be retrieved over a broad range of angles. This sensitivity to weak scattering signals may set the basis for a new generation of cytometry technology, which, in addition to the intensity information, will extract the structural details encoded in the phase of the optical field.

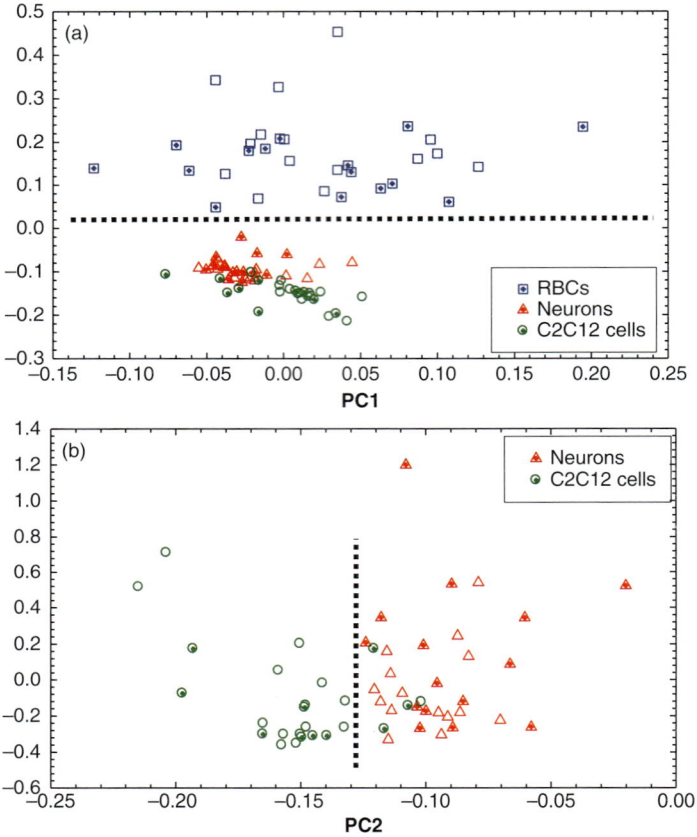

FIGURE **13.8**   PCA of the experimental data for the three cell types, as indicated. The symbols with "+" sign in the middle are the testing measurements for each sample, respectively. (*a*) PC2 vs. PC1. (*b*) PC3 vs. PC2. [IEEE, used with permission, *IEEE Journal of Selected Topics in Quantum Electronics*, H. Ding, E. Berl, Z. Wang, L.J. Millet, M.U. Gillette, J. Liu, M. Boppart, and G. Popescu, "Fourier transform light scattering of biological structures and dynamics," (2010). (*From Fig. 6, Ref. 5.*)]

13.3.3   Dynamic Light Scattering of Cell Membrane Fluctuations

Dynamic properties of cell membrane components such as actin and microtubules have been the subject of intense scientific interest.[68-72] In particular, it has been shown that actin filaments play an important role in various aspects of cell dynamics, including cell motility.[69,70] In this section, we briefly discussed the application of FTLS to study the fluctuating membranes of RBCs.[1] To determine how the cell membrane flickering contributes to the dynamic light scattering of cells, RBCs from healthy volunteer sandwiched with two glass coverslips were imaged via DPM by acquiring 256 frames, at 20 frames/s.

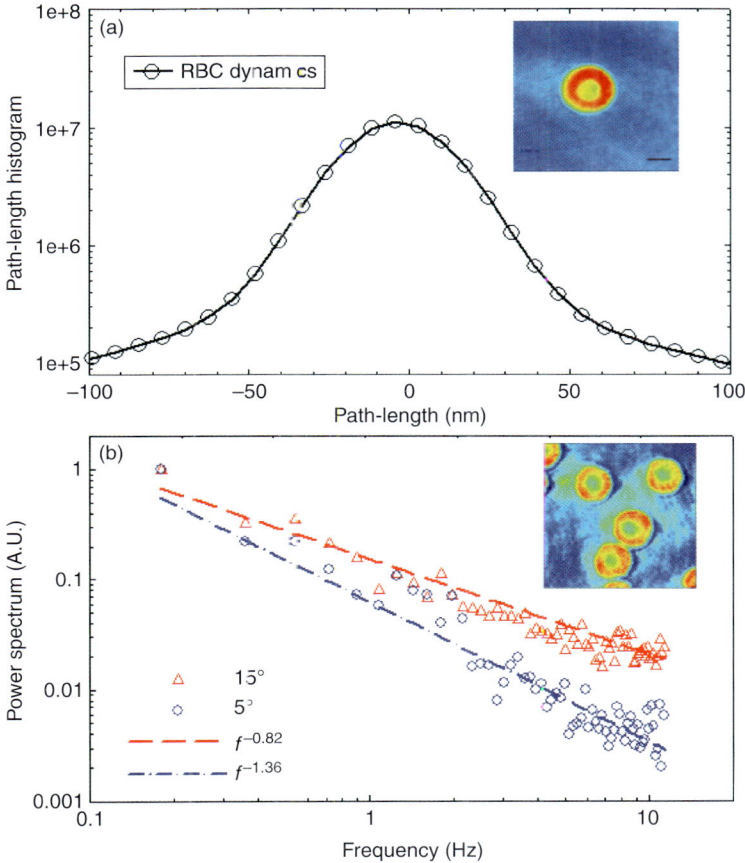

FIGURE 13.9 (a) Histogram of the path-length displacements of a red blood cell. The inset is the phase image. (b) Dynamics FTLS of red blood cells: log–log power spectra at 5° and 15° with the respective power law fits, as indicated. The inset shows one RBC phase image from the time sequence. [APS, used with permission, *Physical Review Letters*, H.F. Ding, Z. Wang, F. Nguyen, S.A. Boppart & G. Popescu, "Fourier transform light scattering of inhomogeneous and dynamic structures," 101 (2008)." (*From Fig. 4, Ref. 1.*)]

Figure 13.9a shows the membrane displacement histograms of a RBC. The power spectrum in Fig. 13.9b follows power laws with different exponents in time for all scattering angles (or, equivalently, wave numbers). As expected, the slower frequency decay at larger q-values indicates a more solid behavior, i.e., the cell is more compliant at longer spatial wavelengths. Notably, the exponent of −1.36 of the longer wavelength (5° angle), is compatible with the −1.33 value predicted by Brochard et al. for the fluctuations at each point on the cell.[73] This is expected, as at each point the motions are dominated by long wavelengths.[74] The dynamic FTLS studies of RBC rheology can be performed on many cells simultaneously, which is an advantage over the previous flickering studies.[51,73,75]

We believe that these initial results are extremely promising and that the label-free approach proposed here for studying cell dynamics will complement very well the existing fluorescence studies. In the next section, we extend the dynamic FTLS study to more complex cells, i.e., cells with a 3D cytoskeleton, and extract information from the temporal light-scattering signals associated with the activity of individual cell component such as filamentous actin and microtubules. The polymerization and depolymerization of these components are highly dynamic processes and have important functional roles.[44]

13.3.4 Dynamic Light Scattering of Cell Cytoskeleton Fluctuations

Recently, dynamic properties of cytoskeleton have been the subject of intense scientific interest.[68-72,76-81] In particular, it has been shown that actin filaments play an important role in various aspects of cell dynamics, including cell motility.[69,70,76] Previously, actin polymerization has been studied in real time by total reflection fluorescence microscopy.[71,77] In this section, we demonstrate that FTLS is capable of sensing the spatio-temporal behavior of active (ATP-consuming) dynamics due to f-actin in single glial cells as addressed previously.[82] This activity is mediated by motor protein myosin II and underlies diverse cellular processes, including cell division, developmental polarity, cell migration, filopodial extension, and intracellular transport.

We used FTLS to study the slow active dynamics of Enteric glial cytoskeleton. During the FTLS measurement, the cells were maintained under constant temperature at 37°C via the incubation system that equips the microscope. The sensitivity of FTLS to actin dynamics was tested by controlling its polymerization activity. In order to inhibit actin polymerization, Cytochalasin-D (Cyto-D), approximately 5 μm in Hibernate-A, was added to the sample dishes. Cyto-D is a naturally occurring fungal metabolite known to have potent inhibitory action on actin filaments by capping and preventing polymerization and depolymerization at the rapidly elongating end of the filament. By capping this "barbed" end, the increased dissociation at the pointed end continues to shorten the actin filament.

The established single particle-tracking method[40] was first applied to test the efficacy of Cyto-D as actin inhibitor. With 1-μm diameter beads attached to the cell membrane as probes, the dynamic property of the cell membrane was investigated. Sets of 512 quantitative-phase images of cells with beads attached were acquired at 1/5 frames/s, with a total acquisition time of 45 minutes. The x and y coordinates of the tracked beads were recorded as a function of time and the trajectories were used to calculate the mean squared displacement (MSD),[83]

$$MSD(\Delta t) = \left\langle [x(t+\Delta t) - x(t)]^2 + [y(t+\Delta t) - y(t)]^2 \right\rangle \qquad (13.10)$$

where <...> indicates the average over time and also over all the tracked particles. We recorded the displacements of the attached beads before and after treatment with Cyto-D. The MSD results are summarized in Fig. 13.10. Each curve is the result of an average over all the beads ($N > 10$) tracked under the same experimental condition. The data for normal (before treatment) cells exhibit a power law trend over two distinct temporal regions, separated at approximately $\tau_0 = 215$ s. This change in slope reflects a characteristic life time τ_0 associated with actin polymerization, which is known to be in the minute range.[84] For both curves, the fit at $t > \tau_0$ gives a power law of exponent approximately 1, indicating diffusive motion. However, the membrane displacements for the untreated cells are a factor of 1.3 larger, consistent with the qualitative observation of the particle trajectories. Thus, at long time, the effect of actin on membrane dynamics is to enhance its motions without changing its temporal statistics.

To determine how the actin cytoskeleton contributes to the dynamic light scattering of cells alone, cells without beads were imaged via DPM by acquiring 512 frames, at 0.2 frames/s, over ~45 minutes, prior to and after Cyto-D application, respectively (Fig. 13.11*a* and *b*).

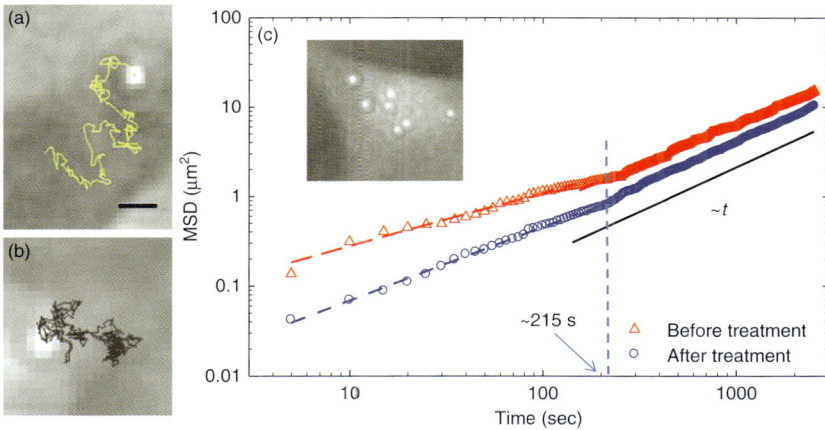

FIGURE 13.10 Tracking beads attached to the cell membrane: (*a*) bead trajectories before Cyto-D treatment and (*b*) after drug treatment, scale bar for 2 μm. (*c*) Corresponding mean-square displacement of the tracked beads before (red) and after (blue) treatment. The fitting with a power law function over two different time windows is indicated. The inset shows a quantitative-phase image of an EGC with 1-μm beads attached to the membrane. [IEEE, used with permission, *IEEE Journal of Selected Topics in Quantum Electronics*, H. Ding, E. Berl, Z. Wang, L.J. Millet, M.U. Gillette, J. Liu, M. Boppart, and G. Popescu, "Fourier transform light scattering of biological structures and dynamics," (2010). (*From Ref. 7.*)]

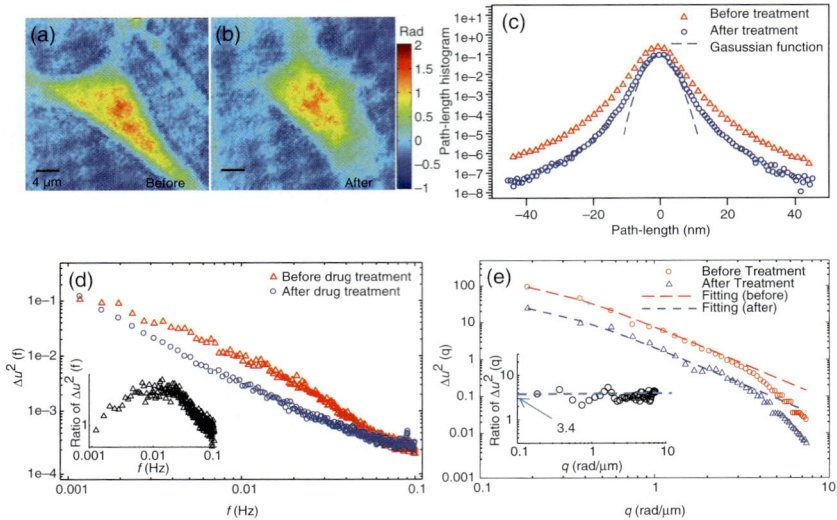

FIGURE 13.11 (*a* and *b*) Quantitative-phase images of glial cell before and after Cyto-D treatment. (*c*) Histogram of the path length displacements of a glial cell before and after drug treatment, as indicated. The blue dash line indicates the fit with a Gaussian function. (*d*) Spatially-averaged power spectrum of glial cells before and after drug treatment, as indicated. The inset shows the ratio between these two power spectrums. (*e*) Temporally averaged power spectrum before and after drug treatment. The dash lines are the fittings with (7), and the inset shows the ratio of the two spectrums. [IEEE, used with permission, *IEEE Journal of Selected Topics in Quantum Electronics*, H. Ding, E. Berl, Z. Wang, L.J. Millet, M.U. Gillette, J. Liu, M. Boppart, and G. Popescu, "Fourier transform light scattering of biological structures and dynamics," (2010). (*From Ref. 7.*)]

Figure 13.11*c* shows a comparison between the membrane displacement histograms of a cell before and after the actin inhibitor. It is evident from this result that the polymerization phenomenon is a significant contributor to the overall cell membrane dynamics, as indicated by the broader histogram distribution. Further, both curves exhibit non-Gaussian shapes at displacements larger than 10 nm, which suggests that the cell motions both before and after actin inhibition, are characterized by non-equilibrium dynamics. Figure 13.11*d* presents the comparison of the spatially-averaged power spectra associated with the FTLS signal for a single cell, before and after treatment with the actin-blocking drug. The broader power spectrum of the untreated cell membrane motions are consistent with the histogram distribution in Fig. 13.11*c* and also the particle tracking results. Further, both frequency-averaged (statics) curves shown in Fig. 13.11*e* indicates similar functional dependence on the wave vector, $\mathbf{q}$, but with enhanced fluctuations for the normal cell, by a factor of ~3.4.

In summary, we reviewed FTLS as a new approach for studying static and dynamic light scattering with unprecedented sensitivity, granted by the image-plane measurement of the optical phase and amplitude. In FTLS, spatial resolution of the scatterer positions is well preserved. These features of FTLS are due to the interferometric experimental geometry and the reliable phase retrieval. FTLS has been applied to study the tissue optical properties, cell type characterization, and dynamic structure of cell membrane. We anticipate that this type of measurement will enable new advances in life sciences, due to the ability to detect weak scattering signals over broad temporal (milliseconds to days) and spatial (fraction of microns to centimeters) scales. Very recent efforts in FTLS applied to tissue optics will be presented in Chap. 15.

References

1. H. F. Ding, Z. Wang, F. Nguyen, S. A. Boppart and G. Popescu. "Fourier transform light scattering of inhomogeneous and dynamic structures." *Physical Review Letters*, 101, (2008).
2. Z. Wang, H. Ding and G. Popescu. "Scattering-phase theorem." *Optics Letters* (under review).
3. H. Ding, X. Liang, Z. Wang, S. A. Boppart, and G. Popescu. "Tissue scattering parameters from organelle to organ scale." *Optics Letters* (under review).
4. H. Ding, Z. Wang, F. Nguyen, S. A. Boppart, L. J. Millet, M. U. Gillette, J. Liu, M. Boppart, and G. Popescu. "Fourier transform light scattering (FTLS) of cells and tissues." *Journal of Computational and Theoretical Nanoscience*, 7, 1546 (in press).
5. H. Ding, E. Berl, Z. Wang, L. J. Millet, M. U. Gillette, J. Liu, M. Boppart, and G. Popescu. "Fourier transform light scattering of biological structures and dynamics." *IEEE Journal of Selected Topics in Quantum Electronics*, 16, 909–918 (2010).
6. Y. Park, M. Diez-Silva, D. Fu, G. Popescu, W. Choi, I. Barman, S. Suresh, and M. S. Feld. "Static and dynamic light scattering of healthy and malaria-parasite invaded red blood cells." *Journal of Biomedical Optics*, 15, 020506 (2010).
7. H. Ding, L. J. Millet, M. U. Gillette, and G. Popescu. "Actin-driven cell dynamics probed by Fourier transform light scattering." *Biomedical Optics Express*, 1, 260 (2010).
8. H. Ding, F. Nguyen, S. A. Boppart, and G. Popescu. "Optical properties of tissues quantified by Fourier transform light scattering." *Optics Letters*, 34, 1372 (2009).
9. H. C. van de Hulst. *Light Scattering by Small Particles* (Dover Publications, New York, 1981).
10. B. J. Berne and R. Pecora. *Dynamic Light Scattering with Applications to Chemistry, Biology and Physics* (Wiley, New York, 1976).
11. D. J. Pine, D. A. Weitz, P. M. Chaikin, and E. Herbolzheimer. "Diffusing-wave spectroscopy." *Physical Review Letters*, 60, 1134–1137 (1988).
12. T. G. Mason and D. A. Weitz. "Optical measurements of frequency-dependent linear viscoelastic moduli of complex fluids." *Physical Review Letters*, 74, 1250–1253 (1995).
13. D. Mizuno, C. Tardin, C. F. Schmidt, and F. C. MacKintosh. "Nonequilibrium mechanics of active cytoskeletal networks." *Science*, 315, 370–373 (2007).
14. T. G. Mason, K. Ganesan, J. H. vanZanten, D. Wirtz, and S. C. Kuo. "Particle tracking microrheology of complex fluids." *Physical Review Letters*, 79, 3282–3285 (1997).
15. J. C. Crocker, M. T. Valentine, E. R. Weeks, T. Gisler, P. D. Kaplan, A. G. Yodh, and D. A. Weitz. "Two-point microrheology of inhomogeneous soft materials." *Physical Review Letters*, 85, 888–891 (2000).

16. A. P. Y. Wong and P. Wiltzius. "Dynamic light-scattering with a CCD camera." *Review of Scientific Instruments*, 64, 2547–2549 (1993).
17. F. Scheffold and R. Cerbino. "New trends in light scattering." *Current Opinion in Colloid & Interface Science*, 12, 50–57 (2007).
18. R. Dzakpasu and D. Axelrod. "Dynamic light scattering microscopy. A novel optical technique to image submicroscopic motions. II: Experimental applications." *Biophysical Journal*, 87, 1288–1297 (2004).
19. M. S. Amin, Y. K. Park, N. Lue, R. R. Dasari, K. Badizadegan, M. S. Feld, and G. Popescu. "Microrheology of red blood cell membranes using dynamic scattering microscopy." *Optics Express*, 15, 17001 (2007).
20. J. Neukammer, C. Gohlke, A. Hope, T. Wessel, and H. Rinneberg. "Angular distribution of light scattered by single biological cells and oriented particle agglomerates." *Applied Optics*, 42, 6388–6397 (2003).
21. M. Giglio, M. Carpineti, and A. Vailati. "Space intensity correlations in the near field of the scattered light: A direct measurement of the density correlation function g(r)." *Physical Review Letters*, 85, 1416–1419 (2000).
22. D. Brogioli, A. Vailati, and M. Giglio. "A schlieren method for ultra-low-angle light scattering measurements." *Europhysics Letters*, 63, 220–225 (2003).
23. F. Croccolo, D. Brogioli, A. Vailati, M. Giglio, and D. S. Cannell. "Use of dynamic schlieren interferometry to study fluctuations during free diffusion." *Applied Optics*, 45, 2166–2173 (2006).
24. R. Drezek, A. Dunn, and R. Richards-Kortum. "Light scattering from cells: finite-difference time-domain simulations and goniometric measurements." Applied Optics, 38, 3651–3661 (1999).
25. J. R. Mourant, J. M. Canpolat, C. Brocker, O. Esponda-Ramos, T. M. Johnson, A. Matanock, K. Stetter, and J. P. Freyer. "Light scattering from cells: The contribution of the nucleus and the effects of proliferative status." *Journal of Biomedical Optics*, 5, 131–137 (2000).
26. C. S. Mulvey, A. L. Curtis, S. K. Singh, and I. J. Bigio. "Elastic scattering spectroscopy as a diagnostic tool for apoptosis in cell cultures." *IEEE Journal of Selected Topics in Quantum Electronics*, 13, 1663–1670 (2007).
27. H. Ding, J. Q. Lu, R. S. Brock, T. J. McConnell, J. F. Ojeda, K. M. Jacobs, and X. H. Hu. "Angle-resolved Mueller matrix study of light scattering by R-cells at three wavelengths of 442, 633, and 850 nm." *Journal of Biomedical Optics*, 12, 034032 (2007).
28. M. T. Valentine, A. K. Popp, D. A. Weitz, and P. D. Kaplan. "Microscope-based static light-scattering instrument." *Optics Letters*, 26, 890–892 (2001).
29. W. J. Cottrell, J. D. Wilson, and T. H. Foster. "Microscope enabling multimodality imaging, angle-resolved scattering, and scattering spectroscopy." *Optics Letters*, 32, 2348–2350 (2007).
30. A. Wax, C. H. Yang, V. Backman, K. Badizadegan, C. W. Boone, R. R. Dasari, and M. S. Feld. "Cellular organization and substructure measured using angle-resolved low-coherence interferometry." *Biophysical Journal*, 82, 2256–2264 (2002).
31. V. Backman, M. B. Wallace, L. T. Perelman, J. T. Arendt, R. Gurjar, M. G. Muller, Q. Zhang, G. Zonios, E. Kline, T. McGillican, S. Shapshay, T. Valdez, K. Badizadegan, J. M. Crawford, M. Fitzmaurice, S. Kabani, H. S. Levin, M. Seiler, R. R. Dasari, I. Itzkan, J. Van Dam, M. S. Feld. "Detection of preinvasive cancer cells." *Nature*, 406, 35–36 (2000).
32. D. F. Abbott, P. D. Kearney, and K. A. Nugent. "3-Dimensional imaging using an optical microscope." *Journal of Mod. Optics*, 37, 1887–1893 (1990).
33. J. A. Newmark, W. C. Warger, C. Chang, G. E. Herrera, D. H. Brooks, C. A. DiMarzio, and C. M. Warner. "Determination of the number of cells in preimplantation embryos by using noninvasive optical quadrature microscopy in conjunction with differential interference contrast microscopy." *Microscopy and Microanalysis*, 13, 118–127 (2007).
34. F. Charriere, N. Pavillon, T. Colomb, C. Depeursinge, T. J. Heger, E. A. D. Mitchell, P. Marquet, and B. Rappaz. "Living specimen tomography by digital holographic microscopy: morphometry of testate amoeba." *Optics Express*, 14, 7005–7013 (2006).

35. F. J. Blonigen, A. Nieva, C. A. DiMarzio, S. Manneville, L. Sui, G. Maguluri, T. W. Murray, and R. A. Roy. "Computations of the acoustically induced phase shifts of optical paths in acoustophotonic imaging with photorefractive-based detection." *Applied Optics*, 44, 3735–3746 (2005).

36. F. Gittes, B. Schnurr, P. D. Olmsted, F. C. MacKintosh, and C. F. Schmidt. "Microscopic viscoelasticity: Shear moduli of soft materials determined from thermal fluctuations." *Physical Review Letters*, 79, 3286–3289 (1997).

37. V. Pelletier, N. Gal, P. Fournier, and M. L. Kilfoil. "Microrheology of microtubule solutions and actin-microtubule composite networks." *Physical Review Letters*, 102, 188303, (2009).

38. L. Ji, J. Lim and G. Danuser "Fluctuations of intracellular forces during cell protrusion." *Nature Cell Biology*, 10, 1393–U38 (2008).

39. M. L. Gardel, M. T. Valentine, J. C. Crocker, A. R. Bausch, and D. A. Weitz. "Microrheology of entangled F-actin solutions." *Physical Review Letters*, 91 (2003).

40. A. J. Levine and T. C. Lubensky. "One- and two-particle microrheology." *Physical Review Letters*, 85, 1774–1777 (2000).

41. A. W. C. Lau, B. D. Hoffman, A. Davies, J. C. Crocker, and T. C. Lubensky. "Microrheology, stress fluctuations, and active behavior of living cells." *Physical Review Letters*, 91 (2003).

42. C. P. Brangwynne, G. H. Koenderink, E. Barry, Z. Dogic, F. C. MacKintosh, and D. A. Weitz. "Bending dynamics of fluctuating biopolymers probed by automated high-resolution filament tracking." *Biophysical Journal*, 93, 346–359 (2007).

43. M. Schlosshauer. "Decoherence, the measurement problem, and interpretations of quantum mechanics." *Reviews of Modern Physics*, 76, 1267–1305 (2004).

44. A. Caspi, R. Granek, and M. Elbaum. "Diffusion and directed motion in cellular transport." *Physical Review E*, 66 (2002).

45. N. Lue, J. Bewersdorf, M. D Lessard, K. Badizadegan, K. Dasari, M. S. Feld, and G. Popescu. "Tissue refractometry using Hilbert phase microscopy." *Optics Letters*, 32, 3522 (2007).

46. B. Rappaz, P. Marquet, E. Cuche, Y. Emery, C. Depeursinge, and P. J. Magistretti. "Measurement of the integral refractive index and dynamic cell morphometry of living cells with digital holographic microscopy." *Optics Express*, 13, 9361–9373 (2005).

47. W. Choi, C. C. Yu, C. Fang-Yen, K. Badizadegan, R. R. Dasari, and M. S. Feld. "Field-based angle-resolved light-scatteiring study of single live COS." *Optics Letters*, 33, 1596–1598 (2008).

48. G. Depetris, A. Dimarzio, and F. Grandinetti. "H2no2+ ions in the gas-phase—a mass-spectrometric and post-Scf Abinitio study." *Journal of Physical Chemistry*, 95, 9782–9787 (1991).

49. G. Popescu, T. Ikeda, R. R. Dasari, and M. S. Feld. "Diffraction phase microscopy for quantifying cell structure and dynamics." *Optics Letters*, 31, 775–777 (2006).

50. W. Choi, C. C. Yu, C. Fang-Yen, K. Badizadegan, R. R. Dasari, and M. S. Feld. "Field-based angle-resolved light-scattering study of single live cells." *Optics Letters*, 33, 1596–1598 (2008).

51. G. Popescu, T. Ikeda, K. Goda, C. A. Best-Popescu, M. Laposata, S. Manley, R. R. Dasari, K. Badizadegan, and M. S. Feld. "Optical measurement of cell membrane tension." *Physical Review Letters*, 97, 218101 (2006).

52. T. Ikeda, G. Popescu, R. R. Dasari, and M. S. Feld. "Hilbert phase microscopy for investigating fast dynamics in transparent systems." *Optics Letters*, 30, 1165–1168 (2005).

53. E. Abbe. "Beiträge zur theorie des mikroskops und der mikroskopischen Wahrnehmung." *Arch. Mikrosk. Anat.*, 9, 431 (1873).

54. V. V. Tuchin and Society of Photo-optical Instrumentation Engineers. *Tissue optics:light scattering methods and instruments for medical diagnosis* (SPIE/International Society for Optical Engineering, Bellingham, Wash., 2007).

55. Y. K. Park, M. Diez-Silva, G. Popescu, G. Lykotrafitis, W. Choi, M. S. Feld, and S. Suresh. "Refractive index maps and membrane dynamics of human red blood cells parasitized by *Plasmodium falciparum*." *Proceedings of the National Academy of Sciences*, 105, 13730 (2008).

56. A. Karlsson, J. P. He, J. Swartling, and S. Andersson-Engels. "Numerical simulations of light scattering by red blood cells." *IEEE Transactions on Biomedical Engineering*, 52, 13–18 (2005).

57. M. Hunter, V. Backman, G. Popescu, M. Kalashnikov, C. W. Boone, A. Wax, G. Venkatesh, K. Badizadegan, G. D. Stoner, and M. S. Feld. "Tissue self-affinity and light scattering in the Born approximation: A new model for precancer diagnosis." *Physical Review Letters*, 97, 138102 (2006).

58. H. F. Ding, F. Nguyen, S. A. Boppart, and G. Popescu. "Optical properties of tissues quantified by Fourier-transform light scattering." *Optics Letters*, 34, 1372–1374 (2009).

59. G. Popescu, in Methods in Cell Biology, 87-115, Jena, B. P., ed. (Academic Press, San Diego, CA, 2008).

60. K. Fujita, M. Kobayashi, S. Kawano, M. Yamanaka, and S. Kawata. "High-resolution confocal microscopy by saturated excitation of fluorescence." *Physical Review Letters*, 99, 228105 (2007).

61. J. M. Schmitt and G. Kumar. "Optical scattering properties of soft tissue: A discrete particle model." *Applied Optics*, 37, 2788–2797 (1998).

62. D. O. Hogenboom, C. A. DiMarzio, T. J. Gaudette, A. J. Devaney, and S. C. Lindberg. "Three-dimensional images generated by quadrature interferometry." *Optics Letters*, 23, 783–785 (1998).

63. J. R. Mourant, J. P. Freyer, A. H. Hielscher, A. A. Eick, D. Shen, and T. M. Johnson. "Mechanisms of light scattering from biological cells relevant to noninvasive optical-tissue diagnostics." *Applied Optics*, 37, 3586–3593 (1998).

64. G. Popescu and A. Dogariu. "Scattering of low coherence radiation and applications." invited review paper, *E. Phys. J.*, 42, 73–93 (2005).

65. S. A. Alexandrov, T. R. Hillman, and D. D. Sampson. "Spatially resolved Fourier holographic light scattering angular spectroscopy." *Optics Letters*, 30, 3305–3307 (2005).

66. I. T. Jolliffe. *Principal Component Analysis, 2nd ed.* (Springer; New York, 2002).

67. T. W. Loong. "Understanding sensitivity and specificity with the right side of the brain." *British Medical Journal*, 327, 716–719 (2003).

68. M. L. Gardel, J. H. Shin, F. C. MacKintosh, L. Mahadevan, P. Matsudaira, and D. A. Weitz. "Elastic behavior of cross-linked and bundled actin networks." *Science*, 304, 1301–1305 (2004).

69. T. D. Pollard and G. G. Borisy. "Cellular motility driven by assembly and disassembly of actin filaments." *Cell*, 112, 453–465 (2003).

70. T. J. Mitchison and L. P. Cramer. "Actin-based cell motility and cell locomotion." *Cell*, 84, 371–379 (1996).

71. T. D. Pollard. "The cytoskeleton, cellular motility and the reductionist agenda." *Nature*, 422, 741–745 (2003).

72. J. A. Theriot and T. J. Mitchison. "Actin microfilament dynamics in locomoting cells." *Nature*, 352, 126–131 (1991).

73. F. Brochard and J. F. Lennon. "Frequency spectrum of the flicker phenomenon in erythrocytes." *J. Physique*, 36, 1035–1047 (1975).

74. N. Gov, A. G. Zilman, and S. Safran. "Cytoskeleton confinement and tension of red blood cell membranes." *Physical Review Letters*, 90, 228101 (2003).

75. A. Zilker, M. Ziegler, and E. Sackmann. "Spectral-analysis of erythrocyte flickering in the 0.3-4-mu-m-1 regime by microinterferometry combined with fast image-processing." *Physical Review A*, 46, 7998–8002 (1992).

76. J. A. Cooper and D. A. Schafer. "Control of actin assembly and disassembly at filament ends." *Current Opinion in Cell Biology*, 12, 97–103 (2000).

77. J. R. Kuhn and T. D. Pollard. "Real-time measurements of actin filament polymerization by total internal reflection fluorescence microscopy." *Biophysical Journal*, 88, 1387–1402 (2005).

78. K. J. Amann and T. D. Pollard. "Direct real-time observation of actin filament branching mediated by Arp2/3 complex using total internal reflection fluorescence microscopy." *Proceedings of the National Academy of Sciences of the United States of America*, 98, 15009–15013 (2001).

79. J. A. Theriot, T. J. Mitchison, L. G. Tilney, and D. A. Portnoy. "The rate of actin-based motility of intracellular listeria-monocytogenes equals the rate of actin polymerization." *Nature*, 357, 257–260 (1992).
80. D. Uttenweiler, C. Veigel, R. Steubing, C. Gotz, S. Mann, H. Haussecker, B. Jahne, and R. H. A. Fink "Motion determination in actin filament fluorescence images with a spatio-temporal orientation analysis method." *Biophysical Journal*, 78, 2709–2715 (2000).
81. C. C. Wang, J. Y. Lin, H. C. Chen, and C. H. Lee. "Dynamics of cell membranes and the underlying cytoskeletons observed by noninterferometric widefield optical profilometry and fluorescence microscopy." *Optics Letters*, 31, 2873–2875 (2006).
82. L. J. M. Huafeng Ding, Martha U. Gillette, and Gabriel Popescu. "Spatio-temporal cytoskeleton fluctuations probed by Fourier transform light scattering." *Physical Review Letters* (submitted).
83. W. C. Warger, G. S. Laevsky, D. J. Townsend, M. Rajadhyaksha, and C. A. DiMarzio. "Multimodal optical microscope for detecting viability of mouse embryos *in vitro*." *Journal of Biomedical Optics*, 12, 044006 (2007).
84. N. Watanabe and T. J. Mitchison. "Single-molecule speckle analysis of actin filament turnover in lamellipodia." *Science*, 295, 1083–1086 (2002).

CHAPTER 14

Current Trends in Methods

C urrently, QPI is rapidly evolving, with efforts devoted to both method development and applications. Thus, in the last two chapters of the book, we will review very recent progress and current trends in both these areas. As already discussed, several efficient methods for measuring quantitative-phase images, i.e., $\phi(x, y)$, are already in place. Naturally, current research is mainly focused on expanding the QPI data to other dimensions, as follows. The real part of refractive index is most generally a function of three spatial coordinates and one temporal coordinate $n(x, y, z; t)$ or, equivalently, of the four respective frequencies, $n(k_x, k_y, k_z; \omega)$. However, as discussed so far, a phase image only gives access to the integrated refractive index along z dimension. Therefore, current trends in QPI methods are aimed at measuring the third spatial dimension, z, i.e., refractive index *tomography,* and wavelength, i.e., refractive index *dispersion.* While the first type of measurement offers full, 3D view of transparent structures (e.g., live cells), the wavelength dependence may provide a mean for achieving *chemical specificity*.

14.1 Tomography via QPI

Imaging cells in 3D has been largely limited to fluorescence confocal microscopy, in which often the specimen is fixed.[1] Deconvolution fluorescence microscopy is another method for 3D reconstruction, and an alternative to confocal microscopy.[2] For confocal microscopy, much of the out-of-focus light is rejected by a pinhole in front of the detector, whereas in deconvolution, this light is numerically propagated back to where it originated from.

Recent advances in QPI have enabled optical tomography of transparent structures, without fluorescent labels. This is a significant achievement because it allows imaging live cells in 3D noninvasively and indefinitely. The two main approaches used for 3D imaging via QPI are *computed tomography (CT)* and *diffraction tomography (DT).* CT uses the Radon-transform-based reconstruction algorithms borrowed from X-ray-computed imaging, as developed by von Laue.[3] In CT, the ray-like propagation or geometrical optics approximation is used. By contrast, DT takes into account the scattering properties of the object via a certain model, e.g., the first Born approximation (see Sec. 2.5), as first described by Wolf in 1969.[4] The relationship between CT and DT was investigated theoretically by Gbur and Wolf[5], and, under certain conditions, CT was proven to be the short-wavelength approximation of DT. Below, we present these two approaches separately.

14.1.1   Computed Tomography (CT) Using Digital Holographic Microscopy

Advances in phase-sensitive measurements enabled optical tomography of transparent structures, following reconstruction algorithms borrowed from X-ray CT, in which scattering and diffraction effects are assumed to be negligible.[5-9] QPI-based projection tomography has been applied to live cells.[10-12] Thus, in 2006, the Depeursinge group from Ecole Polytechnique Federale de Lausanne reported, "Cell refractive index tomography by digital holographic microscopy."[11] Digital holographic microscopy (DHM), described in Sec. 9.1, was combined with a sample holder (micropipette) that allowed the specimen to be rotated around an axis perpendicular to the illumination direction (Fig. 14.1). A diode laser of 635-nm wavelength was used as illumination for the typical DHM setup, which uses a 63×, 0.85NA objective and 512- × 512-pixel CCD, which can record up to 25 frames/s. The off-axis geometry allows for single-shot phase images, with a field of view of 80×80 μm^2 and a transverse resolution of approximately 1 μm.[11]

A pollen grain was used as proof of principle specimen. The grain was introduced in a glass micropipette of internal diameter 100 μm and immersed in glycerol. To minimize strong light refraction by the micropipette, which acts as a cylindrical lens with regard to the illuminating light, the volume between a glass coverslip and the MO is filled with an index matching fluid that suppresses the refraction at the air–glass interface. Like in a standard CT reconstruction, 2D phase-image distributions were recorded for different sample orientations covering an angle range of 180°. To reconstruct one tomogram, 90 images were acquired with a 2° step at a rate of 1 Hz. Each quantitative-phase image can be written as

$$\phi(x,y) = \frac{2\pi}{\lambda} \int \Delta n(x,y,z) dz \tag{14.1}$$

The desired 3D refractive index contrast, $\Delta n(x,y,z)$, was reconstructed from the measured *sinograms* via an inverse Radon transform. The *Fourier slice theorem* (Chap. 3 in Ref. 3) relates the Fourier transform of a projection to the Fourier transform of the object in a

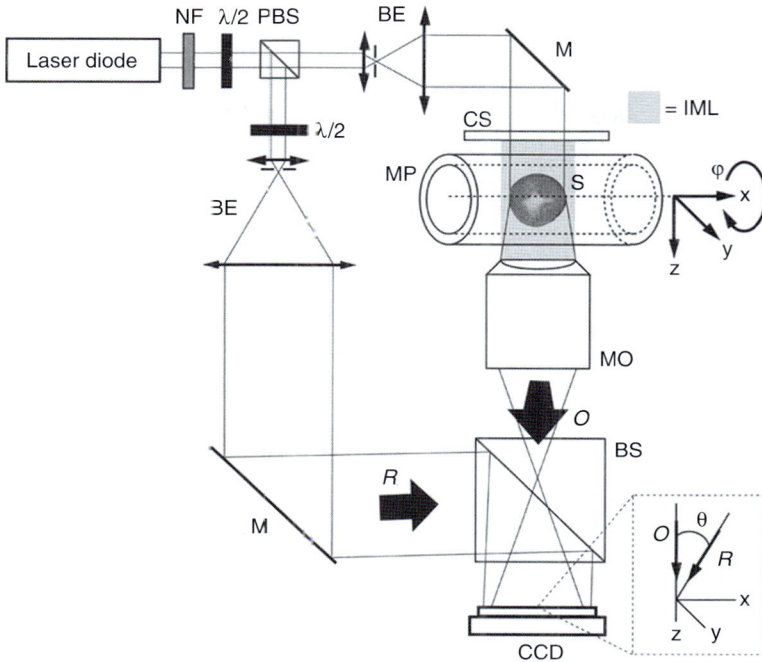

FIGURE 14.1   Holographic microscope for transmission imaging: NF, neutral-density filter; PBS, polarizing beam splitter; BE, beam expander with spatial filter; /2, half-wave plate; MO, microscope objective; M, mirror; BS, beam splitter; O, object wave; R, reference wave; MP, micropipette; S, specimen; IML, index matching liquid. Inset: detail showing the off-axis geometry at the incidence on the CCD. [The Optical Society, used with permission, *Optics Letters*, F. Charriere A. Marian, F. Montfort, J. Kuehn, T. Colomb, E. Cuche, P. Marquet, & C. Depeursinge. "Cell refractive index tomography by digital holographic microscopy," 31, 178–180 (2006). (*From Fig. 1, Ref. 11.*)]

cross-section plane. However, practical implementations typically use the filtered *backprojection algorithm,* in which a weighting function is used to "filter" each Fourier slice.[3]

The proof of principle reconstruction of the pollen grain using DHM is shown in Fig. 14.2, in which different cuts through the object are illustrated. With the known refractive index of the surrounding glycerol of 1.473, a refractive index contrast value of $\Delta n = 0.06 \pm 0.01$ was measured in the nucleus. The wall thickness is comparable with the resolution of the instrument and, thus, its refractive index could not be measured. Because the overall optical thickness of the pollen grain is in some areas larger than 2π, phase unwrapping becomes an issue and, more importantly, the ray optics approximation becomes questionable. Nevertheless, the reconstruction in Fig. 14.2 demonstrates optical sectioning in biological structures using QPI and the CT principle. Later, this approach was used to render tomography of live amoebas.[10]

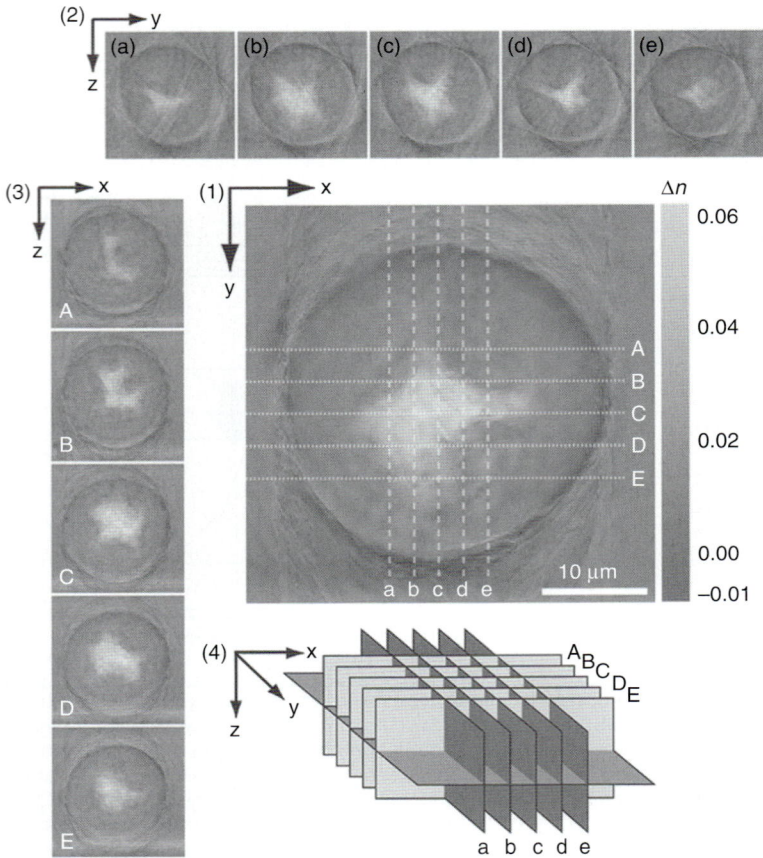

FIGURE 14.2 Tomography of a pollen cell refractive index: (1) cut along the *xy* plane in the middle of the pollen cell; (2) cuts at different positions in the cell along the *yz* plane, and (3) the *xz* plane; and (4) schematic of the presented cuts. [The Optical Society, used with permission, *Optics Letters*, F. Charriere, A. Marian, F. Montfort, J. Kuehn, T. Colomb, E. Cuche, P. Marquet, & C. Depeursinge, "Cell refractive index tomography by digital holographic microscopy," 31, 178–180 (2006). (*From Fig. 1, Ref. 11.*)]

In 2007, the Feld group at Massachusetts Institute of Technology applied the same measurement principle, achieving the variable-angle illumination by scanning the beam itself rather than rotating the specimen.[12] In this case, the angular range of illumination is limited by the numerical aperture of the objective to (−60°, 60°). Still, the reconstruction via the filtered backprojection algorithm demonstrates 3D imaging in live cells (Fig. 14.3).

Clearly, the geometrical optics approximation used in CT fails for high numerical aperture imaging, where diffraction and scattering effects are essential. This aspect was discussed by Gbur and Wolf, who concluded that CT is the zero-wavelength approximation to DT,

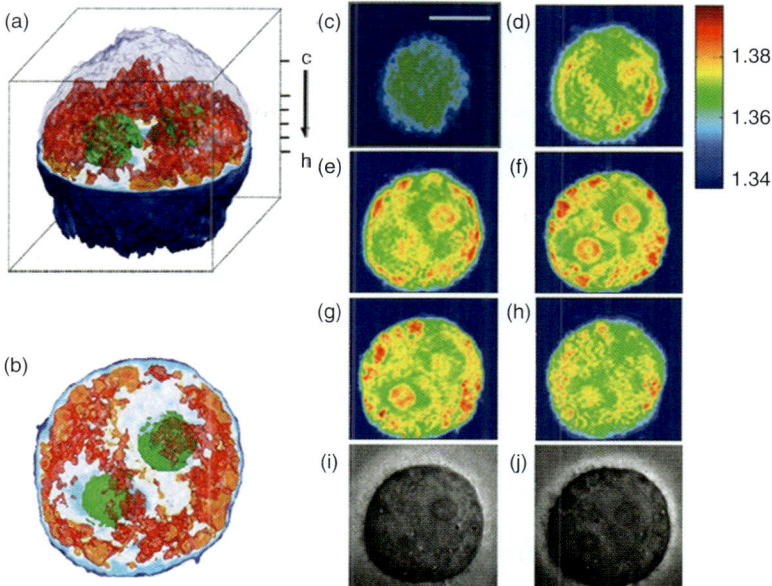

FIGURE **14.3** Refractive index tomogram of a HeLa cell. (*a*) A 3D rendered image of a HeLa cell. The outermost layer of the upper hemisphere of the cell is omitted to visualize the inner structure. Nucleoli are colored green and parts of cytoplasm with refractive index higher than 1.36 are colored red. The side of the cube is 20 mm. (*b*) Top view of *a*. (*c–h*) Slices of the tomogram at heights indicated in *a*. Scale bar, 10 mm. Colorbar, refractive index at λ = 633 nm. (*i*, *j*) Bright field images for objective focus corresponding to *e* and *f*, respectively. [*Nature Methods*, used with permission, W. Choi, C. Fang-Yen, K. Badizadegan, S. Oh, N. Lue, R.R. Dasari, & M.S. Feld, "Tomographic phase microscopy," 4, 717–719 (2007). (*From Fig. 2, Ref. 12.*)]

provided that the propagation angles are small.[5] In other words, when applied to high numerical aperture imaging, CT reconstructions suffer from limited depth of field. More recently Choi et al. have discussed extending the depth of field in CT measurements via numerical field propagation[13].

14.1.2 Diffraction Tomography (DT) via Spatial Light Interference Microscopy

Spatial light interference microscopy (SLIM), a white light QPI technique, is described in more detail in Sec. 12.2 (see Ref. 14). Here we show that by combining white light illumination, high numerical aperture, and phase-resolved detection, SLIM provides optical sectioning. This ability is essentially due to the micron-range coherence window, which overlaps axially with the plane of focus. To obtain a tomographic image of the specimen, we performed axial scanning by translating the sample through the focal plane in step sizes of less than half the Rayleigh range and with an accuracy of 20 nm

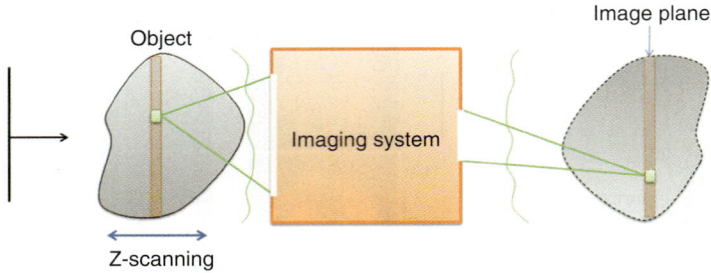

FIGURE 14.4   Experimental arrangement for diffraction tomography using SLIM.

(Fig. 14.4). For a transparent object, such as a live cell, the 3D distribution of the real field measured, U, is the result of the convolution between the scattering potential of the specimen and the Green's function (point-spread function), P, of the microscope,

$$U(\mathbf{r}) = \iiint_{V\,V\,V} F(\mathbf{r}')P(\mathbf{r}-\mathbf{r}')d^3\mathbf{r}' \qquad (14.2)$$

where $F(\mathbf{r}) \propto n^2(\mathbf{r})-1$ is the scattering potential and $\mathbf{r} = (x,\, y,\, z)$. We retrieved P experimentally by measuring a set of axially-resolved quantitative-phase images of a point scattered. In order to retrieve F, we performed the inverse operation in the spatial frequency domain as

$$\tilde{F}(\mathbf{q}) = \tilde{U}(\mathbf{q})/\tilde{P}(\mathbf{q}) \qquad (14.3)$$

where ~ indicates Fourier transformed functions. We implemented this operation numerically by employing the Wiener regularization method, and generated a refractive index map of a single neuron at two different depths, separated by 5.6 μm (Fig. 14.5a-b). The *nucleolus* (arrow, Fig. 14.5b) has the highest refractive index, n ~ 1.46. Figure 14.5c shows a 3D rendering of the same hippocampal neuron generated from 71 images separated by 14 μm. For comparison, we used fluorescence confocal microscopy to obtain a similar view of a different hippocampal neuron cultured under identical conditions. This neuron was stained with anti-polysialic acid IgG #735. The numerical aperture of the confocal microscope objective was NA = 1.2, higher than the NA = 0.75 objective used in SLIM, which explains the higher resolution of the confocal image. Nevertheless the 3D *label-free* imaging by SLIM is qualitatively similar to that obtained by *fluorescence* confocal microscopy.

Thus, in contrast to confocal microscopy, QPI-based tomography enables non-invasive imaging of living cells over long periods of time, with substantially lower illumination power density. For example, the high refractive index associated with segregating chromosomes

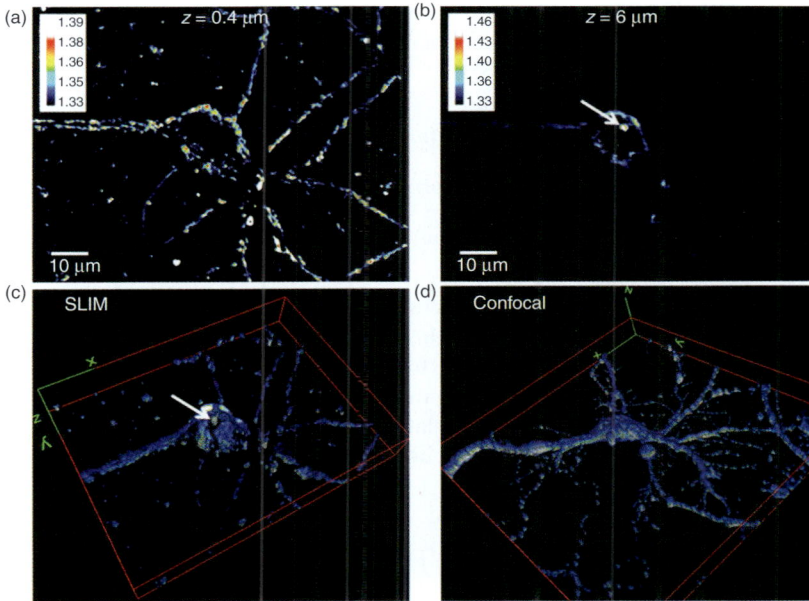

FIGURE 14.5 Tomography capability. (*a* and *b*) Refractive index distribution through a live neuron at position $z = 0.4$ μm (*a*) and 6.0 μm (*b*). The soma and nucleolus (arrow) are clearly visible; scale bars, 10 μm. (*c*) 3D rendering of the same cell. The field of view is 100 μm × 75 μm × 14 μm and NA = 0.75. (*d*) confocal microscopy of a stained neuron with same field of view and NA = 1.2. Neurons were labeled with anti-polysialic acid IgG #735. The 3D rendering in (*c*) and (*d*) was done by Imagej 3D viewer. (*From Ref. 14.*)

allows their imaging during cell mitosis.[14] This type of 4D (x, y, z, time) imaging may yield new insights into cell division, motility, differentiation, and growth.

14.2   Spectroscopic QPI

Phase measurements at multiple wavelengths have been used in the past to overcome issues related to phase wrapping[15-20] and phase noise.[21] Further, multicolor digital holography has been implemented for imaging large fields of view.[22-24] However, more recently multi-wavelength QPI has been developed for a different purpose: to obtain *spectroscopic* information about the biological structure.

Spectroscopic QPI, which deals with the wavelength dependence of the real part (n') of the refractive index, is fully equivalent to the *absorption spectroscopy*, which exploits the imaginary part (n'') of the refractive index (absorption) vs. wavelength. The dependence between the real and imaginary parts of the material response function is referred to as the *dispersion relation* and is captured by the so called

Kramers-Kronig relationship (see, for example, Refs. 25-27), as follows. First, we start with the induced polarization, **P**, associated with the medium (see Chap. 2 in Ref. 28),

$$\mathbf{P}(\omega) = \varepsilon_0 \chi(\omega)\mathbf{E}$$

$$\chi(\omega) = \varepsilon_r(\omega) - 1 \qquad (14.4)$$

$$= n^2(\omega) - 1$$

In Eq. (14.4), χ is the *dielectric susceptibility*, generally a *tensor* quantity, ε_0 is the vacuum dielectric permittivity, ε_r is the dielectric permittivity of the medium relative to that of vacuum, n is the refractive index. Assuming a refractive index that is close to unity, i.e., *weakly scattering* specimens, the relationship between the response function, χ, and refractive index, n, becomes linear,

$$\chi(\omega) \simeq 2[n(\omega) - 1]$$

$$= 2[n'(\omega) - 1] + i2n''(\omega) \qquad (14.5)$$

The Kramers-Kronig relationship establishes a transformation by which the real and imaginary parts of $\chi(\omega) = \chi'(\omega) + i\chi''(\omega)$ can be obtained from one another.

$$\chi'(\omega) = \frac{1}{\pi} P \int_{-\infty}^{\infty} \frac{\chi''(\omega')}{\omega' - \omega} d\omega'$$

$$n'(\omega) = 1 + \frac{1}{\pi} P \int_{-\infty}^{\infty} \frac{n''(\omega')}{\omega' - \omega} d\omega' \qquad (14.6)$$

The relationships in Eq. (14.6) are the result of the time-domain response function, $\chi(t)$ being zero for $t < 0$, which establishes that the system is *causal*. Note that Eq. (14.6a-b) represent Hilbert transforms, as encountered in the context of *complex analytic signals* (App. A). This is not surprising, as in both cases a vanishing function over the negative values of one domain had a Fourier transform with the real and imaginary parts connected via a Hilbert transform.

Equation (14.6b) states that the same information is retrieved by measuring either the real part (n') or imaginary part (n'') of the refractive index, as one can be obtained from the other through a Hilbert transform. Figure 14.6 illustrates the relationship between refractive index and absorption for oxygenated hemoglobin (HbO_2). An optical field passing through a specimen of thickness, L, characterized by such a refractive index can be written as

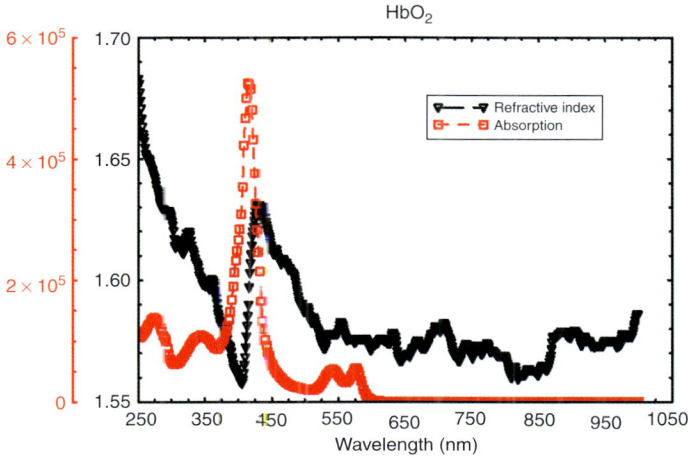

FIGURE 14.6 Wavelength dependence of absorption (a.u.) and refractive index for oxyhemoglobin.

$$U(\omega) = U_0(\omega)e^{-n''(\omega)k_0 L}e^{in'(\omega)k_0 L} \qquad (14.7)$$

where the first exponential indicates an attenuation and the second phase delay; $k_0 = \omega/c$. Traditional spectroscopy deals with measuring the absorption (sometimes emission) coefficient of materials, i.e., n'', because this quantity is readily accessible via amplitude measurements. However, such measurements are impractical in thin specimens such as cells, unless the absorption coefficient (or n'') for the species under investigation is very high.

14.2.1 Spectroscopic Diffraction-Phase Microscopy

In 2009, Park et al. reported a spectroscopic QPI method for measuring hemoglobin concentration in RBCs.[29] The experimental setup combines diffraction-phase microscopy (DPM) (see Sec. 11.2. and Refs. 30, 31) and multiwavelength illumination, as illustrated in Fig. 14.7. The multiwavelength illumination field is obtained using white light lamp and color filters. In Ref. 29, seven different color filters were used: 440 ± 20, 546 ± 10, 560 ± 20, 580 ± 25, 600 ± 20, 655 ± 20, and 700 ± 20 nm. These fields were also spatially filtered to achieve spatial coherence. Due to the common-path geometry associated with DPM, the system can be aligned to be essentially dispersion free, such that the respective images for each color share the same focal plane.

As described in Sec. 11.2., the instrument uses a diffraction grating for achieving a compact Mach-Zender interferometer and the

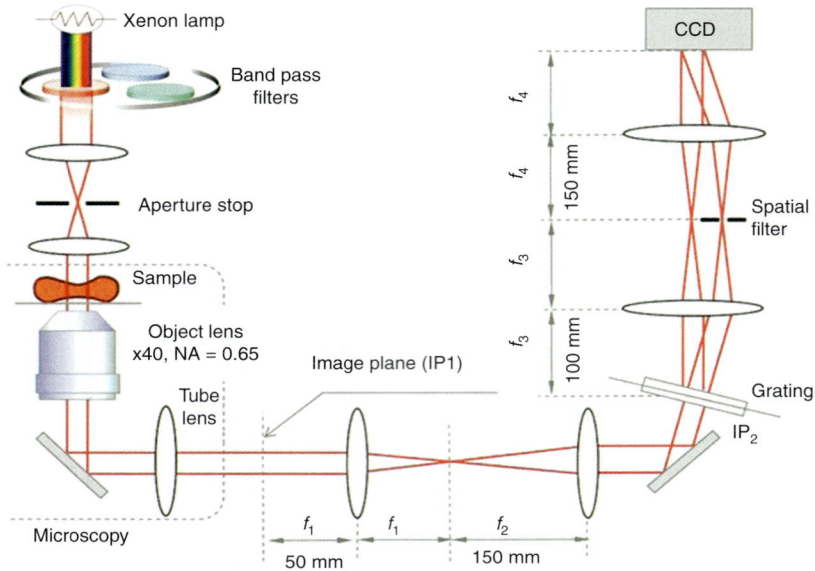

FIGURE 14.7   Spectroscopic DPM experimental setup. [The Optical Society, used with permission, *Optics Letters*, Y. Park, T. Yamauchi, W. Choi, R. Dasari, & M.S. Feld, "Spectroscopic phase microscopy for quantifying hemoglobin concentrations in intact red blood cells," 34, 3668–3670 (2009)—this article had footnotes in article that cite other, external information—P. Schiebener, J. Straub, J. Sengers, and J. Gallagher, *J. Phys. Chem. Ref. Data* 19, 677 (1990). (*From Fig. 1, Ref. 29.*)]

phase-reconstruction algorithm involves a spatial Hilbert transform, as follows. A specimen, located at the sample plane of an inverted microscope, is projected onto image plane IP1. The image is further magnified and delivered to IP_2, where a grating is placed to generate multiple diffraction orders, out of which only the 0th- and 1st-order beams are isolated. The 0th-order beam is spatially low-pass filtered by using a pinhole in a 4f lens system; the beam then approximates a plane wave reference field at the camera plane. The 1st-order beam served as sample beam. Both beams interfere and generate a spatially modulated interference image, which is captured by a CCD camera. Since both beams share almost the same beam path, common-mode phase noise is canceled out on interference. The electric field was extracted from the recorded interferogram by a spatial Hilbert transform (Sec. 11.2). The grating period, 30 μm, was set to be smaller than the diffraction-limited spot of the microscopic imaging system at the grating plane. All the lenses were achromatic to minimize chromatic dispersion.

The Hb refractive index is linearly proportional with the concentration,

$$n(\lambda, C) = \alpha(\lambda)C + n_w \qquad (14.8)$$

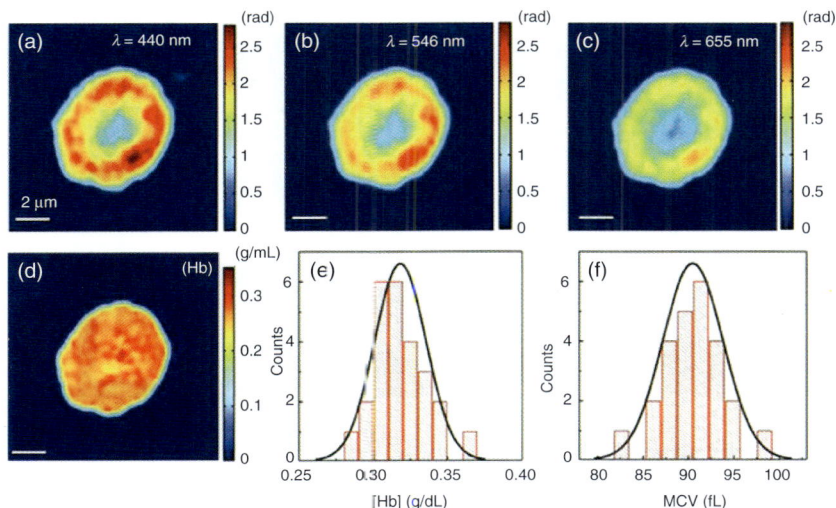

FIGURE 14.8 (*a-c*) Quantitative-phase maps of an RBC at three different wavelengths. (*d*) Retrieved Hb concentration. Histogram of (*e*) Hb concentration and (*f*) mean cell volumes (N = 25). The Optical Society, used with permission, *Optics Letters*, Y. Park, T. Yamauchi, W. choi, R. Dasari, & M.S. Feld, "Spectroscopic phase microscopy for quantifying hemoglobin concentrations in intact red blood cells," 34, 3668–3670 (2009)—this article had footnotes in article that cite other, external information—P. Schiebener J. Straub, J. Sengers, and J. Gallagher, *J. Phys. Chem. Ref. Data*, 19, 677 (1990). (*From Fig. 2, Ref. 29.*)]

where C is the concentration, α is the refractive index increment (a function of wavelength), and n_w the refractive index of water. First, a calibration measurement was performed to obtain the refractive index increment of hemoglobin at various wavelengths: $\alpha(440\text{ nm}) = 0.240 \pm 0.007$ mL/g, $\alpha(560\text{ nm}) = 0.227 \pm 0.004$ mL/g, and $\alpha(660\text{ nm}) = 0.221 \pm 0.005$ mL/g. It was also found that albumin has a very similar dispersion curve to water over the entire visible range. Further, the method was employed to measure Hb concentration and mean cell volume (MCV) for individual RBCs. In principle, measurements at three colors allow for extracting these two parameters independently. Figure 14.8 illustrates the results obtained for N = 25 RBCs.[29] The average Hb concentration and MCV were 0.318+/−0.17 g/mL and 90.5+/−3.3. fL, respectively.

14.2.2 Instantaneous Spatial Light Interference Microscopy (iSLIM)

In 2010, we reported a different approach for spectroscopic QPI.[32] The method, referred to as iSLIM to emphasize its single-shot capability with respect to SLIM (Sec 12.2.), combines the principle of DPM with a commercial (white light) phase-contrast microscope and a RGB

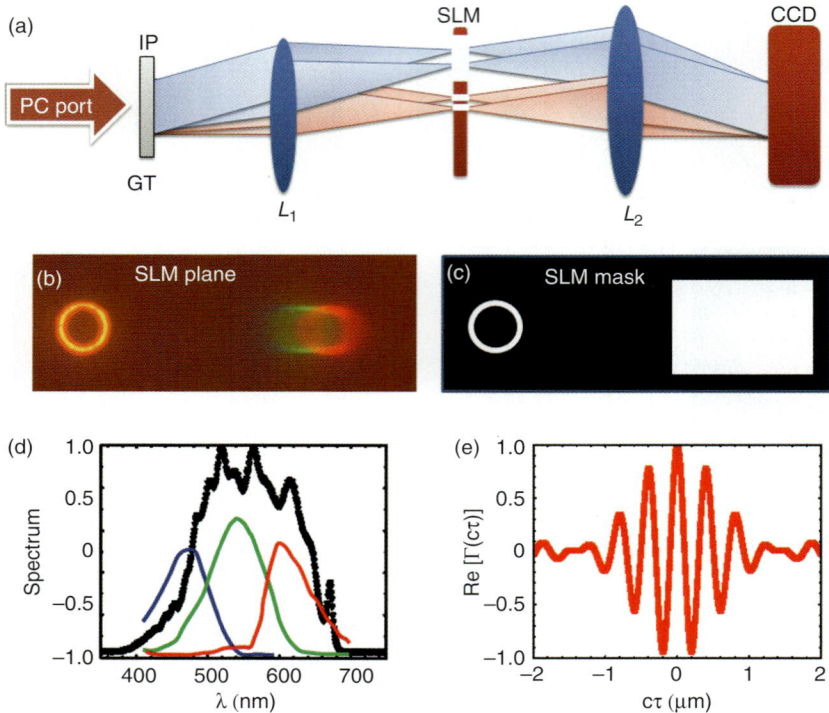

FIGURE **14.9**   (*a*) iSLIM experimental setup: IP, image plane outputted at the phase-contrast (PC) port; L_1-L_2 lenses; SLM, spatial light modulator; CCD, charged-coupled device. (*b*) Intensity distribution at the SLM plane imaged by color camera. (*c*) SLM transmission mask: white represents maximum transmission and black minimum. (*d*) Spectrum of halogen lamp (black symbols) and spectral sensitivity for each of the red, green, and blue channels of the RGB camera. (*e*) The temporal autocorrelation function of the illumination field; τ is the temporal delay, and *c* the speed of light in water. (*From Ref. 32.*)

camera as both filter and multiwavelength detector. The experimental setup is shown in Fig. 14.9. As shown in Fig. 14.9*a*, iSLIM employs spatially coherent white light illumination, commonly obtained from a halogen lamp and spatially filtered through the condenser annulus (Axio Observer Z1, Zeiss). With this modification, the common-path geometry specific to DPM provides not only high-phase sensitivity and diffraction-limited transverse resolution, but also high contrast to noise. Like in DPM (Sec. 11.2 and Ref. 30), at the image plane (IP) of the inverted microscope, we place an amplitude diffraction grating, which generates distinct diffraction orders. We separate these orders in the Fourier plane generated by lens 1 (L_1), where only the 0th and 1st orders are allowed to pass. Figure 14.9*b* shows this Fourier plane as imaged by a color camera, where the 1st diffraction order shows the color spread due to the grating dispersion. Note that the lens system L1-L2 forms a highly stable Mach-Zehnder interferometer. In order to generate the reference beam for this interferometer, the 0th order is spatially low-passed filtered by the spatial light modulator (SLM). The SLM filter is designed to match identically the actual image of the condenser annulus, such that only the DC component of the 0th order is passed, as shown in Fig. 14.9c. Finally, at the CCD plane, we obtain an interferogram that has high-contrast due to the intrinsic coherence matching and comparable power levels of the two beams, and is extremely stable due to the common-path geometry. The spectrum of the illumination field is shown in Fig. 14.9d. Using the Wiener-Kintchin theorem, which establishes the Fourier relationship between the power spectrum and autocorrelation function, we obtained the temporal correlation function, Γ, the real part of which is shown in Fig. 14.9e.

We demonstrate an effortless implementation of quantitative phase imaging at three different colors by simply recording the interferogram with an RGB camera (Zeiss Axiocam MRc). The spectral response for each of the channels is presented in Fig. 14.9d. The central wavelengths for each of the red, green, and blue channels are $\lambda_R = 620$ nm, $\lambda_G = 530$ nm, and $\lambda_B = 460$ nm. Thus, from a single RGB frame, we simultaneously reconstruct quantitative-phase images at all three colors. We used this approach to image both beads (Fig. 14.10a-d) and RBCs (Fig. 14.10e-h). The wavelength-dependent phase in this case can be written as $\phi(\lambda) = \frac{2\pi}{\lambda}\Delta n(\lambda)h$ where Δn is the refractive index contrast with respect to the surrounding fluid. We note that for beads (Fig. 14.10d), $\phi(\lambda) \propto 1/\lambda$, which indicates that the dispersion of polystyrene and water is small in the visible spectrum, as expected. The similar curves associated with RBCs show qualitative differences, as exemplified by data from three points across the cell (Fig. 14.10h). This result suggests that the hemoglobin and surrounding plasma may exhibit measurable dispersion in the visible spectrum, which iSLIM can quantify via a simple RGB measurement. One interesting problem to be addressed is to find out whether hemoglobin is homogeneously distributed inside the cell. The RGB measurement will provide insight into the mean cell hemoglobin concentration (MCHC), mean cellular oxygen saturation level, or even report on these parameters at subcellular scale.

In summary, spectroscopic DPM and iSLIM are simple and robust methods that can provide simultaneous measurements of both volume and Hb concentration at a single red blood cell scale. These parameters can report on a number of diseases, including various anemias and malaria. With appropriate throughput, these methods may offer an advantage over existing technology in the clinic where only statistical, ensemble-averaged values are available. Further, this type of measurement based on QPI may provide a means for globally affordable blood testing, i.e., become a solution for geographical areas where expensive equipment and trained personnel are not available. The fact that the QPI output is entirely digital should represent an opportunity for remote diagnosis and *telepathology*.

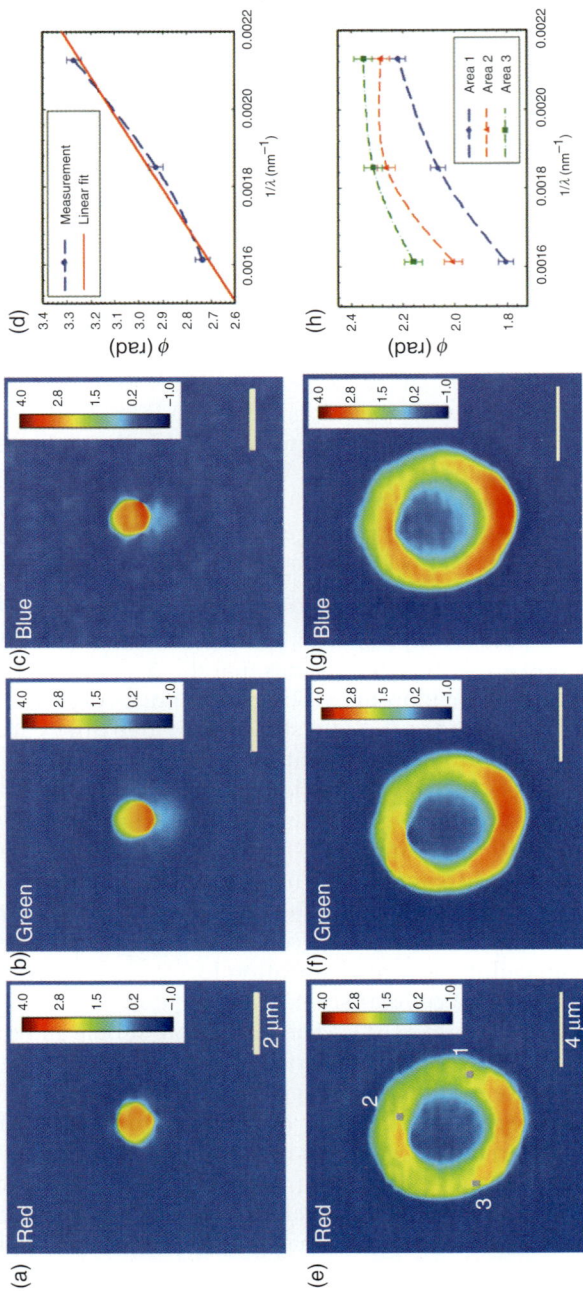

FIGURES 14.10   Phase spectroscopy data (RGB) for 1 micron bead (*a-c*) and red blood cell (*e-g*). The measured phase values are plotted as a function of wave number for the bead (*d*) and the RBC (*h*). The squares in *d* show the three areas chosen for the plotting in *h*. (*From Ref. 32.*)

References

1. J. B. Pawley. *Handbook of biological confocal microscopy.* (Springer, New York, 2006).

2. J. G. McNally, T. Karpova, J. Cooper. and J. A. Conchello. "Three-dimensional imaging by deconvolution microscopy." *Methods—A Companion to Methods in Enzymology,* 19, 373–385 (1999).

3. A. C. Kak and M. Slaney. *Principles of computerized tomographic imaging.* (Society for Industrial and Applied Mathematics, Philadelphia, 2001).

4. E. Wolf. "Three-dimensional structure determination of semi-transparent objects from holographic data." *Optics Communications,* 1, 153 (1969).

5. G. Gbur and E. Wolf. "Relation between computed tomography and diffraction tomography." *Journal of the Optical Society of America A,* 18, 2132–2137 (2001).

6. B. Q. Chen and J. J. Stamnes. "Validity of diffraction tomography based on the first Born and the first Rytov approximations." *Applied Optics,* 37, 2996–3006 (1998).

7. P. S. Carney, E. Wolf, and G. S. Agarwal. "Diffraction tomography using power extinction measurements." *JOSA a-Optics Image Science and Vision,* 16, 2643–2648 (1999).

8. V. Lauer. "New approach to optical diffraction tomography yielding a vector equation of diffraction tomography and a novel tomographic microscope." *Journal of Microscopy-Oxford,* 205, 165–176 (2002).

9. A. M. Zysk, J. J. Reynolds. D. L. Marks, P. S. Carney, and S. A. Boppart. "Projected index computed tomography." *Optics Letters,* 28, 701–703 (2003).

10. F. Charriere, N. Pavillon, T. Colomb, C. Depeursinge, T. J. Heger, E. A. D. Mitchell, P. Marquet, and B. Rappaz. "Living specimen tomography by digital holographic microscopy: morphometry of testate amoeba." *Optics Express,* 14, 7005–7013 (2006).

11. F. Charriere, A. Marian, F. Montfort, J. Kuehn, T. Colomb, E. Cuche, P. Marquet, and C. Depeursinge. "Cell refractive index tomography by digital holographic microscopy." *Optics Letters,* 31, 178–180 (2006).

12. W. Choi, C. Fang-Yen, K. Badizadegan, S. Oh, N. Lue, R. R. Dasari, and M. S. Feld. "Tomographic phase microscopy." *Nature Methods,* 4, 717–719 (2007).

13. W. S. Choi, C. Fang-Yen, K. Badizadegan, R. R. Dasari, and M. S. Feld. "Extended depth of focus in tomographic phase microscopy using a propagation algorithm." *Optics Letters,* 33, 171–173 (2008).

14. Z. Wang, M. Mir, L. J. Millet, H. Ding, S. Unarunotai, J. A. Rogers, R. Bashir, M. Bednarz, I. Golding, Z. Shen, S. G. Prasanth, M. U. Gillette, and G. Popescu. "Spatial light interference microscopy (SLIM)." *Nature Methods,* under review.

15. J. Gass, A. Dakoff, and M. K. Kim. "Phase imaging without 2 pi ambiguity by multiwavelength digital holography." *Optics Letters,* 28, 1141–1143 (2003).

16. C. H. Yang, A. Wax, R. R. Dasari, and M. S. Feld. "2 pi ambiguity-free optical distance measurement with subnanometer precision with a novel phase-crossing low-coherence interferometer." *Optics Letters,* 27, 77–79 (2002).

17. S. De Nicola, A. Finizio, G. Pierattini, D. Alfieri, S. Grilli, L. Sansone, and P. Ferraro. "Recovering correct phase information in multiwavelength digital holographic microscopy by compensation for chromatic aberrations." *Optics Letters,* 30, 2706–2708 (2005).

18. N. Warnasooriya and M. K. Kim. "LED-based multi-wavelength phase imaging interference microscopy." *Optics Express,* 15, 9239–9247 (2007).

19. A. Khmaladze, M. Kim, and C. M. Lo. "Phase imaging of cells by simultaneous dual-wavelength reflection digital holography." *Optics Express,* 16, 10900–10911 (2008).

20. H. C. Hendargo, M. T. Zhao, N. Shepherd and J. A. Izatt. "Synthetic wavelength based phase unwrapping in spectral domain optical coherence tomography." *Optics Express,* 17, 5039–5051 (2009).

21. A. Ahn, C. H. Yang, A. Wax, G. Popescu, C. Fang-Yen, K. Badizadegan, R. R. Dasari, and M. S. Feld. "Harmonic phase-dispersion microscope with a Mach-Zehnder interferometer." *Applied Optics,* 44, 1188–1190 (2005).

22. J. Kato, I. Yamaguchi, and T. Matsumura. "Multicolor digital holography with an achromatic phase shifter." *Optics Letters*, 27, 1403–1405 (2002).

23. I. Yamaguchi, T. Matsumura, and J. Kato. "Phase-shifting color digital holography." *Optics Letters*, 27, 1108–1110 (2002).

24. P. Ferraro, L. Miccio, S. Grilli, M. Paturzo, S. De Nicola, A. Finizio, R. Osellame, and P. Laporta. "Quantitative phase microscopy of microstructures with extended measurement range and correction of chromatic aberrations by multiwavelength digital holography." *Optics Express*, 15, 14591–14600 (2007).

25. M. Born and E. Wolf. *Principles of Optics: Electromagnetic Theory of Propagation, Interference and Diffraction of Light.* (Cambridge University Press, Cambridge; New York, 1999).

26. G. R. Fowles. *Introduction to Modern Optics.* (Holt, New York, 1975).

27. P. Drude. *The Theory of Optics.* (Dover Publications, New York, 1959).

28. Popescu, G. *Nanobiophotonics.* (McGraw-Hill, New York, 2010).

29. Y. Park, T. Yamauchi, W. Choi, R. Dasari. and M. S. Feld. "Spectroscopic phase microscopy for quantifying hemoglobin concentrations in intact red blood cells." *Optics Letters*, 34, 3668–3670 (2009).

30. G. Popescu, T. Ikeda, R. R. Dasari, and M. S. Feld. "Diffraction phase microscopy for quantifying cell structure and dynamics." *Optics Letters*, 31, 775–777 (2006).

31. Y. K. Park, G. Popescu, K. Badizadegan, R. R. Dasari, and M. S. Feld. "Diffraction phase and fluorescence microscopy." *Optics Express*, 14, 8263 (2006).

32. H. F. Ding and G. Popescu. "Instantaneous spatial light interference microscopy." *Optics Express*, 18, 1569–1575 (2010).

CHAPTER 15

Current Trends in Applications

As evidenced throughout the book, much of the past efforts in QPI have been devoted to technology development. Nevertheless, it is clear that the impact of QPI as a new field of study will stem from its ability to enable new biological discoveries. In this chapter, the current trends in biological applications are presented, with the understanding that, most likely, many exciting QPI uses are still to be unraveled. These applications can be categorized as *basic,* i.e., pertaining to basic biological research, and *clinical,* i.e., of medical relevance. For basic studies, applications in cell dynamics (Sec. 15.1.), cell growth (Sec. 15.2.), and tissue optics (Sec. 15.3.) are among the most promising. To achieve clinical significance, QPI methods must satisfy an additional set of constraints, such as high-throughput and cost-effectiveness, which are specific for a clinical setting. New investigations in blood screening and tissue diagnosis are likely to satisfy these requirements (Sec. 15.3).

15.1   Cell Dynamics

15.1.1   Background and Motivation

The live cell is a machine where the *random* and *deterministic* motions of its constitutive components coexist in harmoy. In cells, particles (i.e., mass bits) undergo these fundamentally different types of transport: random (equilibrium, thermally-driven, Brownian) and deterministic (out-of-equilibrium, ATP-consuming, directed). For example, the dominant mechanism for moving certain molecules across the lipid bilayer might be thermal diffusion, while the opposite is true for a vesicle sliding along a dendrite. Clearly, a life form is generated when deterministic behavior emerges from thermal noise, while cell death results in the return of the system to thermal equilibrium.

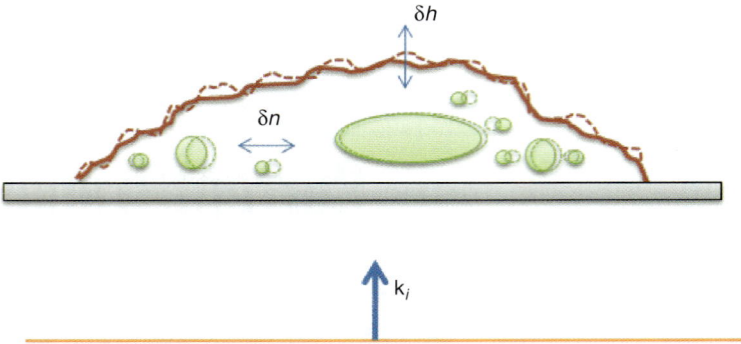

Figure 15.1 Plane-wave incident on a live cell (k_i incident wave vector). The phase fluctuations are due to out-of-plane membrane motions (δh) and in-plane mass transport, which generates local refractive index fluctuations (δn).

Approaching this problem experimentally can be reduced fundamentally to quantifying spatially heterogeneous dynamics at the microscopic scale. QPI methods can provide the necessary ultrasensitive, quantitative methods for cell imaging, to retrieve information about the intricate mechanisms of cell function, various abnormalities, and death. The dynamic information provided by QPI can be divided into two contributions: local cell thickness fluctuations and local refractive index fluctuations (Fig. 15.1). Thus, small optical pathlength fluctuations can be expressed as

$$\delta s(x, y) = \delta[n(x, y)h(x, y)]$$
$$\simeq \bar{n}(x, y)\delta h(x, y) + \bar{h}(x, y)\delta n(x, y) \tag{15.1}$$

In Eq. (15.1), s is the pathlength, n is the refractive index, and h is the thickness. Thus, Eq. (15.1) indicates that the measurable fluctuations can be due to out-of-plane thickness changes (h contributions) at constant refractive index and in-plane particle transport (n contributions), which gives rise to refractive index changes at constant cell thickness. Note that typically, these two phenomena are characterized by very different time scales: milliseconds for membrane fluctuations and seconds-minutes for intracellular transport. Further, the relationship between the spatial and temporal behavior of these fluctuations, i.e., the *dispersion relations,* are specific to each type of motion. This is extremely useful in practice as it allows to experimentally decouple the contributions of n and h, which is a well-known challenge in QPI (as also stated in the introduction of this book). Below we apply this new QPI dynamics concept to studying active (ATP-dependent) membrane fluctuations in RBCs (h contributions) and intracellular mass transport (n contributions).

15.1.2   Active Membrane Fluctuations

The RBC membrane cortex consists of two coupled membranes: a lipid bilayer and a two-dimensional spectrin network.[1] This membrane is remarkably soft and elastic. and thus exhibits fluctuations, observed as "flickering", with amplitudes of the order of tens of nanometers. These fluctuations have been studied for a long time to better understand the interaction between the lipid bilayer and the cytoskeleton.[2-6]

Although RBC membrane dynamics has been explored extensively, no definitive experiment has determined whether flickering is purely thermally driven or contains active contributions. First observed a century ago,[7] flickering is generally believed to stem from thermal noise.[2,8] Different interference microscopic techniques have been employed to study membrane fluctuations and mechanical properties assuming Brownian dynamics.[3,5] In contrast, a technique that qualitatively measured the local fluctuations of RBC membranes, reported a correlation between the ATP concentration and the fluctuation amplitude.[9] However, more recent experimental work, in which only edge shapes of RBCs were probed, showed no relation between ATP depletion and membrane fluctuations.[10] Theoretically, RBC membrane fluctuations were traditionally studied using models of thermally driven equilibrium systems.[2,3] A more recent theoretical model,[11,12] validated by simulation,[13,14] showed that local breaking and reforming of the spectrin network can result in enhanced fluctuations.

In Ref. 15, we studied ATP effects on RBC membrane morphology and fluctuations through diffraction-phase microscopy[16,17] (see Sec. 11.2). By extracting the optical path-length shifts produced across the cell, we measured cell thickness with nanometer sensitivity and millisecond temporal resolution. During this time scale, the cell can be assumed a closed system, such that the variations in refractive index are negligible, i.e., we can assume only h-contributions (see Eq. [15.1]).

RBC samples were prepared under four different conditions: healthy RBCs, and RBCs with irreversibly depleted ATP, metabolically depleted ATP, and repleted ATP groups. After collection, a group of healthy RBCs was minimally prepared. For RBCs in the irreversibly depleted ATP group, the cytoplasmic pool of ATP was depleted by inosine and iodoacetamide. For the metabolically depleted ATP group, healthy RBCs were incubated in a glucose-free medium for 24 hours. For RBCs in the ATP repleted group, cytoplasmic ATP was first metabolically depleted, and then regenerated through the addition of D-glucose. We first address the effects of ATP on the morphologies of RBC membranes From the measured cell thickness profiles at a given time t, $h(x, y, t)$, we calculated (Fig. 15.2a-d) time-averaged heights $\langle h(x, y) \rangle$ and observed the characteristic biconcave shape for healthy RBCs. When ATP was depleted, for both the irreversibly and the metabolically depleted groups, we observed loss of biconcave shape and echinocyte shape transformation. Reintroducing ATP

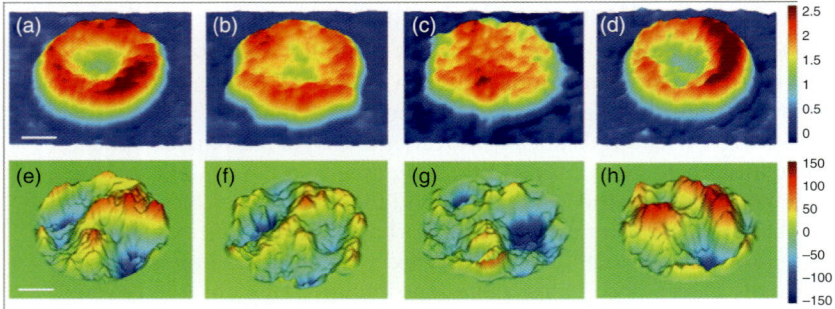

FIGURE **15.2** Effects of ATP on morphology and dynamic fluctuation in RBC membrane. (*a-d*) Topography of a healthy RBC (*a*), of an ATP-depleted RBC (irreversible -ATP group) (*b*), of an ATP-depleted RBC (metabolic -ATP group) (*c*), and of a RBC with recovered ATP level (+ATP group) (*d*), respectively. (*e-h*) Instantaneous displacement maps of membrane fluctuation in the Fig. 15.2*a-d*, respectively. The scale bar is 2 μm. The colorbar scales are in micrometer and nanometer, respectively. [*Proc. Nat. Acad Sci,* used with permission, Y.K. Park, C.A. Best, T. Auth, N. Gov, S.A. Safran, G. Popescu, S. Suresh & M.S. Feld, "Metabolic remodeling of the human red blood cell membrane," 107, 1289 (2010). (*From Fig. 1, Ref. 15.*)]

resulted in the recovery of biconcave shape. This shows that ATP is crucial to maintaining biconcave shape of RBCs.[18]

To probe dynamic membrane fluctuations, we analyzed the membrane displacement map by subtracting the averaged shape from the cell thickness map, $\Delta h(x,y,t) = h(x,y,t) - \langle h(x,y) \rangle$ (Fig. 15.2*e-h*). Compared to healthy RBCs, the fluctuation amplitudes were decreased in both ATP-depleted groups. Reintroducing ATP, however, increased the fluctuation amplitudes to healthy RBC levels. We calculated the root mean squared (RMS) displacement of membrane fluctuations, $\sqrt{\langle \Delta h^2 \rangle}$, which covers the entire cell area for 2 s at 120 frame/s (Fig. 15.3*a*). The RMS displacement of healthy RBCs is 41.5 ± 5.7 nm. Fluctuations significantly decreased to 32.0 ± 7.8 nm and 33.4 ± 8.7 nm in both the irreversibly metabolically ATP-depleted groups, respectively. However, the fluctuations in the ATP-repleted group returned to the level of healthy RBCs (48.4 ± 10.2 nm). This is in agreement with an earlier report using the point measurement technique.[9]

Although the results in Fig. 15.3*a* show that the membrane fluctuations decrease in the absence of ATP, this result does not yet answer the question of whether ATP drives "active" non-equilibrium dynamics, or simply modifies membrane elastic properties. Of course, the two different situations can give rise to fundamentally different dynamics: (1) fluctuations exhibit out-of-equilibrium, or (2) the equilibrium Gaussian statistics is preserved. In order to answer this question, we calculated the non-Gaussian parameter, κ, for the membrane fluctuations (Fig. 15.3*b-e*),

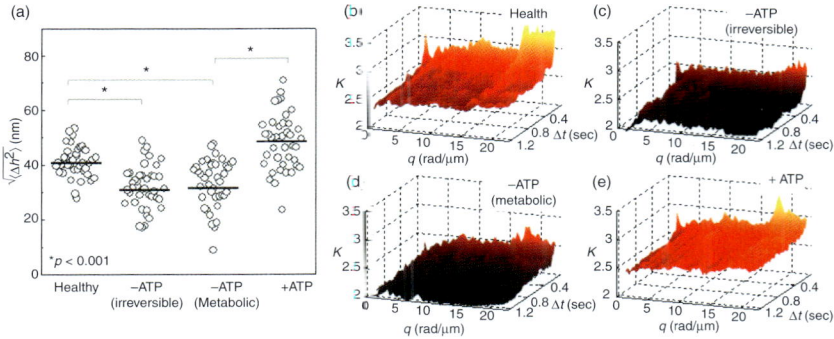

FIGURE 15.3   Non-equilibrium dynamic in RBC membranes. (*a*) RMS displacement of membrane fluctuations for different ATP conditions: healthy RBCs, irreversibly ATP-depleted RBCs, metabolically ATP depleted RBCs, and RBCs in which ATP was reintroduced to metabolically ATP-depleted RBCs. Each symbol represents an individual RBC and the horizontal line is the mean value. (*b-d*) Averaged non-Gaussian parameter vs. a lag time, Δt, and a spatial frequency, q, for membrane fluctuation in healthy RBC (*b*), irreversible ATP depletion group (*c*), metabolic ATP depletion group (*d*), and after reintroducing ATP to metabolic depletion group (*e*), respectively. $N = 40$ RBCs per group. [*Proc. Nat. Acad Sci.* used with permission, Y.K. Park, C.A. Best, T. Auth, N. Gov, S.A. Safran, G. Popescu, S. Suresh & M.S. Feld, "Metabolic remodeling of the human red blood cell membrane," 107, 1289 (2010). (*From Figs. 2-3, Ref. 15.*)]

$$\kappa(q,\tau) = \frac{\left\langle \left| h(q,t+\tau) - h(q,t) \right|^4 \right\rangle_t}{\left\langle \left| h(q,t+\tau) - h(q,t) \right|^2 \right\rangle_t^2} \tag{15.2}$$

From the definition (1), Eq. (15.2), $\kappa = 2$ for purely thermally driven Gaussian motion and increases above two for active non-equilibrium dynamics.[19] For healthy RBCs, the average value of κ that we measured was 2.8, which shows that membrane fluctuations contain non-equilibrium dynamic components, particularly on short-length and time scales ($q > 5$ rad/μm and $\Delta t < 0.5$ s). With depletion of ATP, κ decreased to 2, as expected in purely thermally driven dynamics (the average values of κ were 2.06 and 2.19 for the irreversibly depleted and metabolically depleted ATP groups, respectively). Reintroducing ATP increased κ to healthy RBC levels (average value $\kappa = 2.98$).

These results show that QPI is a valuable tool for studying subtle phenomena, such as active, nanoscale motions in live cells. Our data clearly proves that active, metabolic energy from ATP contributes an enhancement in RMS displacements by 44.9%. This measured value is lower than predicted by a theoretical model, where an increase of at least 100% was expected.[19] Still, this is a measurable effect, which is likely to correlate on cell function. As a result, in this case the *fluctuation-dissipation theorem*,[20] which allowed us to extract membrane rheology (Sec. 11.2.3.2), is expected to introduce errors, because it assumes thermal equilibrium. Overall, we expect the models based on the fluctuation-dissipation theorem to underestimate the mechanical parameters of the membrane. Thus, the contributions to membrane fluctuations due to ATP will be attributed by the equilibrium model to a softer membrane. Nevertheless, the RBC membrane mechanical parameters measured by various methods are within a broad range of one, sometimes two orders of magnitude (see, for example, Tables 5.2. and 5.3 in Ref. 21), which means that the fluctuation dissipation theorem is still a valuable tool for membrane rheology.[22]

15.1.3   Intracellular Mass Transport

Introduction
Cells have developed a complex system to govern the internal transport of materials from single proteins to large complexes such as chromosomes during cell division. These processes are essential for the maintenance of cellular functions. It is now well known that these transport processes do not rely solely on thermal diffusion; a sophisticated system of targeted active transport is essential for the correct distribution of gene products and other intracellular resources.[23] While quantitative measurements of this active, molecular motor driven transport have been made in the past via particle tracking (see, for example, Ref. 24), developing a more global picture of the spatial and temporal distribution of active transport in living cells remains an unsolved problem.

Experimentally, the problem becomes quantifying spatially heterogeneous dynamics at the microscopic scale. Such measurements have been performed, for example, in densely packed colloidal suspensions[25] using multiple particle tracking[26] and in foams using diffusing wave spectroscopy.[27] In Ref. 28, we introduced QPI to map the spatial and temporal distribution of active (*deterministic*) and passive (*diffusive*) transport processes in living cells. We report first on tests of quantitative-phase imaging on dense colloidal systems; these results demonstrate the ability of the technique to measure diffusive motion in a system where the data may be corroborated by particle tracking methods. We then apply this technique to living cells showing that one may extract transport data across the cell body with high spatial and temporal resolution. These data show that transport in cells is highly organized spatially.

As already mentioned, Abbe recognized image formation in a microscope as being the result of light scattering by the specimen, followed by interference of the scattered waves at the image plane.[29] Thus, a microscope is a powerful scattering instrument; the momentum transfer available for measurement is limited only by the numerical aperture of the objective, which can essentially cover scattering over the entire forward hemisphere around the incident wavevector.

In Eq. (15.1), the path-length fluctuation can be expressed to the first order as $\delta s(x,y) \simeq \bar{n}(x,y)\delta h(x,y) + \bar{h}(x,y)\delta n(x,y)$, where the horizontal bars indicate time averaging. Thus, $\delta s(\mathbf{r},t)$ contains information about both thickness (h) and refractive index (n) fluctuations. The *h-contribution* to path-length fluctuations has been shown to reveal new behavior in red blood cells, where n is well approximated by a constant.[15,22,30-32] Studies of refractive index fluctuations have been limited to spatially-averaged measurements in the context of cell growth[33] and cell refractometry.[34]

Principle

We used QPI to study refractive index fluctuations due to mass transport within live cells. These data are characterized by the effective dispersion relation $\omega(q) \sim a^{\alpha}$ to describe the relaxation rate of index of refraction fluctuations. In particular, we expect to observe $\alpha = 1, 2$, respectively, for *advective* or *diffusive* mass transport. We employed spatial light interference microscopy (SLIM)[28,35-37], a quantitative-phase imaging method described in Sec. 12.2, to extract cell mass distributions over broad spatiotemporal scales. Because the phase image acquisition rate in SLIM (< 1 frames/s) is much slower than the decay time associated with the bending and tension modes of membrane fluctuations, we can safely ignore thickness fluctuations.[31] The path-length fluctuations in this case report on the dry mass transport within the cell.[33] Without loss of generality, we will consider that the refractive index fluctuations are generated by a system of moving particles,

$$\delta s(\mathbf{r},t) \simeq \bar{h} \cdot \int \Delta n_0(\mathbf{r}') \sum_j \delta[\mathbf{r} - \mathbf{r}_j(t) - \mathbf{r}']d^2\mathbf{r}' \qquad (15.3)$$

Equation (15.3) represents a spatial convolution, which establishes that particles of refractive index contrast, Δn_0, exist at time-variable positions, $\mathbf{r}_j$, throughout the field of view, $d^2\mathbf{r}$. Taking the spatial Fourier transform of Eq. (15.3), we obtain

$$\delta\tilde{s}(\mathbf{q},t) = \bar{h} \cdot \tilde{n}(\mathbf{q}) \sum_j e^{i\mathbf{q}\cdot\mathbf{r}_j(t)} \qquad (15.4)$$

where ~ indicates Fourier transform. The temporal autocorrelation function yields

$$\Gamma(\mathbf{q},\tau) = \langle \delta s(\mathbf{q},t)\delta s(\mathbf{q},t+\tau) \rangle_t$$
$$= N^2\bar{h}^2 \cdot n(\mathbf{q})^2 \langle e^{i\mathbf{q}\cdot\Delta r(\tau)} \rangle_t \qquad (15.5)$$

where N is the total number of particles. We assume that the process is *ergodic*, so that the time average in Eq. (15.5) can be evaluated via an ensemble average,

$$<e^{iq\cdot\Delta r(\tau)}> = \int e^{iq\cdot\Delta r(\tau)} \cdot \psi(\Delta r, \tau) d^2\Delta r$$

$$= \tilde{\psi}(q, \tau) \qquad (15.6)$$

Thus, the ensemble average $\langle e^{iq\cdot\Delta r(\tau)}\rangle$, i.e., the *dynamic structure function*, is determined by the spatial Fourier transform, $\tilde{\psi}$, of the probability density ψ for finding a particle at point r at time t. Generally ψ satisfies an advection-diffusion equation

$$\frac{\partial}{\partial t}\psi + \mathbf{v}\cdot\nabla\psi = D\nabla^2\psi \qquad (15.7)$$

where D is the coefficient of the particles, and $\mathbf{v}$ is the advection velocity due to flows in the sample cell. In general, when both advective and diffusive transport occur, we expect that the dynamic structure factor defined above decays exponentially as

$$\tilde{\psi}(\mathbf{q}, \tau) = \tilde{\psi}(\mathbf{q}, 0)e^{-i\mathbf{v}\cdot\mathbf{q}\tau + Dq^2\tau} \qquad (15.8)$$

The experimental data, however, is averaged over a broad spectrum of local advection velocities, so that we may write the time dependence of the advection-velocity-averaged dynamic structure factor as

$$\tilde{\Psi}(\mathbf{q}, \tau) = \tilde{\psi}(\mathbf{q}, 0)e^{-Dq^2t}\int P(|\mathbf{v} - \mathbf{v}_0|)e^{-i\mathbf{v}\cdot\mathbf{q}\tau}d^2\mathbf{v} \qquad (15.9)$$

where the probability distribution, P, of local advection velocities remains to be determined. In order to gain insight into the local distribution of advection speeds, we note that maximal kinesis speeds are approximately $v = 0.8$ μm/s[38]. Given the varying loads on such motors, we expect that, the typical advection speeds will be distributed below this limit. The average advection speed may in fact be significantly lower than this value as there must be transport in a variety of directions. Consequently, we propose that $P(v)$ is a Lorentzian of width Δv and that the mean advection velocity averaged over the scattering volume is significantly smaller than this velocity, $v_0 \ll \Delta v$. From this simple model we find that the integral in Eq. (15.9) can be evaluated as

$$\tilde{\Psi}(\mathbf{q}, \tau) = \tilde{\Psi}(\mathbf{q}, 0)e^{-i v_0\cdot\mathbf{q}\tau + Dq^2\tau + q\Delta v\tau} \qquad (15.10)$$

Equation (15.10) establishes a relationship between the decay rate, ω, of a spatial mode and its spatial frequency, q, (i.e., *a dispersion relationship*), which has the form

$$\omega(q) = \Delta v q + D q^2 \qquad (15.11)$$

This dispersion relationship is the foundation for QPI-based intracellular transport measurements. Thus, from a time series of quantitative-phase images, one can compute the 3D (x, y, and time) Fourier transform of the data and obtain the frequency bandwidth for each spatial frequency (q_x, q_y). The fit with a polynomial function then provides the width of the velocity distribution from the quadratic behavior, and the diffusion coefficient from the linear one.

Measurements on Brownian Systems

In order to mimic the cell environment conditions, we used SLIM to image the Brownian motion of 1 μm polyethylene spheres in highly concentrated (99%) glycerol. Figure 15.4 shows the SLIM phase image of the sample with a field of view of 73×73 μm². We acquired SLIM images for 10 minutes with an acquisition rate of 1 frame/s. The expected diffusion coefficient can be calculated via the Stokes-Einstein equation. However, since the effective viscosity of the concentrated particulate suspension in glycerol has a highly nonlinear dependence on the concentration, we estimated the diffusion coefficient by tracking individual particles in the phase image. Figure 15.4b shows the mean-squared displacements obtained by averaging the trajectories of 160 particles. The experimental curve has the expected linear trend, with two regions of slightly different slopes, as shown in Fig. 15.4b. The existence of these two diffusive regimes can be understood by taking into account the collective motion of the beads. Thus, at short times and displacements roughly less that the particle diameter,

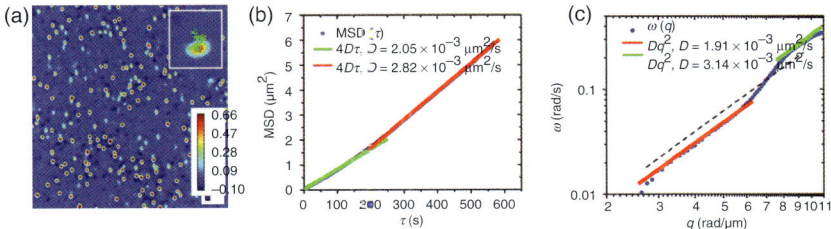

FIGURE 15.4 (a) Quantitative-phase image of 1-μm polyethylene beads in glycerol. Inset illustrates trajectory of a single bead. (b) Mean-squared displacements obtained by tracking individual beads in a. The fits with a linear function yields two diffusion coefficients, as indicated. (c) Dispersion curve associated with the beads in a. The fits with the quadratic function yield two diffusion coefficients, as shown. (*From Ref. 28.*)

particles diffuse more slowly due to their interaction with neighbors. The linear fit yields $D = 2.05 \times 10^{-3}$ $\mu m^2/s$ and $D = 2.82 \times 10^{-3}$ $\mu m^2/s$. From the SLIM data, we calculated the dispersion relation, $\omega(\mathbf{q})$, and performed the angular average to obtain the radial function, $\omega(q)$. The experimental curve exhibits the expected q^2 dependence, as illustrated in the log-log plot of Fig. 15.4c. We found that the two distinct diffusion regimes reveal themselves in this plot as well. Note that the transition between the two regimes appears at the same scale of approximately $2\pi/q = 1$ μm. The diffusion coefficients obtained are $D = 1.91 \times 10^{-3}$ $\mu m^2/s$ and $D = 3.14 \times 10^{-3}$ $\mu m^2/s$ and compare very well with those measured via our particle tracking. Later, we measured lower concentrations of beads, where the particle-particle interaction is insignificant and obtained a single diffusive regime.[28]

Measurements on Live Cells

Note that measuring diffusion via the dispersion relation eliminates the need for tracking individual particles and, more importantly, applies to particles that are not resolved in the image, i.e. are smaller than the diffraction spot of the microscope. This is largely the case in studying mass transport in live cells. We performed experiments on various types of live cells including glia and microglia cells, which are common in the central nervous system. In these cells, individual, unlabeled particles cannot be easily resolved or tracked by light microscopy. The cells are imaged in culture medium under physiological conditions, 37°C and 5% CO_2 controls. Figure 15.5 shows results obtained on a glial cell. The $\omega(q)$ curve exhibits a dominant quadratic shape, which yields a diffusion coefficient $D = 9.6 \times 10^{-3}$ $\mu m^2/s$. Such a small diffusion coefficient is due to the crowded cytosol space that limits the motility of particles within the cell. Further, at the low-wavenumber end of the measurement range, we found a distinct power law of q. From the linear term in fitting the dispersion curve, we extracted the advection velocity distribution width, obtaining a value of $\Delta v = 1.3$ nm/s.

Further, we studied the mass transport in 1D cellular sub-domains. Figure 15.6 shows the SLIM images of various cells in culture and the $\omega(q)$ curves associated with the respective regions. Note that the signal from the strips is associated with fluctuations along a single row of pixels, which explains the relatively higher noise level. These data exhibit a diversity of behavior, from purely diffusive in the microglia culture (Fig. 15.6d) to purely directed, in the dendrite of a neuron (Fig. 15.6f). The entirely directed transport measured along the dendrite is in line with what is generally known about that ATP-consuming cargo transfer along microtubules via protein motors. Examining a narrow strip whose long axis is oriented radially with respect to the cell nucleus (Fig. 15.6b), we found that the transport is diffusive at short scales (below approximately $2\pi/q = 2$ μm) and directed at large scales.

(a)

(b)

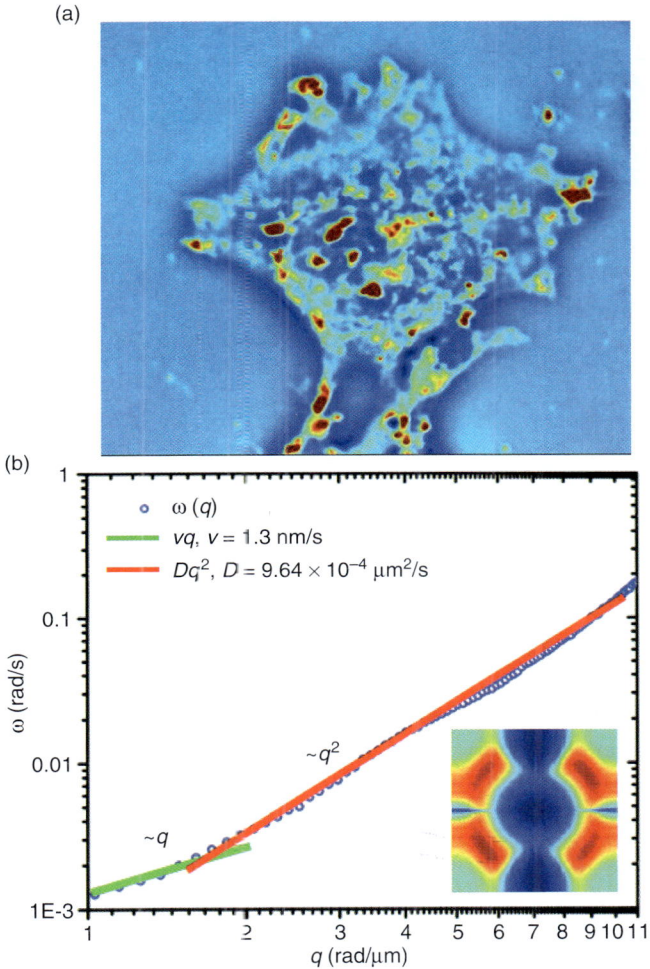

FIGURE **15.5**   (*a*) Quantitative-phase image of a glial cell. (*b*) Dispersion curve measured for the cell in *a*. The green and red lines indicate directed motion and diffusion, respectively. Inset shows the $\Delta\omega$ (q_x, q_y) map. (*From Ref. 28.*)

In sum, we developed a QPI-based approach to study the intracellular transport that relies on measuring the dispersion curves from spatiotemporal refractive index fluctuations. Our experiments show that continuous or completely transparent systems can be studied successfully by this approach, in a label-free manner. Not surprisingly, our results demonstrate that the transport within cells consists of a combination of short-distance diffusion and long-distance deterministic motions.

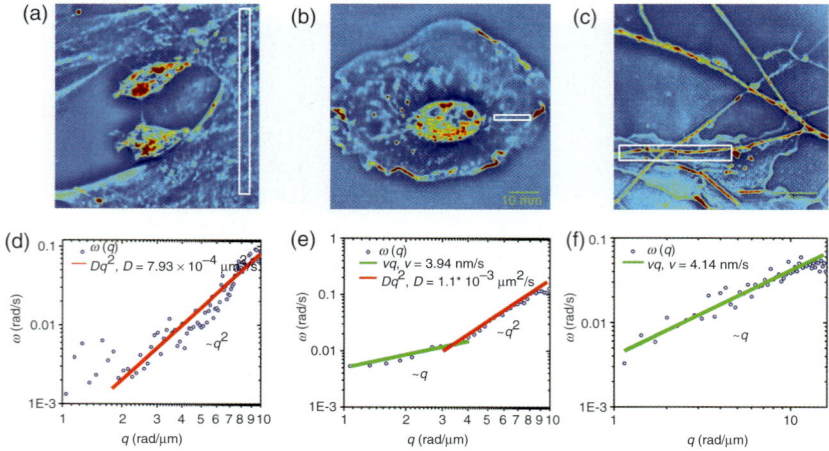

FIGURE 15.6   Quantitative-phase image of a culture of microglia (*a*), glia (*b*), and hippocampal neurons (*c*). (*d-f*) Dispersion curves associated with the regions in *a-c*. The corresponding fits and resulting *D* and Δ*v* values are indicated. (*From Ref. 28.*)

15.2   Cell Growth

15.2.1   Background and Motivation

The question of how single cells regulate their growth has been described as "one of the last big unsolved problems in cell biology."[39] The reason that this debate has persisted despite decades of effort is primarily for the following reasons: cells are small, difficult to measure, and on average only double in size between divisions. Until recently, the state of the art method to assess a single cell growth curve was using Coulter counters to measure the volume of a large number of cells, in combination with careful mathematical analysis.[40] For relatively simple cells such as *Escherichia coli*, traditional microscopy techniques have also been used to assess growth in great detail.[41] These types of methods make the assumption that volume is a good surrogate for mass, although it is known that the volume can change disproportionately to mass, for example, as a result of osmotic response.[42] Recently, shifts in the resonant frequency of vibrating microchannels have been used to quantify the buoyant mass of cells flowing through the structures.[43,44] Using this approach, Godin et al. have shown that several cell types grow exponentially, i.e., heavier cells grow faster than lighter ones.[43] Later, Park et al. extended this principle to allow mass measurements on adherent cells.[45]

An ideal method will perform parallel growth measurements on an ensemble of cells simultaneously and continuously over more than one cell cycle, quantify possible cell-cycle phase-dependent growth, apply equally well to adherent and non-adherent cells, and work in a fully biocompatible environment. Here we demonstrate that SLIM (see Sec. 12.2.) approaches these ideals.[35] As already discussed in Sec. 11.1.3.2., the principle behind using interferometry to measure cell dry mass was established in the early 1950s.[46,47] Recently, it has been shown that the surface integral of the cell phase map is invariant to small osmotic changes.[42] which establishes that quantitative-phase imaging methods can be used for dry mass measurements. The dry mass density at each pixel is calculated as $\rho(x,y) = \dfrac{\lambda}{2\pi\gamma}\phi(x,y)$,

where λ is the centre wavelength, γ is the average refractive increment of protein (0.2 mL/g),[33] and $\phi(x, y)$ is the measured phase. The total dry mass is then calculated by integrating over the region of interest in the dry mass density map. Thus, SLIM's phase stability translates into spatial and temporal sensitivities of 1.5 fg/μm^2 and 0.15 fg/μm^2, respectively.

15.2.2   Cell Cycle-Resolved Cell Growth

By employing SLIM/fluorescence multimodal imaging, we studied cell cycle-dependent growth measurements[48]. We imaged YFP-PCNA (yellow fluorescent protein—proliferating cell nuclear antigen)—transfected human Osteosarcoma (U2OS) cells, which enabled us to monitor PCNA activity via a fluorescence channel.[35] This activity is greatest during the DNA synthesis of the cell cycle and is observed in the localization of the fluorescence signal (its granular appearance), which reveals the S-phase of the cell cycle (Fig. 15.7a-b). Using the fluorescence signal as one marker and the onset of mitosis as the second, it is possible to study cell growth in each phase of the cell cycle separately. We measured a culture of U2OS cells for 51 hours, scanning a 1.2- × 0.9-mm^2 area every 15 minutes and acquiring fluorescence data every 60 minutes as described in Ref. 35. Figure 15.7c shows typical growth curves measured from a single cell as it divides into two cells and then its daughters into four. This ability to differentiate between two daughter cells growing very close together and measure their dry mass independently is a major advantage of SLIM over contemporary methods such as micro-resonators, where such measurements are impossible to perform.

Due to the cell-cycle phase discrimination provided by the YFP-PCNA we can numerically synchronize our population a posteriori (Fig. 15.7d). To our knowledge this is the first time such ensemble measurements have been achieved. Figure 15.7e clearly illustrates the differences in the growth rate between the G1, S, and G2 phases of the cell cycle. It can be seen that during G2, U2OS cells exhibit clear exponential growth. These results establish that SLIM provides a number of advances with respect to existing methods for quantifying cell growth: (1) SLIM can perform parallel growth measurements on an

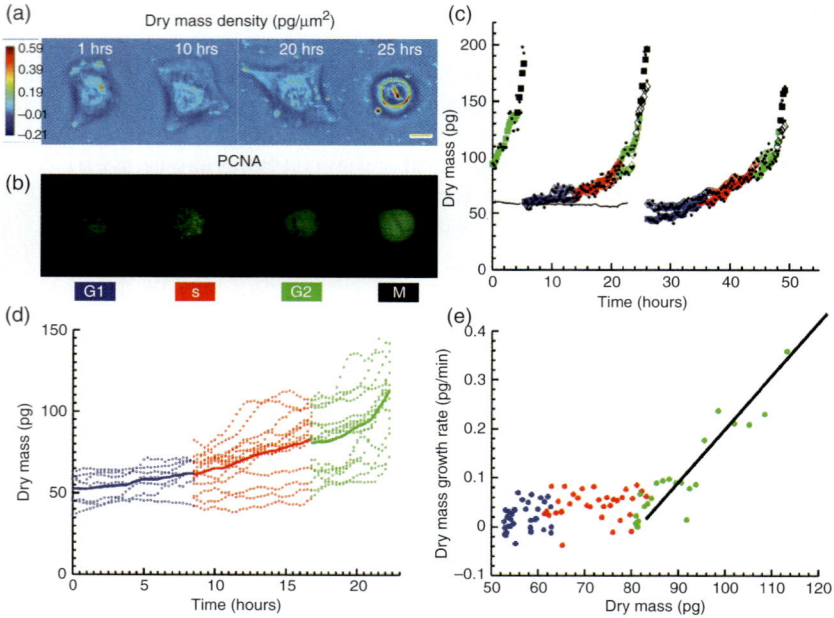

FIGURE 15.7   SLIM measurement of U2OS growth over 2 days. (*a*) Dry mass density maps of a single U2OS cell over its entire cycle at the times indicated, yellow scale bar is 25 µm, colorbar indicates dry mass density in pg/µm^2. (*b*) Simultaneously acquired GFP fluorescence images indicating PCNA activity, the distinct GFP signal during *S* phase, and the morphological changes during mitosis allow for determination of the cell cycle phase. (*c*) Dry mass vs. time for a cell family (i.e., 1->2->4 cells), the two different daughter cell lineages are differentiated by the filled and empty markers, small black markers show raw data, only one daughter cell from each parent is shown for clarity. Different colors indicate the cell cycle as reported by the GFP-PCNA fluorescence. The dotted black line shows measurements from a fixed cell, which has a standard deviation of 1.02 pg. (*d*) *A posteriori* synchronization combination of PCNA stain for S-phase determination and the visual determination of the onset of mitosis allows for the study of cell growth dependence on cell-cycle phase, in an asynchronous culture. The figure shows G1, S, and G2 dependent mass growth as indicated by color, the cycles of the individual cells were aligned as described above, the *x*-axis indicates the average time spent in the respective cell cycle phase. Open circles indicate single cells data and solid lines indicate ensemble averages by cell cycle phase. It can clearly be seen that the cell growth is dependent on both the cell cycle phase and the current mass of the cell. (*e*) Dry mass growth rate vs. dry mass for the ensemble averages, it can be seen that G2 exhibits a distinct exponential growth rate compared to the relatively low growth measured in G1 and S phases. (*From Ref. 35.*)

ensemble of cells simultaneously; (2) the measurement applies equally well to adherent and non-adherent cells; (3) spatial and temporal effects, such as cell-cell interactions can be explored; (4) in combination with fluorescence, specific chemical processes may be probed simultaneously; (5) the environment is fully biocompatible and identical to that in widely used equipment; (6) the imaging nature of SLIM offers a direct look at the cells, which can reveal artifacts, cell morphology, etc; and (7) a lineage study is possible, i.e., a cell and its progeny may be followed.

QPI is very likely to make a significant impact in biology by revealing knowledge about cell growth in isolated cells and populations, in healthy and diseased cells. Diseases such as cancer are characterized by completely different growth kinetics. In combination with fluorescence, QPI can be used to study these differences and perhaps identify drugs that can modulate the growth rate.

15.3 Tissue Optics

15.3.1 Background and Motivation

Upon propagation through inhomogeneous media, optical fields suffer modifications in terms of irradiance, phase, spectrum, direction, and polarization, which, in turn, can reveal information about the sample of interest.[49,50] Light scattering from tissues has evolved as a dynamic area of study and attracted extensive research interest, especially due to the potential it offers for *in-vivo* diagnosis.[51-63] Despite all these efforts, light scattering-based techniques currently have limited use in the clinic. A great challenge is posed by the insufficient knowledge of the tissue optical properties. This information about tissue optics has been limited by two main factors. First, there are very limited experimental means to measure cell and tissue optical properties. Second, the refractive index varies among different organs and spatially within each organ. Therefore, an ideal measurement will provide the tissue scattering properties from the *organelle to organ* scales, which, to our knowledge, remains to be achieved.

Light-tissue interaction can be modeled by a *radiative transport equation*, in complete analogy to the problem of neutron transport in nuclear reactors.[64] With further simplifying assumptions, a *diffusion model* can be applied to describe the *steady state*[65] and *time-resolved*[66] light transport in tissues. The refractive index of biological structures has been modeled both as *discrete* particle distribution[67] and *continuous* or *fractal*[68]. Light propagation in bulk tissue is described by two statistical parameters: the *scattering mean free path*, l_s, which provides the characteristic length scale of the scattering process, and the *anisotropy factor*, g, which scales l_s to higher values, $l_s/(1 - g)$, to account for forward-peaked (or anisotropic) scattering. The direct measurement of these scattering parameters is extremely challenging and, therefore, iterative simulations, e.g., Monte Carlo[69] or finite difference time domain,[70] are often used instead.

In the next section, we show that performing QPI on thin tissue slices reveals the scattering parameters associated with the bulk. First we derive two equations that relate the phase distribution to the scattering parameters of the tissue. Then, we use this theoretical result to map the tissue in terms of these scattering parameters over broad spatial scales.

15.3.2   Scattering-Phase Theorem

As discussed in Chap. 13, Fourier transform light scattering (FTLS) has been developed as the spatial analog of Fourier transform spectroscopy to provide *angular scattering* information from QPI measurements.[71-75] Thus, FTLS was used to measure l_s from angular scattering associated with tissue slices and the anisotropy parameter, g, was determined by fitting the scattering pattern with Gegenbauer Kernel phase function.[75]

More recently, we showed that *quantitative-phase imaging* of thin slices can be used to *spatially map* the tissue in terms of its scattering properties.[76,77] Specifically, we establish mathematical relationships between the phase map, $\phi(\mathbf{r})$, associated with a tissue *slice* of thickness $L << l_s$, and scattering parameters of the *bulk*, i.e., l_s and g. First, we show that the scattering mean free path, l_s, averaged over a certain area across a tissue slice is directly related to the mean-squared phase (variance of the phase) within that region. Second, we prove that the anisotropy factor, g, relates to the *phase gradient* distribution. These relations, which we refer collectively to as *the scattering-phase theorem*, are expressed as

$$l_s = \frac{L}{\left\langle \Delta\phi^2(\mathbf{r}) \right\rangle_{\mathbf{r}}} \tag{15.12a}$$

$$g = 1 - \left(\frac{l_s}{L}\right)^2 \frac{\left\langle \left|\nabla[\phi(\mathbf{r})]\right|^2 \right\rangle_{\mathbf{r}}}{2k_0^2} \tag{15.12b}$$

In Eq. (15.12a-b), L is the tissue slice thickness, $\left\langle \Delta\phi^2(\mathbf{r}) \right\rangle_{\mathbf{r}} = \left\langle \left[\phi(\mathbf{r}) - \left\langle \phi(\mathbf{r}) \right\rangle_{\mathbf{r}}\right]^2 \right\rangle_{\mathbf{r}}$ is the *spatial variance* of ϕ, with $\langle \ \rangle_{\mathbf{r}}$ denoting spatial average, $k_0 = 2\pi/\lambda$, with λ the wavelength of light, $\left|\nabla[\phi(\mathbf{r})]\right|^2 = (\partial\phi/\partial x)^2 + (\partial\phi/\partial y)^2$, with $\mathbf{r} = (x, y)$.

Proof of the l_s-ϕ Relationship

The starting point in proving Eq. (15.12a) is the definition of l_s as the characteristic length in the medium over which the irradiance I_0' of the unscattered light drops to $1/e$ of the original value I_0, i.e., *the Lambert-Beer's law*,

$$I_0' = I_0 e^{-L/l_s} \tag{15.13}$$

In Eq. (15.13), $I_0 = |U_0|^2$ and $I_0' = |U_0'|^2$, where U_0 and U_0' represent the incident plane wave and the unscattered light that passed through

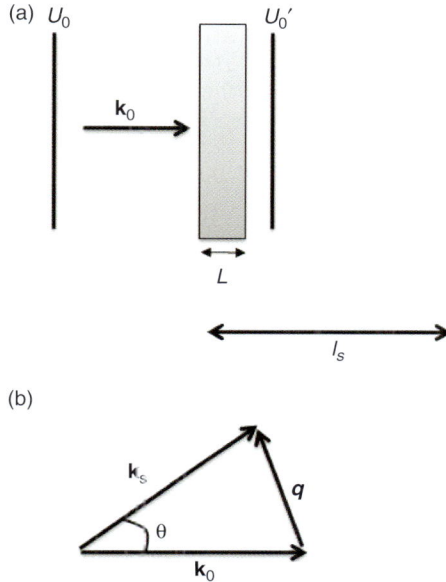

FIGURE 15.8   Diagram for light passing through a thin tissue slice. *L*, thickness of the sample; l_s, mean free scattering path length. K_0 and K_0', wave vector for the incident and scattering field. (*From Ref. 76.*)

the slice, respectively, as illustrated in Fig. 15.8. The field after the tissue slice, U', carries information about the spatial phase distribution, $\phi(\mathbf{r})$, which is available for measurement via QPI, $U'(\mathbf{r}) = U_0 \cdot e^{i\phi(\mathbf{r})}$. The transmitted field can be expressed as the superposition between the scattered and unscattered components,

$$U'(\mathbf{r}) = U_0' + U_1'(\mathbf{r}) \tag{15.14}$$

Note that U_0' is the zero-frequency (unscattered) component of U' and U_1' is the sum of all high-frequency field components. Therefore, U_0' can be expressed as the spatial average of U',

$$U_0' = \left\langle U_0 \cdot e^{i\phi(\mathbf{r})} \right\rangle_\mathbf{r} \tag{15.15}$$

For a *normal distribution* of phase shifts, where the probability density function is a Gaussian function of the form $\exp\left[-\phi^2/2\left\langle\Delta\phi^2\right\rangle_\mathbf{r}\right]/\sqrt{2\pi\left\langle\Delta\phi^2\right\rangle_\mathbf{r}}$, the average in Eq. (15.15) is readily performed as

$$U_0' = \frac{U_0}{\sqrt{2\pi(\delta\phi)^2}} \int_{-\infty}^{\infty} e^{i\phi} e^{-\frac{\phi^2}{2\left\langle\Delta\phi^2\right\rangle_\mathbf{r}}} d\phi$$

$$= U_0 e^{-\frac{\left\langle\Delta\phi^2\right\rangle_\mathbf{r}}{2}} \tag{15.16}$$

In Eq. (15.16), $\langle \Delta\phi^2 \rangle_r$ is the *variance* associated with the phase-shift distribution. Since $U_0'/U_0 = \sqrt{I_0'/I_0}$, combining Eqs. (15.13) and (15.16) yields the expression of the *scattering mean free* path,

$$l_s = \frac{L}{< \Delta\phi^2(\mathbf{r}) >_r}$$

(15.17)

The assumption of Gaussian statistics provides the analytic formula in Eq. (15.17), which is simple and insightful at the same time. However, we note that the average in Eq. (15.15) can be calculated numerically for any non-Gaussian distribution of phase shifts, as long as the quantitative-phase image is known.

Proof of the *g*-ϕ Relationship

By definition, g represents the average-cosine of the scattering angle for *a single scattering event*. Recently, we have extended this concept to continuous distributions of scattering media, including tissues.[75] We showed that, since l_s also means the distance over which, on average, light scatters once, g can be defined by the average cosine of the field transmitted through a slice of thickness l_s,

$$g = < \cos\theta >_\theta$$

(15.18)

As illustrated in Fig. 15.8b, the scattering angle connects the incident wavevector, $\mathbf{k_0}$, the scattered wavevector, $\mathbf{k_s}$, and the *momentum transfer*, $\mathbf{q} = \mathbf{k_s} - \mathbf{k_0}$, as

$$\cos\theta = 1 - \frac{q^2}{2k_0^2}$$

$$q = 2k_0 \sin\frac{\theta}{2}$$

(15.19)

Combining Eqs. (15.18) and (15.19), we can express the average cosine as

$$g = 1 - \frac{1}{2k_0^2} \int q^2 P(q) q \, dq$$

(15.20)

In Eq. (15.20), $P(q)$ is the angular scattering probability distribution of the field exiting a slice of thickness l_s. P is the normalized angular scattering intensity and has the form,

$$P(q) = \frac{\left|\tilde{U}'(q)\right|^2}{\int \left|\tilde{U}'(q)\right|^2 q \, dq}$$

(15.21)

here $\tilde{U}'$ is the spatial Fourier transform of U'. Inserting Eq. (15.21) into Eq. (15.20) we find

$$g = 1 - \frac{1}{2k_0^2} \int \left| q\tilde{U}'(q) \right|^2 q\, dq / \int \left| \tilde{U}'(q) \right|^2 q\, dq \qquad (15.22)$$

Using Parseval's theorem for both the numerator and denominator and applying the *differentiation theorem* to the numerator, we can express g via spatial-domain integrals,[78]

$$g = 1 - \frac{1}{2k_0^2} \int \left| \nabla U'(r) \right|^2 r\, dr / \int \left| \tilde{U}'(r) \right|^2 r\, dr \qquad (15.23)$$

Since the spatial dependence of U' is in the phase only, $U'(\mathbf{r}) = U_0 \cdot e^{i\phi_{ls}(\mathbf{r})}$, the gradient simplifies to

$$\nabla U'(\mathbf{r}) = U'(\mathbf{r})\nabla\phi_{ls}(\mathbf{r}) \qquad (15.24)$$

Thus, combining Eqs. (15.12) and (15.13), we arrive at the final formula for g,

$$g = 1 - \frac{\left\langle \left| \nabla[\phi_{ls}(r)] \right|^2 \right\rangle_{\mathbf{r}}}{2k_0^2} \qquad (15.25)$$

Equation (15.25) expresses the relationship between g and the gradient of the phase shift distribution through a slice of thickness l_s. If the phase image, $\phi(\mathbf{r})$, is obtained over a thickness L, with $L \ll l_s$, then $\phi_{ls} = \phi l_s / L$. Thus, the anisotropy factor depends on the measurable phase image as

$$g = 1 - \left(\frac{l_s}{L} \right)^2 \frac{\left\langle \left| \nabla[\phi(r)] \right|^2 \right\rangle_{\mathbf{r}}}{2k_0^2} \qquad (15.26)$$

In summary, the *scattering-phase theorem* connects the phase image of a thin tissue slice to the scattering properties of the tissue. Note that the tissue can be mapped in terms of l_s and g that are averaged over patches of certain area, S. While this remarkable result may seem counter intuitive, its physical interpretation is straight forward. The l_s-ϕ relationship simply establishes that the attenuation due to scattering is stronger (l_s shorter) as the tissue roughness (*variance*) is larger, i.e., the more inhomogeneous the tissue, the stronger the scattering. For homogeneous tissue, i.e., zero-variance, l_s becomes infinite, which indicates the absence of scattering. On the other hand, the g-ϕ formula contains the spatially-averaged phase gradient intensity, or the variance of the gradient. This phase gradient relates to a tilt in direction of propagation. The modulus squared of the gradient indicates that the angular average is intensity-based. Thus, the higher the squared-averaged gradient, the higher the probability for large scattering angles, i.e., the smaller the g value (Eq. [15.26]). In essence, a thin tissue slice can be assimilated with a (complicated) phase grating, which is characterized by a certain diffraction efficiency (controlled by l_s) and average diffraction angle (reflected in g).

We propose QPI as a direct method for extracting l_s and g, which is likely to have a significant impact in the field of biophotonics. In the experimental results described in the next section, we demonstrate this idea by mapping the scattering properties of tissues over broad spatial scales and also discuss the effects of the limited numerical aperture of the imaging optics.[77] We envision that this approach will facilitate building up a large database, where various tissue types, healthy and diseased, are fully characterized in terms of their scattering properties. Furthermore, these measurements will provide important diagnosis value, as they allow studying both healthy and diseased tissue optics from microscopic (organelle) to macroscopic (organ) scales.

15.3.3 Tissue-Scattering Properties from Organelle to Organ Scale

The *scattering-phase theorem* derived in the previous section[77,79] provides a mathematical relationship between l_s and g on one hand, and the statistics of phase shift distribution associated with a thin tissue slice, as measured via QPI (see Eq. 15.12a-b).

In order to implement this approach experimentally, we used spatial light interference microscopy (SLIM), a new quantitative-phase imaging method developed in our laboratory, which was described in Sec. 12.2.[35] SLIM uses broadband light centered at 532 nm and provides highly sensitive quantitative-phase images, with 0.03-nm path-length sensitivity temporally and 0.3 nm spatially. In order to obtain large fields of view, throughout the experiments presented here, we used a 10×, 0.3 NA objective. This limited numerical aperture effectively acts as a low-pass spatial frequency filter. Thus, the spatial averages performed in deriving Eq. (15.12a-b)[79] are expected to be affected by this cut-off. Still, because tissues scatter strongly forward (g close to unity), we anticipate that the low NA is not a significant error source. However, in order to quantify the effect of the numerical aperture on the overall scattered intensity measured and g estimation, we used the common Henyey-Greenstein angular distribution to calculate the respective error functions,

$$\Delta P(NA, g) = 1 - \int_{\sqrt{1-NA^2}}^{1} P(\cos\theta) d\cos\theta \qquad (15.27a)$$

$$\Delta g(NA, g) = 1 - \int_{\sqrt{1-NA^2}}^{1} \cos\theta P(\cos\theta) d\cos\theta \qquad (15.27b)$$

where P is the Henyey-Greenstein distribution, normalized to unit area, $P(\cos\theta) = const.(1 - g^2)/(1 + g^2 - 2g\cos\theta)^{3/2}$. In Eq. (15.27a-b), ΔP represents the scattered power that is not accounted for due to NA,

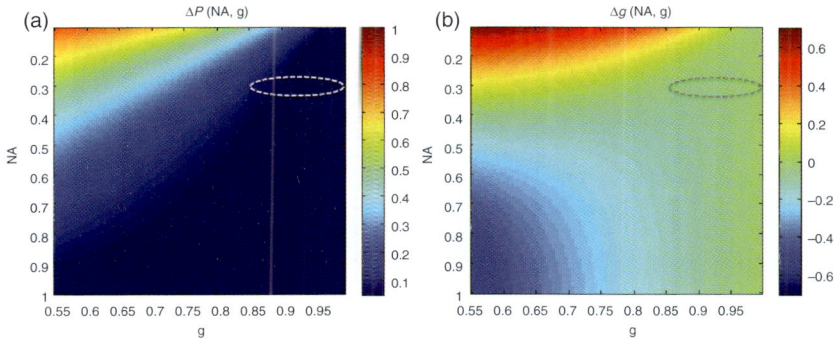

FIGURE 15.9 Error estimation due to limited *NA*. (*a*) Error in power vs. *NA* and *g*. (*b*) Error in *g* vs. *NA* and *g*. The dash ellipses show the regime of our measurements. (*From Ref. 76.*)

and Δg represents the difference between the measured and true average cosine of the scattering angle (i.e., anisotropy factor). Figure 15.9 shows the two error functions. It can be seen that, in the measurement range set by our $NA = 0.3$ and large g values associated with tissues (ellipses in Fig. 15.9*a* and *b*), the errors are below 10% in power and 5% in *g*, and decrease accordingly for higher *NA*.

We acquired quantitative-phase images associated with 5-μ thick tissue slices from rat organs. The specimens were prepared according to a standard procedure under a protocol approved by the Institutional Animal Care and Use Committee at the University of Illinois at Urbana-Champaign. Three slices from each organ of the same rat were cut in succession and imaged by SLIM. The field of view given by the microscope was about 0.4 × 0.3 mm². Therefore, in order to image the entire cross section of the organ (1 cm or larger), the specimen was translated and a mosaic of quantitative phase images were acquired and numerically collaged together. Owing to the 20-nm position accuracy of the translation stage and the in-house–developed code for creating a seamless mosaic, we obtained single quantitative-phase images made of approximately 1000 individual SLIM images. Each of the collages has approximately ~1 *giga-pixels*. Note that these enormous quantitative-phase images cover the entire cross section of a rat organ, with a resolution of ~$\lambda/2NA = 0.9$ μm resolution. Figure 15.10*a* shows one example of quantitative-phase image of a tissue slice cut from a 3-month old rat liver.

Following Eq. (15.12*a-b*), we calculated l_s and *g* in windows of 9×9 μm² across the entire tissue slice. Figure 15.10*b* and *c* shows the maps of l_s and *g*, respectively, for the same rat liver in Fig. 15.10*a*. It is apparent that the tissue scattering parameters exhibit strong inhomogeneities across the organ, mainly due to inclusions which induce refractive index fluctuations. Note that the background l_s values are very high, indicating lack of scattering, as expected. However, the *g*

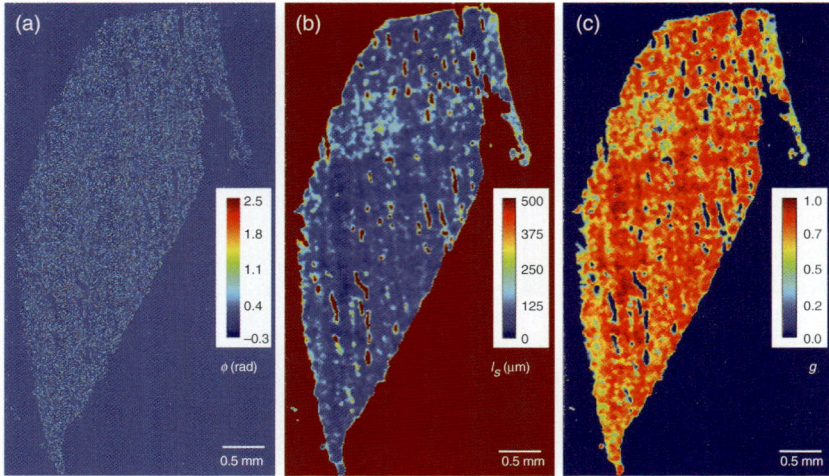

Figure 15.10   Maps of I_s (b) and g (c) for a tissue slice across an entire rat liver. Colorbars show I_s and g, as indicated. (*From Ref. 77.*)

associated with the background appears to be very low, which may seem counter intuitive. The explanation is that, although the background noise is very low, i.e., 2 to 3 orders of magnitude lower than the tissue phase, it has the characteristics of (spatial) white noise, which translates into isotropic scattering.

In summary, following a recent theoretical result, our experimental method provides fast and spatially resolved access to tissue scattering mean free path, l_s, and anisotropy factor, g, from quantitative-phase images of thin tissue slices.[79] This QPI-based approach operates without fitting or iterative procedures. The knowledge of l_s and g has great impact on predicting the outcome of a broad range of scattering experiments on large samples. Our method allows building up an exhaustive database, where various tissue types, healthy and diseased, will be fully characterized in terms of their scattering properties, from microscopic (organelle) to macroscopic (organ) spatial scales.

15.4   Clinical Applications

15.4.1   Background and Motivation

For centuries, light microscopy has been the most widely used tool in biomedicine.[80] During common tissue biopsy or blood smear inspection, the microscope reveals to the pathologist the presence of lesions based upon which a final diagnosis is rendered.[81] The process of histopathology has remained largely unchanged, where tissue biopsied from the patient is typically fixed (e.g., with formalin), embedded (with paraffin or frozen), sectioned in very thin slices (e.g., 3- to 4-μm thick for light microscopy, and 100 nm or less for electron micros-copy), and treated with specific stains for different purposes. Most commonly, the Hematoxylin and Eosin (H&E) stain reveals the baso-philic structures as blue-purple hue (e.g., cell nuclei), and acidic structures in bright pink (e g., most cytoplasm).[81] Effective immuno-chemical staining also relies on the skill of the operator and the exper-imental conditions and methods, which further limited its usage for confirmatory purpose other than a primary diagnostic standard.[82]

Existing clinical technologies used to characterize patient blood such as impedance counters and flow cytometers, though very effec-tive in terms of throughput, offer limited information, are expensive, bulky, costly to maintain, and often require careful calibration. Though, there have been reports of using high throughput cytome-ters to characterize red cell morphology,[83] this approach is limited as it only provides a general description of shape (e.g., ellipsoid vs. spherical) and is unable to provide the resolution required for aiding in differential diagnosis. Automated counters are thus designed to produce accurate measurements of normal blood and to alert the technician with "flags" when numerical abnormalities exist so that a smear may then be prepared and examined.[84] Even though auto-mated blood analyzers have reduced the number of samples that require smears to 15%, the examination of a smear is still an indispen-sible tool in providing differential diagnosis (commonly for anemias and thrombocytopenia), recommending further tests, speedy diagno-sis of certain infections, and the identification of leukemia and lym-phoma.[85] Despite the ability of the automated instruments to measure volume and hemoglobin concentrations, they are unable to accurately measure morphologic abnormalities and variations in shape, at the single cell level, and thus a pathologist is required to manually exam-ine a smear. Other modern methods that can be used for accurately assessing red cell morphology, such as confocal microscopy, suffer from complicated procedures and the need for using specialized exogenous contrast agents.

On the other hand, unstained tissue slices and blood smears exhibit negligible absorption, i.e., they are transparent or *phase objects*, and, as a result, provide no diagnostic information under bright field microscopy. Significant progress has been made in label-free, spectro-scopic imaging of tissues, including infrared[86-88] and Raman[89] spec-troscopic imaging. Spectroscopy methods, although offer information about the tissue chemical composition, are characterized by a low signal-to-noise ratio, which often hinders the data interpretation.[82] The more traditional label-free methods, i.e., phase contrast[90] and dif-ferential interference contrast (DIC)[91] can reveal structures from such transparent structures, without the need for staining. Nonetheless, phase contrast and DIC are plagued by optical artifacts ("halos" and "shadows," respectively), which often conceal details in the struc-tures. Most importantly, the information provided by these methods is *qualitative,* i.e., there is no linear relationship between the intensity in the image and the intrinsic properties of the tissue structures (e.g., mass density). Therefore, common signal-processing techniques, such as spatial correlations, mean, variance, and higher-order moments of potential diagnosis value lose their physical meaning when applied to phase-contrast and DIC images.

As discussed throughout the book, today there are a number of methods capable of providing accurate measurements of quantitative-phase imaging. Nevertheless, the potential of QPI for *label-free pathology* has been explored only recently. In the following we present recent results on using QPI for imaging blood smears and cancer biopsies.

15.4.2   Blood Screening

Recently, we have employed a new QPI on a chip method, *diffraction-phase cytometry* (DPC, see Sec. 14.3. and Ref. 92 for details on its principle), to characterize specific morphological abnormalities in diseased blood.[93] The simplicity and versatility of the DPC technique have been previously demonstrated by combining it with CD-ROM technology, for characterizing red blood cells (RBCs).[94] In order to obtain accurate morphological parameters from the retrieved phase map, we translate it to a thickness map using an index of refraction calculated based on the mean cell hemoglobin concentration of each sample, as measured by the impedance analyzer. Due to the linear dependence on the protein concentration[95] the refractive index can be calculated as: $n_c = n_0 + \beta \otimes MCHC$, where β is the refractive increment of hemoglobin (0.002 dL/g) and MCHC is the concentration of dry protein expressed in g/dL. The phase map $\phi(x, y)$ is then translated to a height map $h(x, y)$ using the contrast in refractive index between the cells and surrounding media, Δn: $h(x,y) = \dfrac{\lambda}{2\pi\Delta n}\phi(x,y)$, where $\lambda = 532$ nm is the wavelength of the illumination and $\Delta n = n_c - n_0$. Once the height information is retrieved, the volume of each cell is calculated by integrating the height map over the projected area as $V = \iint h(x,y)dxdy$. The surface area of individual cells is determined using Monge parameterization,[96] where the area of each pixel element, dA, is calculated as $dA = dxdy\sqrt{1 + h_x^2 + h_y^2}$ where dx and dy are the width and height of each pixel, and h_x and h_y are the gradients along the x and y directions, respectively. The surface area of each cell is then the sum of all the area elements and the projected area, assuming the cell is sitting flat on the cover slip. The sphericity, ψ of RBCs was first determined as an important parameter by Canham and Burton.[97] It is defined as the ratio between the surface area (SA) of a sphere with the same volume as the cell, to the actual surface area of the cell, with values ranging from 0 (for a laminar disk) to 1 for a perfect sphere and is calculated as $\psi = 4.84\dfrac{V^{2/3}}{SA}$. Knowing the surface area and volume, we can calculate parameters such as sphericity (ψ) and minimum cylindrical diameter (MCD).[98] The MCD, also introduced by Canham and Burton, is a theoretical parameter that predicts the smallest capillary diameter that a given RBC can squeeze through. The MCD is obtained by solving the following polynomial equation that defines the cell volume: $V = SA \otimes MCD - \dfrac{\pi MCD^3}{12}$. Overall, for each cell imaged we obtain the following 17 parameters: *perimeter, projected area, circular diameter, surface area, volume, sphericity, eccentricity, minimum, maximum, and mean height, minimum cylindrical diameter, circularity, integrated density, kurtosis, skewness,* and *variance.*

In this study, samples from 32 patients were analyzed using both a clinical Coulter impedance counter and the DPC system; with the DPC system we analyzed an average of 828 cells per sample.[93] In order to evaluate the consistency of the DPC analysis with that of the Coulter counter, we compared the mean corpuscular volumes (MCV) obtained by both methods (Fig. 15.11). Initially the data was analyzed assuming a constant refractive index contrast for all samples, which resulted in a weak correlation (Pearson correlation coefficient, $\rho = 0.52$) between the DPC and CBC volume data (cross symbols in Fig. 15.11). However, once the mean cellular hemoglobin concentration (MCHC) values were taken into account and the refractive index

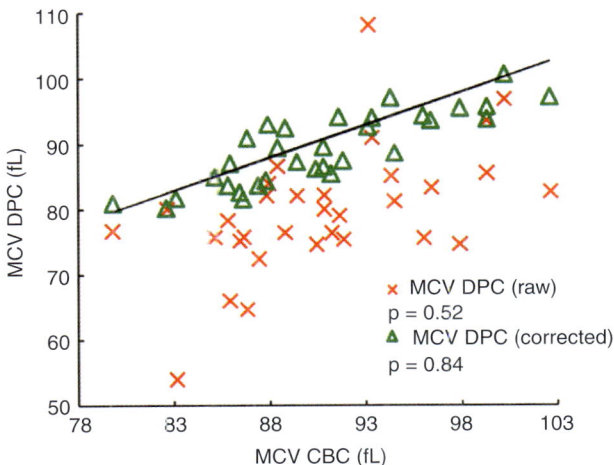

FIGURE 15.11 MCV values measured by DPC vs. impedance counter (complete blood count, CBC). The DPC data is shown before the correction for the refractive index (Raw) and after refractive index correction (Corrected). Pearson correlation coefficients for both data sets are shown in the legend. The straight line, included for comparison, represents the CBC MCV values. [SPIE, used with permission, *J. Biomed Opt.,* M. Mir, M. Ding, Z. Wang, J. Reedy, K. Tangella & G. Popescu, "Blood screening using diffraction phase cytometry," 02016-1-4 (2010). (*From Fig. 1, Ref. 93.*)]

contrast corrected, the correlation improves to $\rho = 0.84$ (triangular symbols in Fig. 15.11).

An important advantage of DPC as an emerging technology is that it recovers all metrics that are familiar and intuitive to pathologists, such as the MCV. One disorder that is fairly common and easy to diagnose using the MCV is *anisocytosis*, which is characterized by large variations in the cell volumes and quantified by the red cell distribution width (RDW). Figure 15.12 shows volume distributions from two patients, one normal and one exhibiting anisocytosis. Again we show images of cells across the distribution to illustrate the information available about each cell. This type of analysis enables the DPC system to accurately identify the morphological abnormalities that are responsible for the anisocytosis. Since anisocytosis could be a result of a variety of disorders such as thalassemia (decreased globin synthesis) and myelodisplastic syndrome (preleukimia),[85] more detailed information on the cause will aid in a quick and early automatic diagnosis of these conditions possible.

The strong dependence of our results on the cell hemoglobin content indicates that an accurate measurement of individual cell protein content needs to be made. A previous method entails measuring the

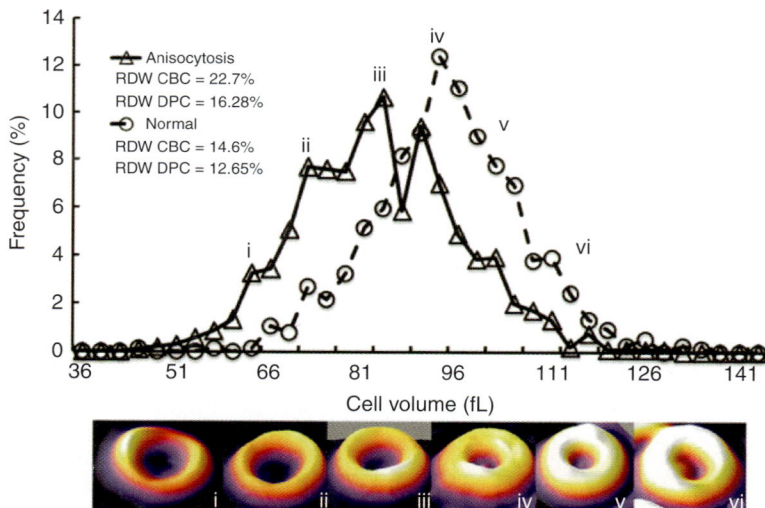

FIGURE 15.12 Comparison of volume distributions of patient exhibiting anisocytosis vs. a normal patient. The DPC measures red cell distribution width (RDW) values of 12.65 and 16.28 for the normal and abnormal patient, respectively. More subpopulations are apparent in the patient with anisocytosis. *i-vi* show examples of cells at the different volume peaks as follows: (*i*) 64 fL (*ii*) 72 fL (*iii*) 84 fL (*iv*) 93 fL (*v*) 102 fL (*vi*) 117 fL. [SPIE, used with permission, *J. Biomed Opt.*, M. Mir, M. Ding, Z. Wang, J. Reedy, K. Tangella & G. Popescu, "Blood screening using diffraction phase cytometry," 02016-1-4 (2010). (*From Fig. 3, Ref. 93.*)]

cells in two solutions with different refractive indices.[99] Though this decoupling method is an effective way to calculate the refractive index, it may be impractical in a clinical setting, due to throughput considerations and because exposing the cells to different solutions may affect their properties. Recently it has been shown that DPC can directly measure single cell hemoglobin concentration by either utilizing a broadband source[100] or by performing DPM at different wavelengths.[101] Both of these techniques rely on the dispersion properties of hemoglobin to infer the protein concentration (see Sec. 14.2). These new methods free the DPC from relying on any external measurements and thus greatly add to both its practical application in a clinic and its power in aiding differential diagnosis.

In conclusion, QPI methods may offer a powerful new blood screening utility that can be used to make differential diagnosis by an experienced pathologist. DPC, a form of QPI on a chip, can be simply added on as a modality to any existing microscopy and no special sample preparation is necessary to integrate it into the clinical workflow. Furthermore, the outputs of DPC are intuitive morphological characteristics, such as sphericity and skewness, meaning that no new specialized knowledge is necessary to take advantage of DPC. Advancements in spectroscopic phase measurements, image processing, and computing power will continue to augment the abilities of DPC, while maintaining its position as a low cost, high throughput, and highly sensitive instrument.

15.4.3 Label-Free Tissue Diagnosis

We implemented SLIM (for details see Sec. 12.2 and Refs. 35, 37) with a programmable scanning stage, which allows for imaging large areas of tissue with high transverse resolution, limited only by the numerical aperture of the objective (i.e., a fraction of a micron). All tissues were handled according to safety regulations by the Institutional Review Board at University of Illinois and Provena Medical Center. The specimen preparation is detailed in Chap. 4 of Ref. 82. Briefly, prostate tissue from patients was fixed with paraffin and sectioned in 4-μm thick slices. Four successive slices were imaged as follows. One unstained slice was de-parafined and placed in xylene solution for SLIM imaging. The other three slices were stained with H&E, K903, and AMACR, respectively, and imaged with the same microscope via the bright field channel equipped with a color camera.

Figure 15.13a shows the bright field (i.e., common intensity) image of the unstained slice. Clearly, little contrast can be observed, which indicates the long-standing motivation for the use of staining in clinical pathology. The H&E stained slice is shown in Fig. 15.13b. The contrast is greatly enhanced as the tissue structures show various shades of color, from dark purple to bright pink. Figure 15.13c shows the optical path-length map rendered by SLIM, which represents a mosaic of 4131

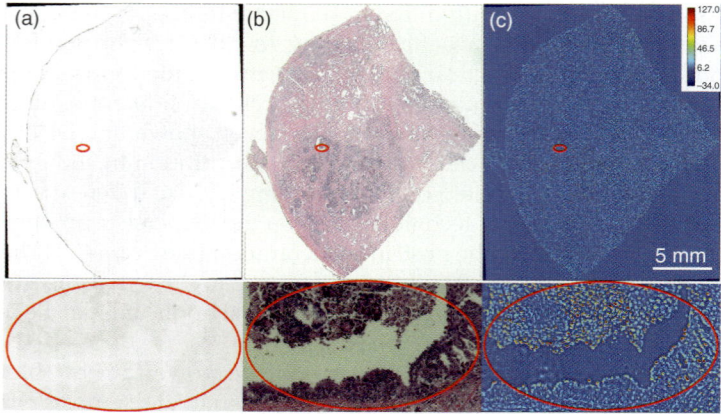

FIGURE 15.13   Multimodal imaging of prostate tissue slices. Objective: 10×/0.3. The field of view is 2.0 cm × 2.4 cm. The size of the blowout area (in red circle) is 630 μm × 340 μm. (*a*) Bright field image of unstained slice; (*b*) bright field image of H&E (hematoxylin and eosin)–stained slice; (*c*) SLIM phase map of the unstained slice. Colorbar indicates optical path length in nanometer. *(From Ref. 35.)* Z. Wang, K. Tangella, A. Balla and G. Popescu, *Tissue refractive index as marker of disease*, Proceedings of the National Academy of Sciences, under review).

individual images. Since the tissue thickness is 4 μm throughout the specimen, the SLIM image quantitatively captures the spatial fluctuations of the refractive index, which fully determines the elastic interaction with the optical field, i.e., its light-scattering properties.[76,77]

Further, we investigated the ability of SLIM's refractive index information to report on breast calcification. Mammogram is an important screening tool for detecting breast cancer.[102] Presence of abnormal calcifications, i.e., calcium phosphate and calcium oxalate,[103] warrants further workup. Calcium oxalate crystals are far more difficult to detect radiologically compared to calcium phosphate, because they do not stain with H&E.[104] Examination of breast tissue under polarized light facilitates detection of calcium oxalate crystals,[105] which is a step that potentially could be missed if the pathologist does not have high index of suspicion. Figure 15.14 illustrates how SLIM may eliminate this challenging task. In Fig. 15.14*b*, the dark H&E staining was identified by pathologists as calcium phosphate. This structure is revealed in the SLIM image as having inhomogeneous refractive index. On the other hand, calcium oxalate is hardly visible in H&E (Fig. 15.14*d*); the faint color hues are due to the well known birefringence of this type of crystal. Clearly, calcium oxalate exhibits a strong refractive index signature, as evidenced by the SLIM image. SLIM's ability to detect calcium oxalate in unstained breast biopsies may become very useful in modern pathology.

We also studied biopsies from *prostate cancer* patients. Ten biopsies from 8 patients were imaged with both SLIM and H&E, as

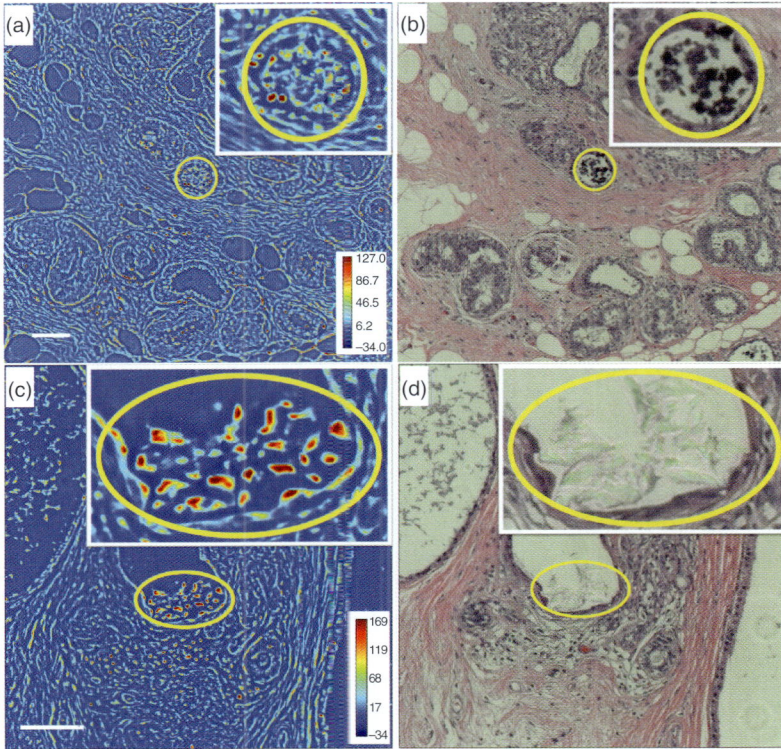

FIGURE 15.14 SLIM imaging signatures for breast tissues. (*a, b*) Breast tissue with calcium phosphate. The whole slice is 2.2 cm × 2.4 cm. The SLIM image is stitched by 4785 images and the H&E is stitched by 925 images. Left: SLIM image, colorbar in nanometer; right: H&E image. Scale bar is 100 μm. (*c, d*) Breast tissue with calcium oxalate. The entire slice is 1.6 cm × 2.4 cm. The SLIM image is stitched by 2840 images and the H&E is stitched by 576 images. Left: SLIM images, colorbar in nanometer; right: H&E image. Scale bar indicates 200 μm. (*From Ref. 35.*)
Z. Wang, K. Tangella, A. Balla and G. Popescu, *Tissue refractive index as marker of disease*, Proceedings of the National Academy of Sciences, under review).

illustrated in Fig. 15.15*a* and *b*. For each biopsy, the pathologist identified regions of normal and malignant tissue. From the SLIM image, we computed the map of phase shift *variance*, $\langle \Delta\phi(\mathbf{r})^2 \rangle$, where the angular brackets denote spatial average (calculated over $32 \times 32 \ \mu m^2$). Fig. 15.15*c* illustrates the map of the scattering mean free path, calculated from the variance as[76,77] $l_s = L/\langle \Delta\phi(\mathbf{r})^2 \rangle$ (see Sec. 15.3.2 for derivation). The spatially resolved scattering map shows very good correlation with cancerous and benign areas. These findings confirm in a direct way the importance of tissue light scattering as means for cancer diagnosis.[53,106–117] It can be easily seen that the regions of high variance, or short scattering mean free path, correspond to the darker staining in H&E, which is associated with cancer. Our

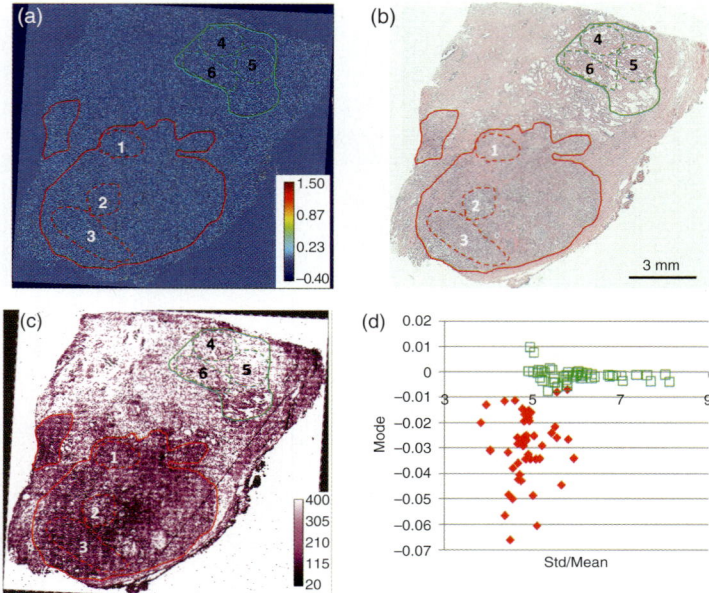

FIGURE 15.15 Multimodal imaging of cancerous prostate tissue slices. Field of view 1.48 cm × 1.44 cm. (*a*) SLIM unstained slice. Colorbar indicates optical path length in radian. The red lines marked the specific cancerous areas and the green lines marked the benign areas by pathologist. Small areas are also picked by pathologist as typical cancerous area (area 1, 2, and 3) and benign area (area 4, 5, and 6). (*b*) H&E–stained slice with the same areas marked as in *a*. (*c*) L_s map of the tissue slice with the same areas marked. Colorbar indicates l_s in μm. (*d*) 2D representation of 56 cancerous areas (red) and 53 benign areas (green) from 11 biopsies. *(From Ref. 35.)* Z. Wang, K. Tangella, A. Balla and G. Popescu, *Tissue refractive index as marker of disease*, Proceedings of the National Academy of Sciences, under review).

measurements indicated that the disease affects the tissue architecture in such a way as to render it more inhomogeneous.

In order to quantitatively analyze the information contained in the refractive index for the cancer and benign regions, we computed statistical parameters of 1st to 4th order via the respective histograms. Based on these distributions, we calculated the mean, standard deviation, mode, skewness, and kurtosis for each of the 56 cancer and 53 benign areas. Thus, we generated a multi-dimensional data space in which we searched for the best separation between the two groups of data points. Figure 15.15*d* illustrates that the mode vs. standard deviation normalized by the mean provides 100% separation (specificity) between the data points.

These promising results will stimulate further studies devoted to using QPI as a label-free method for cancer detection in biopsies. The hope is that the subtle refractive index changes associated with cancer onset and progression can be used as a marker not only for diagnosis but also for prognosis. The prospect of a highly automatic procedure, together with the low cost and high speed associated with the absence of staining, may make a significant impact in pathology at a global scale. We anticipate that this type of imaging will impact the field of biophotonics further by providing direct access to the scattering properties of tissues. Thus, we envision that a database of QPI images associated with various types of tissues, both healthy and diseased, will allow light scattering investigators to look up the scattering mean free path and anisotropy factors of tissues and ultimately predict outcomes of particular experiments, including optical coherence tomography, diffuse backscattering, spectroscopic scattering, enhanced backscattering, etc.

References

1. N. Mohandas and E. Evans, Mechanical Properties of the Red Cell Membrane in Relation to Molecular Structure and Genetic Defects, *Annual Reviews in Biophysics and Biomolecular Structure*, 23, 787–818 (1994).
2. F. Brochard and J. F. Lennon, Frequency spectrum of the flicker phenomenon in erythrocytes, *J. Physique*. 36, 1035–1047 (1975).
3. A. Zilker, H. Engelhardt and E. Sackmann, Dynamic reflection interference contrast (RIC-) microscopy: a new method to study surface excitations of cells and to measure membrane bending elastic moduli, *J. Physique*, 48, 2139–2151 (1987).
4. N. Gov, A. G. Zilman and S. Safran, Cytoskeleton Confinement and Tension of Red Blood Cell Membranes, *Phys. Rev. Lett.*, 90, 228101 (2003).
5. G. Popescu, T. Ikeda, K. Goda, C. A. Best-Popescu, M. Laposata, S. Manley, R. R. Dasari, K. Badizadegan and M. S. Feld, Optical Measurement of Cell Membrane Tension, *Phys. Rev. Lett.*, 97, 218101 (2006).
6. G. Popescu, Y. K. Park, R. R Dasari, K. Badizadegan and M. S. Feld, Coherence properties of red blood cell membrane motions, *Phys. Rev. E*, 76, 31902 (2007).
7. T. Browicz, Further observation of motion phenomena on red blood cells in pathological states, *Zentralbl. Med. Wiss.*, 28, 625–627 (1890).
8. A. Parpart and J. Hoffman, Flicker in erythrocytes. "vibratory movements in the cytoplasm"?, *Journal of Cellular and Comparative Physiology*, 47, 295–303 (1956).
9. S. Tuvia, S. Levin, A. Bitler and R. Korenstein, Mechanical fluctuations of the membrane-skeleton are dependent on F-actin ATPase in human erythrocytes, *J. Cell Biol.*, 141, 1551–1561 (1998).
10. J. Evans, W. Gratzer, N. Mohandas, K. Parker and J. Sleep, Fluctuations of the Red Blood Cell Membrane: Relation to Mechanical Properties and Lack of ATP Dependence, *Biophysical Journal*, 94, 4134 (2008).
11. N. S. Gov and S. A. Safran, Red Blood Cell Membrane Fluctuations and Shape Controlled by ATP-Induced Cytoskeletal Defects, *Biophys. J.*, 88, 1859 (2005).
12. N. S. Gov, Active elastic network: Cytoskeleton of the red blood cell, *Phys. Rev. E*, 75, 11921 (2007).
13. J. Li, G. Lykotrafitis, M. Dao and S. Suresh, Cytoskeletal dynamics of human erythrocyte, *Proceedings of the National Academy of Sciences*, 104, 4937 (2007).
14. R. Zhang and F. Brown, Cytoskeleton mediated effective elastic properties of model red blood cell membranes, *The Journal of Chemical Physics*, 129, 065101 (2008).
15. Y. K. Park, C. A. Best, T. Auth, N. Gov, S. A. Safran, G. Popescu, S. Suresh and M. S. Feld, Metabolic remodeling of the human red blood cell membran, *Proc. Nat. Acad. Sci.*, 107, 1289 (2010).

16. Y. K. Park, G. Popescu, K. Badizadegan, R. R. Dasari and M. S. Feld, Diffraction phase and fluorescence microscopy, *Opt. Express*, 14, 8263–8268 (2006).
17. G. Popescu, T. Ikeda, R. R. Dasari and M. S. Feld, Diffraction phase microscopy for quantifying cell structure and dynamics, *Opt. Lett.*, 31, 775–777 (2006).
18. M. Sheetz and S. Singer, On the mechanism of ATP-induced shape changes in human erythrocyte membranes. I. The role of the spectrin complex, *Journal of Cell Biology*, 73, 638–646 (1977).
19. C. Lawrence, N. Gov and F. Brown, Nonequilibrium membrane fluctuations driven by active proteins, *The Journal of Chemical Physics*, 124, 074903 (2006).
20. R. Kubo, The fluctuation-dissipation theorem, *Rep. Prog. Phys.*, 29, 255–284 (1966).
21. D. H. Boal, Mechanics of the Cell (Cambridge University Press, Cambridge, UK; New York, 2002).
22. Y. K. Park, C. A. Best, K. Badizadegan, R. R. Dasari, M. S. Feld, T. Kuriabova, M. L. Henle, A. J. Levine and G. Popescu, Measurement of red blood cell mechanics during morphological changes, *Proc. Nat. Acad. Sci.*, 107, 6731 (2010).
23. B. Alberts, Essential cell biology : an introduction to the molecular biology of the cell (Garland Pub., New York, 2004).
24. A. Yildiz, J. N. Forkey, S. A. McKinney, T. Ha, Y. E. Goldman and P. R. Selvin, Myosin V walks hand-over-hand: Single fluorophore imaging with 1.5-nm localization, *Science*, 300, 2061–2065 (2003).
25. E. R. Weeks, J. C. Crocker, A. C. Levitt, A. Schofield and D. A. Weitz, Three-dimensional direct imaging of structural relaxation near the colloidal glass transition, *Science*, 287, 627–631 (2000).
26. J. C. Crocker and D. G. Grier, Methods of digital video microscopy for colloidal studies, *Journal of Colloid and Interface Science*, 179, 298–310 (1996).
27. D. J. Durian, D. A. Weitz and D. J. Pine, Multiple Light-Scattering Probes of Foam Structure and Dynamics, *Science*, 252, 686–688 (1991).
28. R. Wang, Z. Wang, L. Millet, M. U. Gillette, A. J. Levine and G. Popescu, Spatiotemporal mass transport in living cells, *Phys. Rev. Lett.* (under review).
29. E. Abbe, Beiträge zur Theorie des Mikroskops und der mikroskopischen Wahrnehmung, *Arch. Mikrosk. Anat.*, 9, 431 (1873).
30. Y. K. Park, M. Diez-Silva, G. Popescu, G. Lykotrafitis, W. Choi, M. S. Feld and S. Suresh, Refractive index maps and membrane dynamics of human red blood cells parasitized by Plasmodium falciparum, *Proc. Natl. Acad. Sci. U S A*, 105, 13730 (2008).
31. G. Popescu, T. Ikeda, K. Goda, C. A. Best-Popescu, M. Laposata, S. Manley, R. R. Dasari, K. Badizadegan and M. S. Feld, Optical measurement of cell membrane tension, *Phys. Rev. Lett.*, 97, 218101 (2006).
32. G. Popescu, Y. Park, W. Choi, R. R. Dasari, M. S. Feld and K. Badizadegan, Imaging red blood cell dynamics by quantitative phase microscopy, *Blood Cells Molecules and Diseases*, 41, 10–16 (2008).
33. G. Popescu, Y. Park, N. Lue, C. Best-Popescu, L. Deflores, R. R. Dasari, M. S. Feld and K. Badizadegan, Optical imaging of cell mass and growth dynamics, *Am. J. Physiol. Cell Physiol.*, 295, C538–44 (2008).
34. N. Lue, G. Popescu, T. Ikeda, R. R. Dasari, K. Badizadegan and M. S. Feld, Live cell refractometry using microfluidic devices, *Opt. Lett.*, 31, 2759 (2006).
35. Z. Wang, K. Tangella, A. Balla and G. Popescu, Tissue refractive index as marker of disease, *Proc. Nat. Acad. Sci.*, (under review).
36. Z. Wang and G. Popescu, Quantitative phase imaging with broadband fields, *Applied Physics Letters*, 96, 051117 (2010).
37. Z. Wang, I. S. Chun, X. L. Li, Z. Y. Ong, E. Pop, L. Millet, M. Gillette and G. Popescu, Topography and refractometry of nanostructures using spatial light interference microscopy, *Opt. Lett.*, 35, 208–210 (2010).
38. D. L. Coy, M. Wagenbach and J. Howard, Kinesin takes one 8-nm step for each ATP that it hydrolyzes, *J. Biol. Chem.*, 274, 3667–3671 (1999).
39. J. B. Weitzman, Growing without a size checkpoint, *J Biol*, 2, 3 (2003).

40. A. Tzur, R. Kafri, V. S. LeElzu, G. Lahav and M. W. Kirschner, Cell growth and size homeostasis in proliferating animal cells, *Science*, 325, 167–171 (2009).
41. G. Reshes, S. Vanounou, I Fishov and M. Feingold, Cell shape dynamics in Escherichia coli, *Biophysical Journal*, 94, 251–264 (2008).
42. G. Popescu, Y. Park, N. Lue, C. Best-Popescu, L. Deflores, R. Dasari, M. Feld and K. Badizadegan, Optical imaging of cell mass and growth dynamics, *Am. J. Physiol.: Cell Physiol.*, 295, C538 (2008).
43. M. Godin, F. F. Delgado, S. Son, W. H. Grover, A. K. Bryan, A. Tzur, P. Jorgensen, K. Payer, A. D. Grossman, M. W. Kirschman and S. R. Manalis, Using buoyant mass to measure the growth of single cells, *Nat. Methods*, 7, 387–390 (2010).
44. A. K. Bryan, A. Goranov, A. Amon and S. R. Manalis, Measurement of mass, density, and volume during the cell cycle of yeast, *Proc Natl Acad Sci U S A*, 107, 999–1004 (2010).
45. K. Park, L. Millet, J. Huan, N Kim, G. Popescu, N. Aluru, K. J. Hsia and R. Bashir, Measurement of Adherent Cell Mass and Growth, *Proc. Nat. Acad. Sci.*, 2010).
46. H. G. Davies and M. H. Wilkins, Interference microscopy and mass determination, *Nature*, 169, 541 (1952).
47. R. Barer, Interference microscopy and mass determination, *Nature*, 169, 366–367 (1952).
48. M. Mir, Z. Wang, Z. Shen, M. Bednarz, R. Bashir, I. Golding, S. G. Prasanth and G. Popescu, Measuring Cell Cycle Dependent Mass Growth, *Proc. Nat. Acad. Sci.* (under review).
49. G. Popescu, Propagation of low-coherence optical fields in inhomogeneous media, *Ph.D. thesis* (CREOL / School of Optics, University of Central Florida, Orlando, 2002).
50. G. Popescu and A. Dogariu, Scattering of low coherence radiation and applications, invited review paper, *E. Phys. J.*, 32, 73–93 (2005).
51. V. Backman, M. B. Wallace, L. T. Perelman, J. T. Arendt, R. Gurjar, M. G. Muller, Q. Zhang, G. Zonics, E. Kline, J. A. McGilligan, S. Shapshay, T. Valdez, K. Badizadegan, J. M. Crawford, M. Fitzmaurice, S. Kabani, H. S. Levin, M. Seiler, R. R. Dasari, I Itzkan, J. Van Dam and M. S. Feld, Detection of preinvasive cancer cells, *Nature*, 406, 35–36 (2000).
52. R. Drezek, A. Dunn and R. Richards-Kortum, Light scattering from cells: finite-difference time-domain simulations and goniometric measurements, *Appl. Opt.*, 38, 3651–3661 (1999).
53. J. R. Mourant, M. Canpolat, C. Brocker, O. Esponda-Ramos, T. M. Johnson, A. Matanock, K. Stetter and J. P. Freyer, Light scattering from cells: the contribution of the nucleus and the effects of proliferative status, *J. Biomed. Opt.*, 5, 131–137 (2000).
54. V. V. Tuchin, Tissue optics (SPIE-The International Society for Optical Engineering, 2000).
55. J. W. Pyhtila, K. J. Chalut, J. D. Boyer, J. Keener, T. D'Amico, M. Gottfried, F. Gress and A. Wax, In situ detection of nuclear atypia in Barrett's esophagus by using angle-resolved low-coherence interferometry, *Gastrointestinal Endoscopy*, 65, 487–491 (2007).
56. C. S. Mulvey, A. L. Curtis, S. K. Singh and I. J. Bigio, Elastic scattering spectroscopy as a diagnostic tool for apoptosis in cell cultures, *IEEE Journal of Selected Topics in Quantum Electronics*, 13, 1663–1670 (2007).
57. H. Ding, J. Q. Lu, R. S. Brock, T. J. McConnell, J. F. Ojeda, K. M. Jacobs and X. H. Hu, Angle-resolved Mueller matrix study of light scattering by R-cells at three wavelengths of 442, 633, and 850 nm, *Journal of Biomedical Optics*, 12, 034032 (2007).
58. W. J. Cottrell, J. D. Wilson and T. H. Foster, Microscope enabling multimodality imaging, angle-resolved scattering, and scattering spectroscopy, *Opt. Lett.*, 32, 2348–2350 (2007).
59. H. Fang, L. Qiu, E. Vitkin, M. M. Zaman, C. Andersson, S. Salahuddin, L. M. Kimerer, P. B. Cipolloni, M. D. Modell, B. S. Turner, S. E. Keates, I. Bigio, I. Itzkan, S. D. Freedman, R. Bansil, E. B. Hanlon and L. T. Perelman, Confocal light absorption and scattering spectroscopic microscopy, *Applied Optics*, 46, 1760–1769 (2007).

60. I. Georgakoudi, B. C. Jacobson, J. Van Dam, V. Backman, M. B. Wallace, M. G. Muller, Q. Zhang, K. Badizadegan, D. Sun, G. A. Thomas, L. T. Perelman and M. S. Feld, Fluorescence, reflectance, and light-scattering spectroscopy for evaluating dysplasia in patients with Barrett's esophagus, *Gastroenterology*, 120, 1620–1629 (2001).

61. I. Itzkan, L. Qiu, H. Fang, M. M. Zaman, E. Vitkin, L. C. Ghiran, S. Salahuddin, M. Modell, C. Andersson, L. M. Kimerer, P. B. Cipolloni, K. H. Lim, S. D. Freedman, I. Bigio, B. P. Sachs, E. B. Hanlon and L. T. Perelman, Confocal light absorption and scattering spectroscopic microscopy monitors organelles in live cells with no exogenous labels, *Proceedings of the National Academy of Sciences of the United States of America*, 104, 17255–17260 (2007).

62. I. J. Bigio, S. G. Bown, G. Briggs, C. Kelley, S. Lakhani, D. Pickard, P. M. Ripley, I. G. Rose and C. Saunders, Diagnosis of breast cancer using elastic-scattering spectroscopy: preliminary clinical results, *J. Biomed. Opt.*, 5, 221–228 (2000).

63. V. Ntziachristos, A. G. Yodh, M. Schnall and B. Chance, Concurrent MRI and diffuse optical tomography of breast after indocyanine green enhancement, *Proceedings of the National Academy of Sciences of the United States of America*, 97, 2767–2772 (2000).

64. J. J. Duderstadt and L. J. Hamilton, Nuclear Reactor Analysis (Wiley, New York, 1976).

65. T. J. Farrell, M. S. Patterson and B. Wilson, A Diffusion-Theory Model of Spatially Resolved, Steady-State Diffuse Reflectance for the Noninvasive Determination of Tissue Optical-Properties Invivo, *Medical Phys.*, 19, 879–888 (1992).

66. M. S. Patterson, B. Chance and B. C. Wilson, Time Resolved Reflectance and Transmittance for the Noninvasive Measurement of Tissue Optical-Properties, *Appl. Opt.*, 28, 2331–2336 (1989).

67. J. M. Schmitt and G. Kumar, Optical scattering properties of soft tissue: a discrete particle model, *Appl. Opt.*, 37, 2788–2797 (1998).

68. M. Hunter, V. Backman, G. Popescu, M. Kalashnikov, C. W. Boone, A. Wax, G. Venkatesh, K. Badizadegan, G. D. Stoner and M. S. Feld, Tissue Self-Affinity and Light Scattering in the Born Approximation: A New Model for Precancer Diagnosis, *Phys. Rev. Lett.*, 97, 138102 (2006).

69. S. T. Flock, M. S. Patterson, B. C. Wilson and D. R. Wyman, Monte-Carlo Modeling of Light-Propagation in Highly Scattering Tissues .1. Model Predictions and Comparison with Diffusion-Theory, *IEEE Transactions on Biomedical Engineering*, 36, 1162–1168 (1989).

70. A. Dunn and R. Richards-Kortum, Three-dimensional computation of light scattering from cells, *IEEE Journal of Selected Topics in Quantum Electronics*, 2, 898–905 (1996).

71. H. Ding, Z. Wang, F. Nguyen, S. A. Boppart and G. Popescu, Fourier transform light scattering of inhomogeneous and dynamic structures, *Phys. Rev. Lett.*, 101, 238102 (2008).

72. H. Ding, Z. Wang, F. Nguyen, S. A. Boppart, L. J. Millet, M. U. Gillette, J. Liu, M. Boppart and G. Popescu, Fourier Transform Light Scattering (FTLS) of Cells and Tissues, *Journal of Computational and Theoretical Nanoscience*, 7, 1546 (in press).

73. H. Ding, E. Berl, Z. Wang, L. J. Millet, M. U. Gillette, J. Liu, M. Boppart and G. Popescu, Fourier transform light scattering of biological structures and dynamics, *IEEE Journal of Selected Topics in Quantum Electronics* (in press).

74. H. Ding, L. J. Millet, M. U. Gillette and G. Popescu, Actin-driven cell dynamics probed by Fourier transform light scattering, *Biomed. Opt. Express*, 1, 260 (2010).

75. H. Ding, F. Nguyen, S. A. Boppart and G. Popescu, Optical properties of tissues quantified by Fourier transform light scattering, *Opt. Lett.*, 34, 1372 (2009).

76. Z. Wang, H. Ding and G. Popescu, Scattering-phase teorem, *Opt. Lett.* (under review).

77. H. Ding, X. Liang, Z. Wang, S. A. Boppart and G. Popescu, Tissue scattering parameters from organnell to organ scale, *Opt. Lett.* (under review).

78. R. N. Bracewell, The Fourier transform and its applications (McGraw Hill, Boston, MA 2000).
79. Z. Wang, H. Ding and G. Popescu, A scattering-phase theorem, *Opt. Lett.* (under review).
80. Milestones in light microscopy, *Nat. Cell Biol.*, 11, 1165–1165 (2009).
81. V. Kumar, A. K. Abbas, N. Fausto, S. L. Robbins and R. S. Cotran, Robbins and Cotran pathologic basis of disease (Elsevier Saunders, Philadelphia, 2005).
82. G. Popescu, *Nanobiophotonics* (2010).
83. M. Piagnerelli, K. Z. Boudjeltia, D. Brohee, A. Vereerstraeten, P. Piro, J.-L. Vincent and M. Vanhaeverbeek, Assessment of erythrocyte shape by flow cytometry techniques, *J. Clin. Pathol.*, 60, 549–554 (2007).
84. B. J. Bain, Blood Cells A Practical Guide (Blackwell Science, London, 2002).
85. B. J. Bain, Diagnosis from the Blood Smear, *The New England Journal of Medicine*, 353, 489–507 (2005).
86. R. Bhargava, Towards a practical Fourier transform infrared chemical imaging protocol for cancer histopathology, *Analytical and Bioanalytical Chemistry*, 389, 1155–1169 (2007).
87. D. C. Fernandez, R. Bhargava, S. M. Hewitt and I. W. Levin, Infrared spectroscopic imaging for histopathologic recognition, *Nature Biotechnology*, 23, 469–474 (2005).
88. O. A. C. Petroff, G. D. Graham, A. M. Blamire, M. Alrayess, D. L. Rothman, P. B. Fayad, L. M. Brass, R. G. Shulman and J. W. Prichard, Spectroscopic Imaging of Stroke In Humans-Histopathology Correlates of Spectral Changes, *Neurology*, 42, 1349–1354 (1992).
89. M. D. Schaeberle, V. F. Kalasinsky, J. L. Luke, E. N. Lewis, I. W. Levin and P. J. Treado, Raman chemical imaging: Histopathology of inclusions in human breast tissue, *Analytical Chemistry*, 68, 1829–1833 (1996).
90. F. Zernike, How I discovered phase contrast, *Science* 121, 345 (1955).
91. G. Nomarski, Differential microinterferometer with polarized waves, *J. Phys. Radium*, 16, 9s–13s (1955).
92. M. Mir, Z. Wang, K. Tangella and G. Popescu, Diffraction Phase Cytometry: blood on a CD-ROM, *Opt Exp.*, 17, 2579 (2009).
93. M. Mir, M. Ding, Z. Wang, J. Reedy, K. Tangella and G. Popescu, Blood screening using diffraction phase cytometry, *J. Biomed. Opt.*, 027016–1–4 (2010).
94. M. Mir, Z. Wang, K. Tangella and G. Popescu, Diffraction Phase Cytometry: blood on a CD-ROM, *Optics Express*, 17, 2579–2585 (2009).
95. G. Popescu, Y. Park, N. Lue, C. Best-Popescu, L. Deflores, R. R. Dasari, M. S. Feld and K. Badizadegan, Optical imaging of cell mass and growth dynamics, *Am. J. Physiol. Cell Physiol*, 295, C538–C544 (2008).
96. S. A. Safran, Statistical Thermodynamics of Surfaces, Interfaces, and Membranes (Addison-Wesley, 1994).
97. P. B. Canham and A. C. Burton, Distribution of Size and Shape in Populations of Normal Human Red Cells, *Circ. Res.*, 22, 405–422 (1968).
98. P. B. Canham, Difference in Geometry of Young and Old Human Erythrocytes Explained by a Filtering Mechanism, *Circ. Res.*, 25, 39–45 (1696).
99. B. Rappaz, A. Barbul, Y. Emery, R. Korenstein, C. Depeursinge, J. M. Pierre and P. Marquet, Comparative Study of Human Erythrocytes by Digital Holographic Microscopy, Confocal Microscopy, and Impedance Volume Analyzer, *Cytometry Part A*, 73A, 895–903 (2008).
100. H. Ding and G. Popescu, Instantaneous Spatial Light Interference Microscopy, *Optics Express*, 18, 1569–1575 (2010).
101. Y. K. Park, T. Yamauchi, W. Choi, R. Dasari and M. S. Feld, Spectroscopy phase microscopy for quantifying hemoglobin concentrations in intact red blood cells, *Opt. Lett.*, 34, 3668–3670 (2009).
102. H. D. Nelson, K. Tyne, A. Naik, C. Bougatsos, B. K. Chan and L. Humphrey, Screening for breast cancer: an update for the U.S. Preventive Services Task Force, *Ann. Intern. Med.*, 151, 727–737, W237–42 (2009).
103. G. M. Tse, P. H. Tan, A. L. Pang, A. P. Tang and H. S. Cheung, Calcification in breast lesions: pathologists' perspective, *J. Clin. Pathol.*, 61, 145–151 (2008).

104. G. M. Tse, P. H. Tan, H. S. Cheung, W. C. Chu and W. W. Lam, Intermediate to highly suspicious calcification in breast lesions: a radio-pathologic correlation, *Breast Cancer Res Treat*, 110, 1–7 (2008).
105. L. Cook, J. Vinding and H. W. Gordon, Polarizing calcifications, *Am. J. Surg. Pathol.*, 21, 255–256 (1997).
106. R. A. Drezek, R. Richards-Kortum, M. A. Brewer, M. S. Feld, C. Pitris, A. Ferenczy, M. L. Faupel and M. Follen, Optical imaging of the cervix, *Cancer*, 98, 2015–2027 (2003).
107. M. G. Muller, T. A. Valdez, I. Georgakoudi, V. Backman, C. Fuentes, S. Kabani, N. Laver, Z. Wang, C. W. Boone, R. R. Dasari, S. M. Shapshay and M. S. Feld, Spectroscopic detection and evaluation of morphologic and biochemical changes in early human oral carcinoma, *Cancer*, 97, 1681–1692 (2003).
108. D. A. Benaron, The future of cancer imaging, *Cancer and Metastasis Reviews*, 21, 45–78 (2002).
109. V. Ntziachristos, J. Ripoll, L. H. V. Wang and R. Weissleder, Looking and listening to light: the evolution of whole-body photonic imaging, *Nature Biotechnology*, 23, 313–320 (2005).
110. H. F. Zhang, K. Maslov, G. Stoica and L. H. V. Wang, Functional photoacoustic microscopy for high-resolution and noninvasive in vivo imaging, *Nature Biotechnology*, 24, 848–851 (2006).
111. R. S. Gurjar, V. Backman, L. T. Perelman, I. Georgakoudi, K. Badizadegan, I. Itzkan, R. R. Dasari and M. S. Feld, Imaging human epithelial properties with polarized light-scattering spectroscopy, *Nature Medicine*, 7, 1245–1248 (2001).
112. L. V. Wang, Multiscale photoacoustic microscopy and computed tomography, *Nature Photonics*, 3, 503–509 (2009).
113. Z. Yaqoob, D. Psaltis, M. S. Feld and C. Yang, Optical phase conjugation for turbidity suppression in biological samples, *Nature Photonics*, 2, 110–115 (2008).
114. B. Brooksby, B. W. Pogue, S. D. Jiang, H. Dehghani, S. Srinivasan, C. Kogel, T. D. Tosteson, J. Weaver, S. P. Poplack and K. D. Paulsen, Imaging breast adipose and fibroglandular tissue molecular signatures by using hybrid MRI-guided near-infrared spectral tomography, *Proceedings of the National Academy of Sciences of the United States of America*, 103, 8828–8833 (2006).
115. A. Cerussi, D. Hsiang, N. Shah, R. Mehta, A. Durkin, J. Butler and B. J. Tromberg, Predicting response to breast cancer neoadjuvant chemotherapy using diffuse optical spectroscopy, *Proceedings of the National Academy of Sciences of the United States of America*, 104, 4014–4019 (2007).
116. S. Srinivasan, B. W. Pogue, S. D. Jiang, H. Dehghani, C. Kogel, S. Soho, J. J. Gibson, T. D. Tosteson, S. P. Poplack and K. D. Paulsen, Interpreting hemoglobin and water concentration, oxygen saturation, and scattering measured in vivo by near-infrared breast tomography, *Proceedings of the National Academy of Sciences of the United States of America*, 100, 12349–12354 (2003).
117. H. Subramanian, P. Pradhan, Y. Liu, I. R. Capoglu, X. Li, J. D. Rogers, A. Heifetz, D. Kunte, H. K. Roy, A. Taflove and V. Backman, Optical methodology for detecting histologically unapparent nanoscale consequences of genetic alterations in biological cells, *Proceedings of the National Academy of Sciences of the United States of America*, 105, 20118–20123 (2008).

APPENDIX **A**

Complex Analytic Signals

$\mathbf{I}$t is very convenient to describe a given real signal (such as an optical field) via an associated complex quantity. This way, many nuisances due to trigonometry are avoided. The key is to realize that a real signal, $U_r(t)$, has a Fourier transform $\tilde{U}(\omega)$ that is *Hermitian*,

$$U_r(t) = \int_{-\infty}^{\infty} \tilde{U}(\omega) \cdot e^{-i\omega t}\, d\omega \tag{A.1a}$$

$$\tilde{U}(-\omega) = \tilde{U}^*(\omega) \tag{A.1b}$$

Thus, the information contained in the negative frequencies is identical with that contained in positive frequencies. Therefore, suppressing the negative frequencies in Eq. (A.1a) does not generate any loss of information. With this operation, we obtain a new function, called the *complex analytic* signal associated with the real signal U_r, defined as

$$U(t) = 2\int_{0}^{\infty} \tilde{U}(\omega) \cdot e^{i\omega t} d\omega \tag{A.2}$$

where the factor of 2 will become clear shortly. Function $U(t)$ is now complex, because its Fourier transform misses the negative frequencies and is, thus, non-Hermitian. In order to find an explicit relationship between $U(t)$ and $U_r(t)$, let us express Eq. (A.2) as

$$U(t) = 2\int_{-\infty}^{\infty} \tilde{U}(\omega) \cdot H(\omega) \cdot e^{-i\omega t}\, dt \tag{A.3a}$$

$$H(\omega) = \begin{vmatrix} 1, & \omega > 0 \\ \tfrac{1}{2}, & \omega = 0 \\ 0, & \omega < 0 \end{vmatrix} \tag{A.3b}$$

where $H(\omega)$ is known as the Heaviside (or step) function and is used to suppress the negative ω. The Fourier transform of the product in Eq. (A.3a) is the convolution of the two Fourier transforms. The inverse Fourier transforms of $\tilde{U}(\omega)$ and $H(\omega)$ are given by

$$\tilde{U}(\omega) \rightarrow U_r(t)$$

$$H(\omega) \rightarrow \frac{1}{2}\delta(t) - \frac{i}{2\pi t} \tag{A.4}$$

As usual, the arrows in Eq. (A.4) indicate Fourier transformation. As a result, Eq. (A.3a) can be rewritten by inserting H and via the *convolution theorem* as

$$U(t) = U_r(t) - \frac{i}{\pi} P \int \frac{U_r(t')}{t-t'} dt' \tag{A.5}$$

where we used the property that a convolution of a function with a Dirac-delta function returns the original function (shifted at the position of the delta function).

Figure A.1 illustrates the operations involved in computing the complex analytic signal associated with a real field. Equation (A.5) reveals very important properties of the *complex analytic signal U*, as follows:

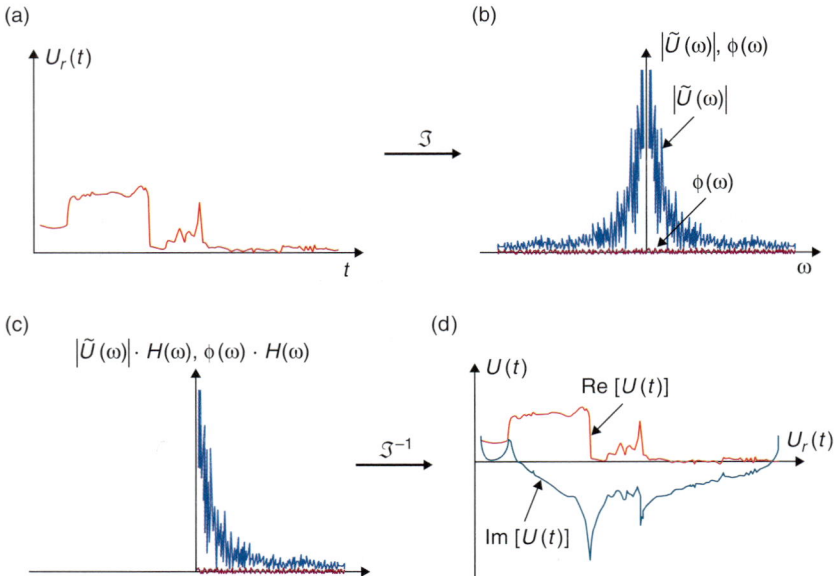

FIGURE A.1   From a real function to its complex analytic signal.

(i)   Its real part recovers the original (real) signal,

$$\text{Re}[U(t)] = U_r(t) \tag{A.6}$$

Note that this relationship holds without scaling factors because the factor of 2 was already incorporated in Eq. (A.2).

(ii) The real part, $\text{Re}[U(t)]$, and imaginary part, $\text{Im}[U(t)]$, are related via

$$U_i(t) = \text{Im}[U(t)]$$
$$= -\frac{1}{\pi} P \int_{-\infty}^{\infty} \frac{U_r(t')}{t - t'} dt' \tag{A.7}$$

The *principal value integral* in Eq. (A.7) is known as the *Hilbert transform*. Mathematically, it can be seen that the Hilbert transform of a function $f(x)$ is equivalent to a convolution with the function $1/x$. This relationship between the real and imaginary parts of U should come as no surprise, because the imaginary part was generated from simply suppressing the negative frequencies of U_r, i.e., U_i does not contain any new information that is not already in U_r.

(iii) The time-averaged modulus squared of the complex analytic signal is proportional to the field *irradiance*,

$$\left\langle \left| U(t) \right|^2 \right\rangle_t = \left\langle U_r^2(t) + U_i^2(t) \right\rangle_t$$
$$= 2 \left\langle U_r^2(t) \right\rangle_t \tag{A.8}$$

With these three properties, we can safely denote a real field by its complex analytic signal, which greatly simplifies the calculations. As a common example, the irradiance associated with the interference between two fields is computed much faster in the complex representation. Thus in the real-field representation, we have (for $a_1, a_2 \in \mathbb{R}$ the two amplitudes, ω the frequency, and φ_1, φ_2 the phases of each field)

$$I = \left\langle [a_1 \cos(\omega t + \varphi_1) + a_2 \cos(\omega t + \varphi_2)]^2 \right\rangle_t$$
$$= \frac{a_1^2}{2} + \frac{a_2^2}{2} + \left\langle 2 a_1 a_2 \cdot \cos(\omega t + \varphi_1) \cdot \cos(\omega t + \varphi_2) \right\rangle_t \tag{A.9}$$
$$= \frac{a_1^2}{2} + \frac{a_2^2}{2} + 2 a_1 a_2 \left[\frac{\cos(\varphi_2 - \varphi_1)}{2} + \left\langle \frac{\cos(2\omega t + \varphi_1 + \varphi_2)}{2} \right\rangle_t \right]$$

The last term, i.e., containing $2\omega t$ as argument, averages to zero such that the final intensity of the sum field reads (we used the fact that the cosine squared averages to $1/2$)

$$I = \frac{a_1^2}{2} + \frac{a_2^2}{2} + a_1 a_2 \cdot \cos(\varphi_2 - \varphi_1) \qquad (A.10)$$

Now, let us calculate the same irradiance via the complex analytic signals (Eq. [A.8])

$$I = \frac{1}{2}\left\langle |U(t)|^2 \right\rangle_t$$

$$= \frac{1}{2}\left\langle \left| a_1 \cdot e^{i\varphi_1} + a_2 \cdot e^{i\varphi_2} \right|^2 \right\rangle_t \qquad (A.11)$$

$$= \frac{1}{2}a_1^2 + \frac{1}{2}a_2^2 + a_1 a_2 \cos(\varphi_2 - \varphi_1)$$

Clearly, the complex representation is far more efficient in describing field superposition, especially when a large number of field components are involved. This formalism will be used throughout the book.

Further Reading

1. M. Born and E. Wolf. *Principles of optic: Electromagnetic Theory of Propagation, Interference and Diffraction of Light* (Cambridge University Press, Cambridge; New York, 1999).
2. L. Mandel and E. Wolf. *Optical Coherence and Quantum Optics* (Cambridge University Press, Cambridge; New York, 1995).
3. E. Wolf. *Introduction to the Theory of Coherence and Polarization of Light* (Cambridge University Press, Cambridge, 2007).
4. A. Papoulis. *The Fourier Integral and its Applications* (McGraw-Hill, New York, 1962).

The Two-Dimensional and Three-Dimensional Fourier Transform

B.1 The 2D Fourier Transform

In studying imaging, the concept of Fourier transforms must be generalized from 1D[1] to 2D and 3D functions.[2-4] For example, diffraction and 2D image formation are treated efficiently via a 2D Fourier transforms, while light scattering and tomographic reconstructions require 3D Fourier transforms.

A 2D function, f, can be reconstructed from its Fourier transform as

$$f(x,y) = \int_{-\infty}^{\infty} \int_{-\infty}^{\infty} \tilde{f}(k_x, k_y) \cdot e^{i(k_x \cdot x + k_y \cdot y)} dk_x dk_y \qquad (B.1)$$

The inverse relationship reads

$$\tilde{f}(k_x, k_y) = \int_{-\infty}^{\infty} \int_{-\infty}^{\infty} f(x,y) \cdot e^{-i(k_x \cdot x + k_y \cdot y)} dx dy \qquad (B.2)$$

Thus $\tilde{f}(k_x, k_y)$, generally a complex function, sets the amplitude and phase associated with the sinusoidal of frequency, $\mathbf{k} = (k_x, k_y)$. The contours of constant phase are given by

$$\phi(x,y) = k_x \cdot x + k_y \cdot y$$
$$= \text{const.} \qquad (B.3)$$

Equation (B.3) can be expressed as

$$\phi(x,y) = k\left(x \cdot \frac{k_x}{k} + y \cdot \frac{k_y}{k}\right)$$

$$= \text{const.}$$
(B.4)

where $k = |\mathbf{k}|$, $k = \sqrt{k_x^2 + k_y^2}$. From Eq. (B.4), it is clear that the direction of the contour makes an angle $\frac{k_x}{|k|}$ with the x-axis and has a wavelength $\Lambda = \frac{2\pi}{|k|}$ (Fig. B.1).

Thus, Eq. (B.1) indicates that the 2D function, f, (e.g., an image) is a superposition of waves of the type shown in Fig. B.1c and d, with appropriate amplitude and phase for each frequency (k_x, k_y). The

(a)

(b)

(c)

(d)

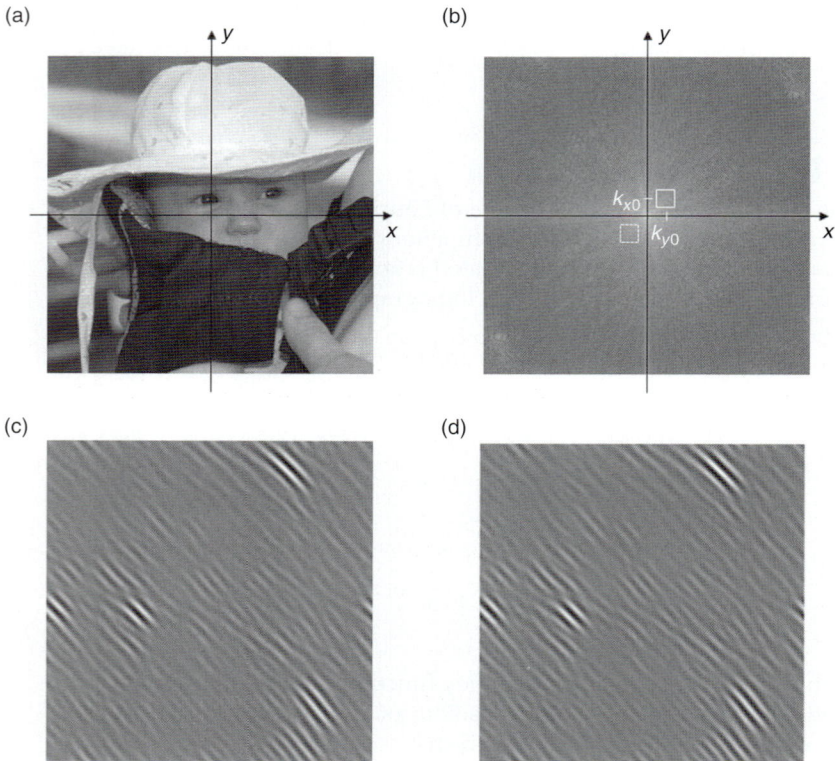

FIGURE B.1   (a) Example of a 2D function $f(x,y)$. (b) The modulus of the Fourier transform (i.e., spectrum). (c) The real part of the image obtained by band-passing the signal around the frequency (k_{x0}, k_{y0}), as indicated by the square region in b. (d) The imaginary counterpart of the image in c. Note that if the filter in b retains also the frequencies symmetric with respect to the origin, i.e., in the dash square, then the resulting image will be real, i.e., the image in d would be identically zero.

Fourier transform $\tilde{f}(k_x, k_y)$ simply assigns these amplitudes and phases for each frequency component (k_x, k_y).

B.2 Two-Dimensional Convolution

The *convolution* operation between two 2D functions, $f(x, y)$ and $g(x, y)$ is defined as

$$f \circledS_{xy} g = \int_{-\infty}^{\infty} \int_{-\infty}^{\infty} f(x', y') g(x - x', y - y') dx' \, dy' \tag{B.5}$$

where the symbol $*_{xy}$ indicates a *convolution* along the x and y coordinates. Note that g is rotated by $180°$ about the origin due to the change of sign in both x' and y', then displaced, and the products integrated over the plane. Often, one encounters one-dimensional convolutions of 2D functions,

$$f \circledS_x g = \int_{-\infty}^{\infty} \bar{f}(x', y') g(x - x', y') dx' \tag{B.6}$$

One example of such convolutions may occur when using cylindrical optics.

The 2D *cross-correlation* integral of f and g is

$$f \otimes_{xy} g = \int_{-\infty}^{\infty} \int_{-\infty}^{\infty} f(x', y') \cdot g(x + x', y + y') dx' dy' \tag{B.7}$$

In Eq. (B.7), $\otimes_{xy}$ stands for the 2D *cross-correlation* operation. Thus, like in the 1D case, the only difference between convolution and correlation is in the sign of g, which establishes whether or not g is rotated around the origin.

B.3 Theorems Specific to Two-Dimensional Functions

Rotation theorem: If $f(x, y)$ is rotated in the (x, y) plane then its Fourier transform is rotated in the (k_x, k_y) plane by the same angle. The rotated coordinates are

$$\begin{pmatrix} u \\ v \end{pmatrix} = \begin{pmatrix} \cos\theta - \sin\theta \\ \sin\theta \quad \cos\theta \end{pmatrix} \begin{pmatrix} x \\ y \end{pmatrix}$$

$$= \begin{pmatrix} x\cos\theta - y\sin\theta \\ x\sin\theta + y\cos\theta \end{pmatrix} \tag{B.8}$$

Thus the rotation theorem states

$$f(x\cos\theta - y\sin\theta, x\sin\theta + y\cos\theta) \rightarrow \tilde{f}(k_x\cos\theta - k_y\sin\theta, k_x\sin\theta + k_y\cos\theta) \tag{B.9}$$

Proof: The Fourier transform of the rotated function is

$$\tilde{f}_\theta(k_x, k_y) = \iint f(x\cos\theta - y\sin\theta, x\sin\theta + y\cos\theta)e^{-i(k_x \cdot x + k_y \cdot y)}dxdy$$

(B.10)

Let us use the change of variables

$$u = x\cos\theta - y\sin\theta$$
$$v = x\sin\theta + y\cos\theta$$
$$x = u\cos\theta + v\sin\theta$$
$$y = -u\sin\theta + v\cos\theta$$

(B.11)

It follows that

$$dudv = \begin{vmatrix} \dfrac{du}{dx} & \dfrac{du}{dy} \\ \dfrac{dv}{dx} & \dfrac{dv}{dy} \end{vmatrix} dxdy$$

(B.12)

$$= dxdy$$

Equation (B.10) becomes

$$f_\theta(k_x, k_y) = \iint f(u,v) \cdot e^{-i[k_x(u\cos\theta + v\sin\theta) + k_y(-u\sin\theta + v\cos\theta)]}dudv$$

$$= \iint f(u,v) \cdot e^{-i[u(k_x\cos\theta - k_y\sin\theta) + v(k_x\sin\theta + k_y\cos\theta)]}dudv$$

(B.13)

$$= \tilde{f}(k_x\cos\theta - k_y\sin\theta, k_x\sin\theta + k_y\cos\theta) \qquad (q.e.d.)$$

B.4   Generalization of 1D Theorems

Central ordinate theorem

$$\int_{-\infty}^{\infty}\int_{-\infty}^{\infty} f(x,y)dxdy = \tilde{f}(0,0)$$

(B.14)

Shift theorem

$$f(x - a, y - b) \rightarrow e^{-i(k_x \cdot a + k_y \cdot b)} \cdot \tilde{f}(k_x, k_y)$$

(B.15)

Similarity theorem

$$f(ax, by) \rightarrow \frac{1}{|ab|} \cdot \tilde{f}\left(\frac{k_x}{a}, \frac{k_y}{b}\right)$$

(B.16)

Convolution theorem

$$f(x,y) \otimes_{xy} g(x,y) \rightarrow \tilde{f}(k_x,k_y) \cdot \tilde{g}(k_x,k_y) \qquad \text{(B.17)}$$

Correlation theorem

$$f(x,y) \otimes_{xy} g(x,y) \rightarrow \tilde{f}(k_x,k_y) \cdot \tilde{g}^*(k_x,k_y) \qquad \text{(B.18)}$$

Modulation theorem

$$f(x,y)\cos bx \rightarrow \frac{1}{2}\tilde{f}(k_x+b,k_y) + \frac{1}{2}\tilde{f}(k_x-b,k_y) \qquad \text{(B.19)}$$

Parseval's theorem

$$\int_{-\infty}^{\infty}\int_{-\infty}^{\infty} |f(x,y)|^2\, dx\, dy = \int_{-\infty}^{\infty}\int_{-\infty}^{\infty} |F(k_x,k_y)|^2\, dk_x\, dk_y \qquad \text{(B.20)}$$

Differentiation properties

$$\left(\frac{\partial}{\partial x}\right)^m \left(\frac{\partial}{\partial y}\right)^n f(z,y) \rightarrow (ik_x)^m (ik_y)^n\, \tilde{f}(k_x,k_y) \qquad \text{(B.21a)}$$

$$\left[\left(\frac{\partial}{\partial x}\right)^m + \left(\frac{\partial}{\partial y}\right)^n\right] f(x,y) \rightarrow \left[(ik_x)^m + (ik_y)^n\right]\tilde{f}(k_x,k_y) \qquad \text{(B.21b)}$$

First moments

$$\int_{-\infty}^{\infty}\int_{-\infty}^{\infty} xf(x,y)dxdy = i\frac{\partial \tilde{f}}{\partial k_x}(0,0)$$

$$\int_{-\infty}^{\infty}\int_{-\infty}^{\infty} yf(x,y)dxdy = i\frac{\partial \tilde{f}}{\partial k_y}(0,0)$$

$$\text{(B.22)}$$

Center of gravity

$$\langle x \rangle = \frac{\displaystyle\int_{-\infty}^{\infty}\int_{-\infty}^{\infty} xf(x,y)dxdy}{\displaystyle\int_{-\infty}^{\infty}\int_{-\infty}^{\infty} f(z,y)dxdy} = i\frac{\left.\dfrac{\partial \tilde{f}}{\partial k_x}\right/ (0,0)}{\tilde{f}(0,0)}$$

$$\langle y \rangle = \frac{\displaystyle\int_{-\infty}^{\infty}\int_{-\infty}^{\infty} yf(x,y)dxdy}{\displaystyle\int_{-\infty}^{\infty}\int_{-\infty}^{\infty} f(x,y)dxdy} = i\frac{\left.\dfrac{\partial \tilde{f}}{\partial k_y}\right/ (0,0)}{\tilde{f}(0,0)}$$

$$\text{(B.23)}$$

Second moments

$$\int_{-\infty}^{\infty}\int_{-\infty}^{\infty} x^2 f(x,y)dxdy = -\frac{\partial^2 \tilde{f}}{\partial k_x^2}(0,0)$$

$$\int_{-\infty}^{\infty}\int_{-\infty}^{\infty} y^2 f(x,y)dxdy = -\frac{\partial^2 \tilde{f}}{\partial k_y^2}(0,0)$$

$$\int_{-\infty}^{\infty}\int_{-\infty}^{\infty} xyf(x,y)dxdy = -\frac{\partial^2 \tilde{f}}{\partial k_x \partial k_y}(0,0)$$ (B.24)

$$\int_{-\infty}^{\infty}\int_{-\infty}^{\infty} (x^2+y^2)f(x,y)dxdy = -\left[\frac{\partial^2 \tilde{f}}{\partial k_x^2}(0,0) + \frac{\partial^2 \tilde{f}}{\partial k_y^2}(0,0)\right]$$

Equivalent width

$$\frac{\int_{-\infty}^{\infty}\int_{-\infty}^{\infty} f(x,y)dxdy}{f(0,0)} = \frac{\tilde{f}(0,0)}{\int_{-\infty}^{\infty}\int_{-\infty}^{\infty} \tilde{f}(k_x,k_y)dk_x dk_y}$$ (B.25)

B.5   The Hankel Transform

Many optical systems exhibit circular symmetry. Light emitted in 2D by a point source also exhibits this symmetry. In this case the problem simplifies significantly as the only nontrivial variable is the radial coordinate. Thus, changing from cartesian to polar coordinates, we obtain.

$$x = r\cos\theta$$

$$y = r\sin\theta$$

$$r = \sqrt{x^2 + y^2}$$ (B.26)

$$\theta = \tan^{-1}\frac{y}{x}$$

Similarly, using the polar representation of the Fourier domain, we have

$$k_x = k\cos\theta'$$

$$k_y = k\sin\theta'$$

$$k = \sqrt{k_x^2 + k_y^2}$$ (B.27)

$$\theta' = \tan^{-1}\frac{k_y}{k_x}$$

From the rotation theorem, if a function is circularly symmetric, i.e., $f(x,y) = f(r)$, then its Fourier transform is also circularly symmetric, $f(k_x, k_y) = f(k)$.

The 2D Fourier transform is

$$\int_{-\infty}^{\infty}\int_{-\infty}^{\infty} f(x,y)\cdot e^{-i(k_x\cdot x+k_y\cdot y)}\,dxdy = \int_0^{2\pi}\int_0^{\infty} f(r)\cdot e^{-ikr\cos(\theta-\theta')}\,rdr\,d\theta$$

$$= \int_0^{\infty} f(r)\left[\int_0^{2\pi} e^{-ikr\cos\theta}\,d\theta\right] rdr$$

(B.28)

For circular symmetry, we see that the integral in Eq. B.28 does no longer depend on θ'. The integral over θ defines the Bessel function of *zeroth order* and *first kind*,

$$J_0(kr) = \frac{1}{2\pi}\int_0^{2\pi} e^{-ikr\cos\theta}\,d\theta$$

(B.29)

Thus, the resulting Fourier relationships become

$$\tilde{f}(k) = 2\pi\int_0^{\infty} f(r)J_0(kr)rdr$$

$$f(r) = 1/2\pi\int_0^{\infty} \tilde{f}(k)J_0(kr)kdk$$

(B.30)

Equations (B.30a-b) define a *Hankel transform* relationship (of 0th order) between f and $\tilde{f}$. Thus, because of the circular symmetry, the 2D Fourier transfer reduces to a 1D integral, where the $e^{ik\cdot r}$ kernel is replaced by $J_0(kr)$. Figure B.2 illustrates the behavior of the first order Bessel functions of orders $n = 0,1,2$.

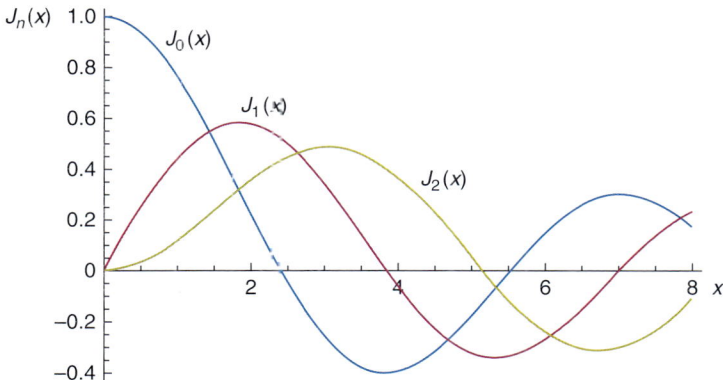

FIGURE B.2   Bessel function of first kind and various orders.

Some useful identities for Bessel functions of first kind are

$$J_n(x) = \frac{1}{2\pi i^n} \cdot \int_0^{2\pi} e^{ix\cos\theta} \cdot e^{in\theta}\, d\theta$$

$$\frac{d}{dx}\left[x^m J_m(x)\right] = x^m J_{m-1}(x) \tag{B.31}$$

$$\int_0^\infty J_n(x)\, dx = 1$$

The Hankel transform satisfies some important theorems, which are analogous to those of 1D and 2D Fourier transforms.[2]

Central ordinate theorem

$$\tilde{f}(0) = 2\pi \int_0^\infty \tilde{f}(r) J_0(0) r\, dr$$

$$= 2\pi \int_0^\infty f(r) r\, dr \tag{B.32}$$

Shift theorem

The circular symmetry is destroyed upon a shift in origin; Hankel transform does not apply.

Similarity theorem

$$f(ar) \to \frac{1}{a^2} \tilde{f}\left(\frac{k}{a}\right) \tag{B.33}$$

Convolution theorem

$$\int_0^\infty \int_0^{2\pi} f(r') g(\rho) r'\, dr'\, d\theta \to \tilde{f}(k) \cdot \tilde{g}(k)$$

$$(\rho^2 = r^2 + r'^2 - 2rr'\cos\theta) \tag{B.34}$$

Parseval's theorem

$$\int_0^\infty |f(r)|^2 r\, dr = \frac{1}{(2\pi)^2} \int_0^\infty |\tilde{f}(k)|^2 k\, dk \tag{B.35}$$

Laplacian

$$\nabla^2 f = \frac{d^2 f}{dr^2} + \frac{1}{r}\frac{df}{dr} \to -k^2 \tilde{f}(k) \tag{B.36}$$

Second moment

$$\int_0^\infty r^2 f(r)r\,dr = \frac{\tilde{f}''(0)}{-4\pi^3} \tag{B.37}$$

Equivalent width

$$\frac{2\pi\int_0^\infty f(r)r\,dr}{f(0)} = \frac{\tilde{f}}{2\pi\int_0^\infty \tilde{f}(k)k\,dk} \tag{B.38}$$

B.6   The 3D Fourier Transform

The Fourier pairs naturally extend to 3D functions as

$$f(x,y,z) = \int_{-\infty}^\infty \int_{-\infty}^\infty \int_{-\infty}^\infty \tilde{f}(k_x,k_y,k_z) \cdot e^{i(k_x \cdot x + k_y \cdot y + k_z \cdot z)} dk_x\,dk_y\,dk_z$$

$$\tilde{f}(k_x,k_y,k_z) = \int_{-\infty}^\infty \int_{-\infty}^\infty \int_{-\infty}^\infty f(x,y,z) \cdot e^{-i(k_x \cdot x + k_y \cdot y + k_z \cdot z)} dx\,dy\,dz \tag{B.39}$$

Next we discuss the 3D Fourier transform in *cylindrical* and *spherical* coordinates.

B.7   Cylindrical Coordinates

In this case,

$$f(x,y,z) = g(r,\theta,z)$$
$$\tilde{f}(k_x,k_y,k_z) = \tilde{g}(k_\perp,\theta',k_z) \tag{B.40}$$

where,

$$x + iy = r \cdot e^{i\theta}$$
$$k_x + ik_y = k_\perp \cdot e^{i\theta'} \tag{B.41}$$

The functions g and $\tilde{g}$ are related by

$$g(r,\theta,z) = \int_{-\infty}^\infty \int_0^{2\pi} \int_0^\infty \tilde{g}(k_\perp,\theta',k_z) \cdot e^{i[k_\perp r\cos(\theta-\theta')+k_z \cdot z]} dk_\perp\,d\theta'\,dk_z \tag{B.42a}$$

$$\tilde{g}(k_\perp,\theta',k_z) = \int_{-\infty}^\infty \int_0^{2\pi} \int_0^\infty g(r,\theta,z) \cdot e^{-i[k_\perp r\cos(\theta-\theta')+k_z \cdot z]} dr\,d\theta\,dz \tag{B.42b}$$

Under *circular symmetry*, i.e., f independent of θ, and, thus, $\tilde{f}$ independent of θ', we have

$$f(x,y,z) = h(r,z)$$

$$\tilde{f}(k_x,k_y,k_z) = \tilde{h}(k_\perp,k_z) \tag{B.43}$$

Therefore, Eq. (B.42*a-b*) becomes (using the properties of the Hankel transform)

$$h(r,z) = \frac{1}{2\pi}\int\limits_{-\infty}^{\infty}\int\limits_{0}^{\infty}\tilde{h}(k_\perp,k_z)\cdot J_0(k_\perp\cdot r)\cdot e^{ik_z\cdot z}\cdot k_\perp dk_\perp dk_z \tag{B.44a}$$

$$\tilde{h}(k_\perp,k_z) = 2\pi\int\limits_{-\infty}^{\infty}\int\limits_{0}^{\infty}h(r,z)\cdot J_0(k_\perp\cdot r)\cdot e^{-ik_z\cdot z}\cdot r\,dr\,dz \tag{B.44b}$$

It can be seen that the integral in Eq. (B.44*a*) represents a 1D Fourier transform along z of the Hankel transform with respect to r.

If the problem has *cylindrical symmetry*, i.e., f is independent of both θ and z, we have

$$f(x,y,z) = p(r)$$

$$\tilde{p}(k_x,k_y,k_z) = \tilde{p}(k_\perp)\cdot\delta(k_z) \tag{B.45}$$

The Fourier transform simplifies to

$$p(r) = \frac{1}{2\pi}\int\limits_{0}^{\infty}\tilde{p}(k_\perp)\cdot J_0(k_\perp r)k_\perp dk_\perp$$

$$\tilde{p}(k_\perp) = 2\pi\int\limits_{0}^{\infty}p(r)\cdot J_0(k_\perp r)r\,dr \tag{B.46}$$

B.8   Spherical Coordinates

In spherical coordinates,

$$f(x,y,z) = g(r,\theta,\phi)$$

$$\tilde{f}(k_x,k_y,k_z) = \tilde{g}(k,\theta',\phi') \tag{B.47}$$

The change of coordinates takes the form

$$x = r\sin\theta\cos\phi,\ y = r\sin\theta\sin\phi,\ z = r\cos\theta$$

$$k_x = k\sin\theta'\cos\phi',\ k_y = k\sin\theta'\sin\phi',\ k_z = k\cos\theta' \tag{B.48}$$

where, again, $k = \sqrt{k_x^2 + k_y^2 + k_z^2}$.

The Fourier integrals become

$$g(r,\theta,\phi) = \int\limits_{-\infty}^{\infty}\int\limits_{0}^{2\pi}\int\limits_{0}^{2\pi} \tilde{g}(k,\theta',\phi') \cdot e^{i[\cos\theta\cos\theta' + \sin\theta\sin\theta'\cos(\phi-\phi')]} \cdot k^2 \sin\theta' \, dk\, d\theta' \, d\phi'$$

$$\tilde{g}(k,\theta',\phi') = \int\limits_{-\infty}^{\infty}\int\limits_{0}^{2\pi}\int\limits_{0}^{2\pi} g(r,\theta,\phi) \cdot e^{-i[\cos\theta\cos\theta' + \sin\theta\sin\theta'\cos(\phi-\phi')]} \cdot r^2 \sin\theta \, dr\, d\theta \, d\phi$$

$$(B.49)$$

Under *circular symmetry*, i.e., $f(x,y,z)$ is independent of ϕ, we have

$$f(x,y,z) = g(r,\theta)$$
$$\tilde{f}(k_r, k_y, k_z) = \tilde{g}(k,\theta')$$

$$(B.50)$$

The Fourier transforms for circular symmetry are

$$g(r,\theta) = 2\pi\int\limits_{0}^{\infty}\int\limits_{0}^{\pi} \tilde{g}(k,\theta')J_0(kr\sin\theta \cdot \sin\theta') \cdot e^{ikr\cos\theta\cos\theta'} k^2 \sin\theta' \, dk\, d\theta'$$

$$(B.51)$$

$$\tilde{g}(k,\theta') = 2\pi\int\limits_{0}^{\infty}\int\limits_{0}^{\pi} g(r,\theta)J_0(kr\sin\theta \cdot \sin\theta') \cdot e^{-ikr\cos\theta\cos\theta'} r^2 \sin\theta \, dr\, d\theta$$

For *spherical symmetry*, the function is independent of both angles, such that we have

$$f(x,y,z) = h(r)$$
$$\tilde{f}(k_x, k_y, k_z) = \tilde{h}(k)$$

$$(B.52)$$

In this case, the integrals reduce to

$$h(r) = \frac{1}{2\pi^2}\int\limits_{0}^{\infty} \tilde{h}(k)\,\mathrm{sinc}(kr)k^2 dk$$

$$(B.53)$$

$$\tilde{h}(k) = 4\pi\int\limits_{0}^{\infty} h(r)\,\mathrm{sinc}(qr)r^2 dr$$

A few examples of 3D Fourier pairs, relevant for scattering problems, are shown in the table below.

$f(x, y, z)$	$\tilde{f}(k_x, k_y, k_z)$
$\delta(x-a, y-b, z-c)$	$e^{i(k_x \cdot a + k_y \cdot b + k_z \cdot c)}$
$\Pi(x, y, z)$ (cube)	$\dfrac{1}{(2\pi)^3} \sin\left(\dfrac{k_x}{2\pi}\right) \cdot \sin\left(\dfrac{k_y}{2\pi}\right) \cdot \sin\left(\dfrac{k_z}{2\pi}\right)$
$\Pi(x, y)$ (bar)	$\dfrac{1}{(2\pi)^2} \sin\left(\dfrac{k_x}{2\pi}\right) \cdot \sin\left(\dfrac{k_y}{2\pi}\right) \cdot \delta(k_z)$
$\Pi(x)$ (slab)	$\dfrac{1}{2\pi} \cdot \sin\left(\dfrac{k_x}{2\pi}\right) \cdot \delta(k_y) \cdot \delta(k_z)$
$\Pi\left(\dfrac{r}{2}\right)$ (ball)	$\dfrac{\sin k - k \cdot \cos k}{2k^3}$
$\Pi(x)\Pi\left(\sqrt{y^2 + z^2}\right)$ (disk)	$\dfrac{1}{(2\pi)^3} \sin\left(\dfrac{k_x}{2\pi}\right) \cdot \dfrac{J_1\left(\sqrt{k_y{}^2 + k_z{}^2}\middle/ 2\right)}{\sqrt{k_y{}^2 + k_z{}^2}\middle/ \pi}$
$\dfrac{e^{-\alpha r}}{4\pi r}$	$\dfrac{1}{\alpha^2 + k^2}$
$\dfrac{e^{-r/R}}{\frac{4}{3}\pi R^3}$	$\dfrac{1}{(2\pi)^3} \dfrac{6}{(1 + k^2 R^2)^2}$
$e^{\frac{-\alpha r^2}{2}}$	$e^{\frac{-k^2}{2\alpha}}$

References

1. A. Papoulis. *The Fourier Integral and Its Applications* (McGraw-Hill, New York, 1962).
2. R. N. Bracewell. *The Fourier Transform and Its Applications* (McGraw-Hill, Boston, MA, 2000).
3. J. D. Gaskill. *Linear Systems, Fourier Transforms, and Optics* (Wiley, New York, 1978).
4. J. W. Goodman. *Introduction to Fourier Optics.* (McGraw-Hill, New York, 1996).

APPENDIX C
QPI Artwork

Plate 1. Red blood cells (acquired using FPM).

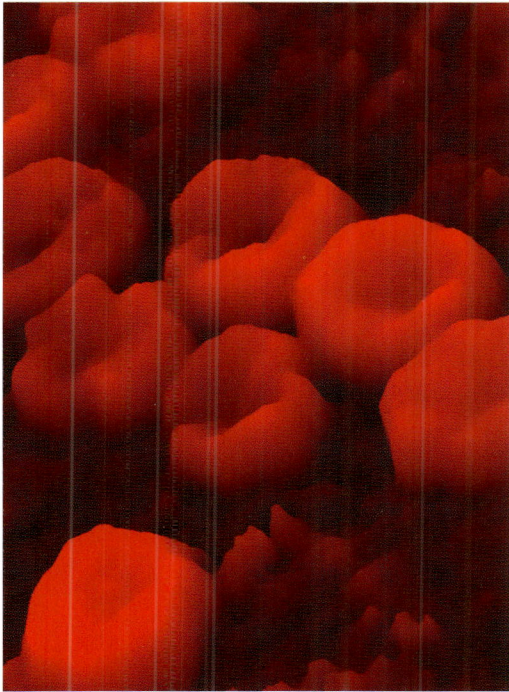

Plate 2. Neuron (acquired by SLIM-red channel and fluorescence-green and blue channels).

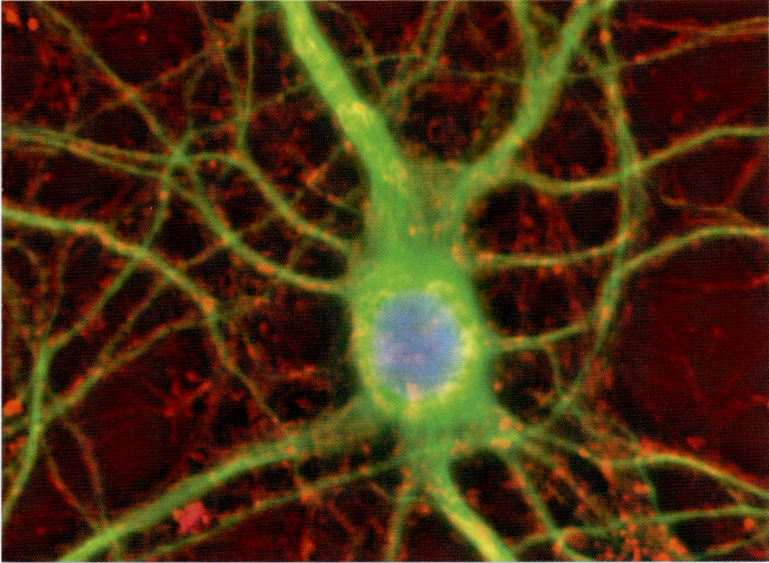

Plate 3. Red blood cells and putative macrophage (acquired using HPM).

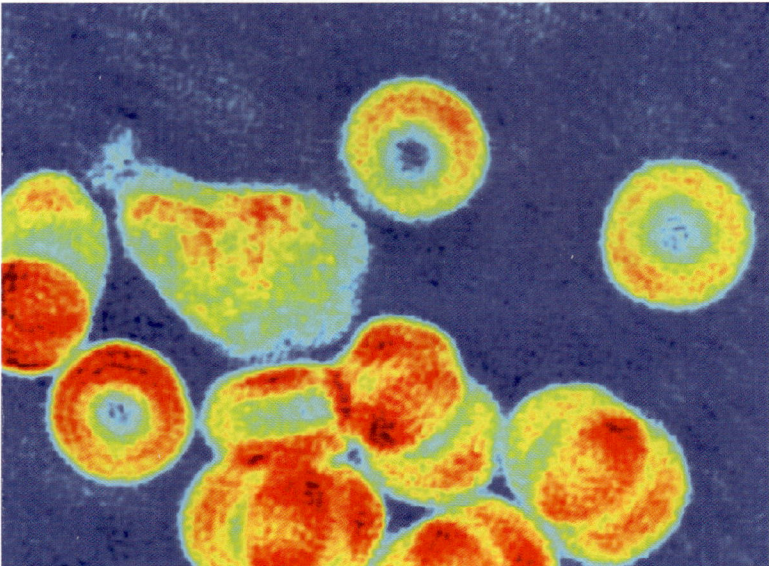

Plate 4. Glial cell (acquired by SLIM).

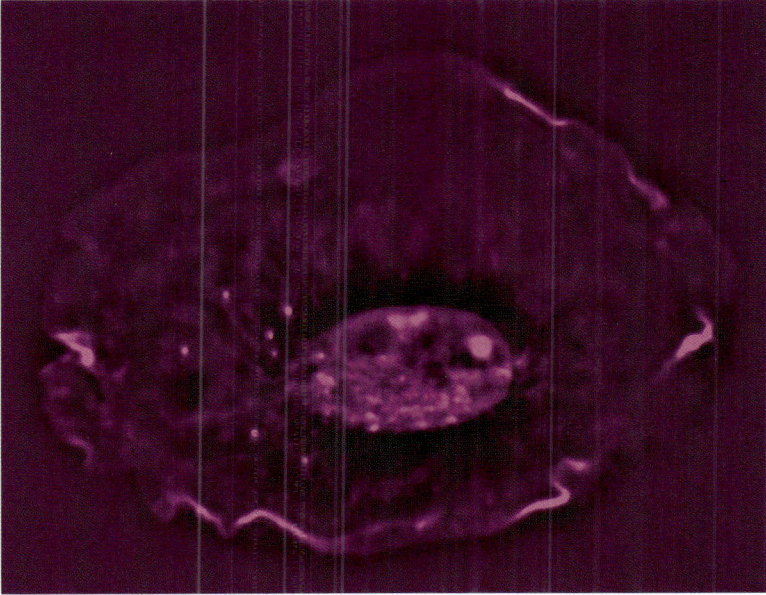

Plate 5. HeLa cells (acquired by FPM).

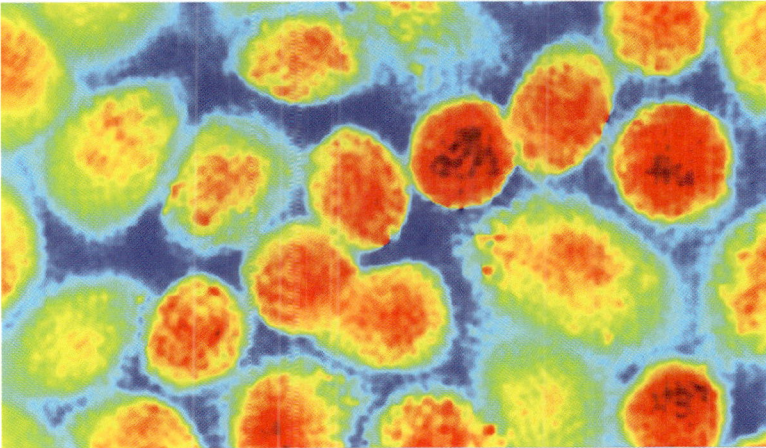

Plate 6. RBC squished by macrophage (acquired using HPM).

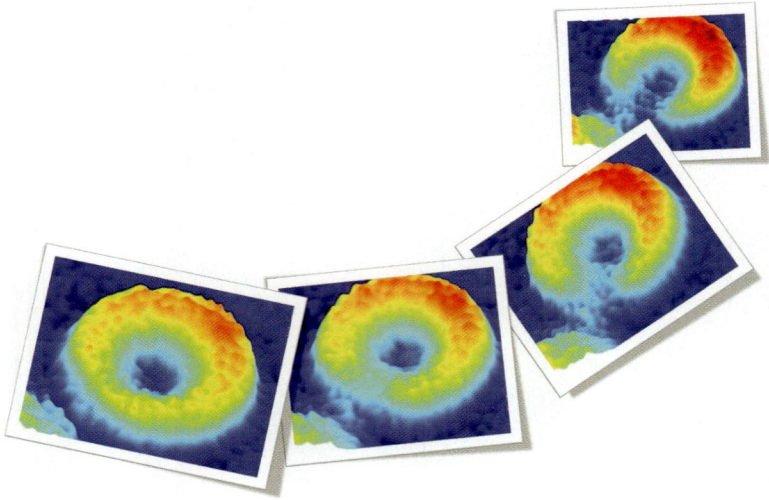

Plate 7. Neuron (acquired by SLIM).

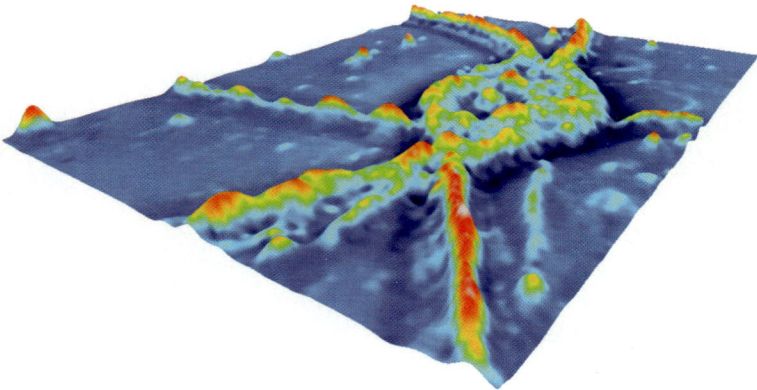

Plate 8. Malaria-infected cell (acquired by DPM).

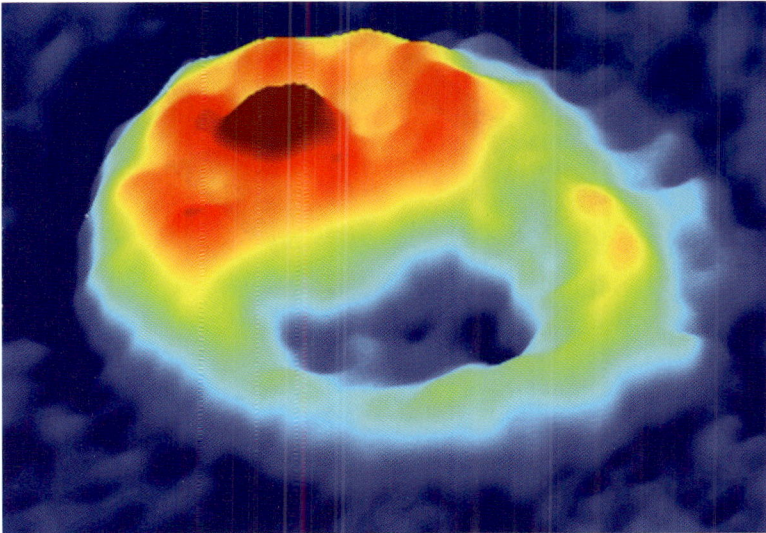

Plate 9. Neuron network: SLIM (red), Laplacian of SLIM (green).

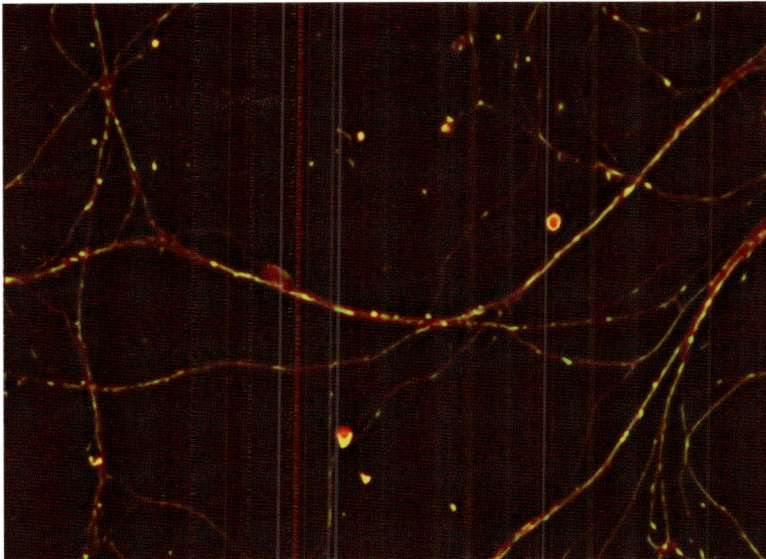

Plate 10. Discocyte, echinocyte, spherocyte (acquired using stabilized HPM).

(a)

(b)

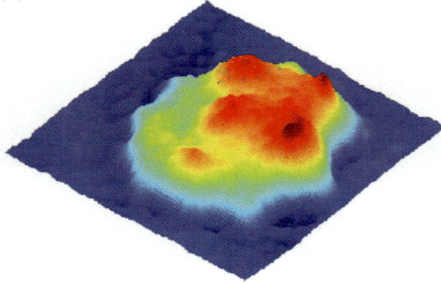

(c)

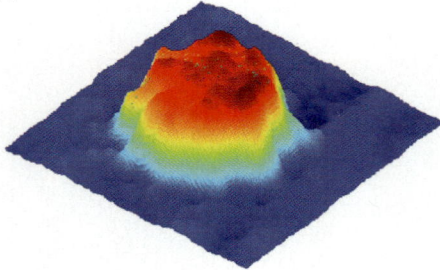

Plate 11. Eosinophil marching by a red blood cell (acquired by HPM).

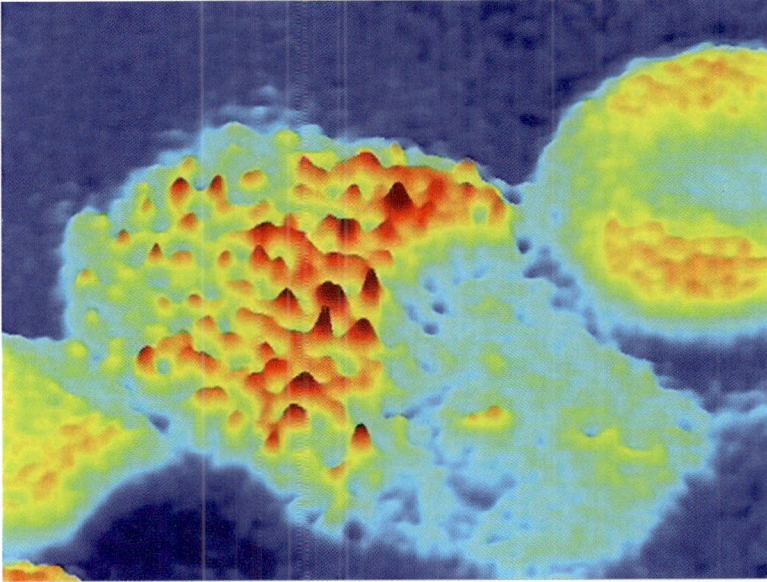

Plate 12. Neuron (acquired by SLIM).

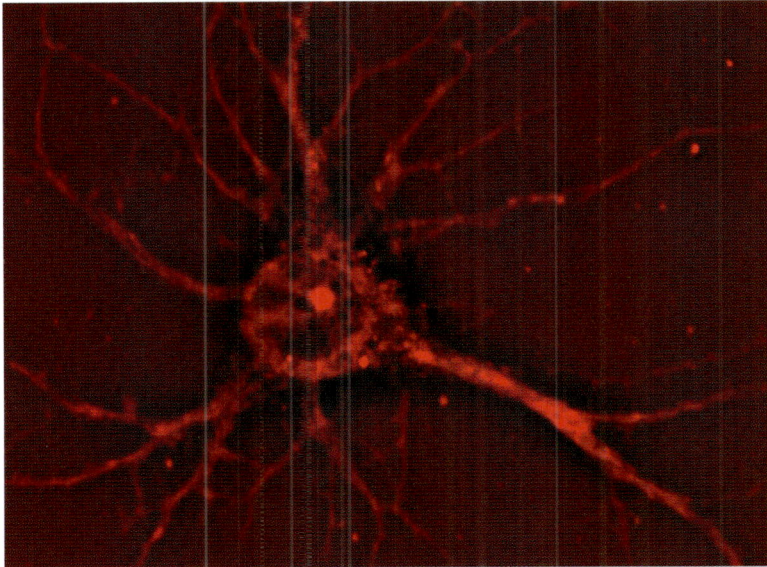

Plate 13. Dendritic structure (acquired by SLIM).

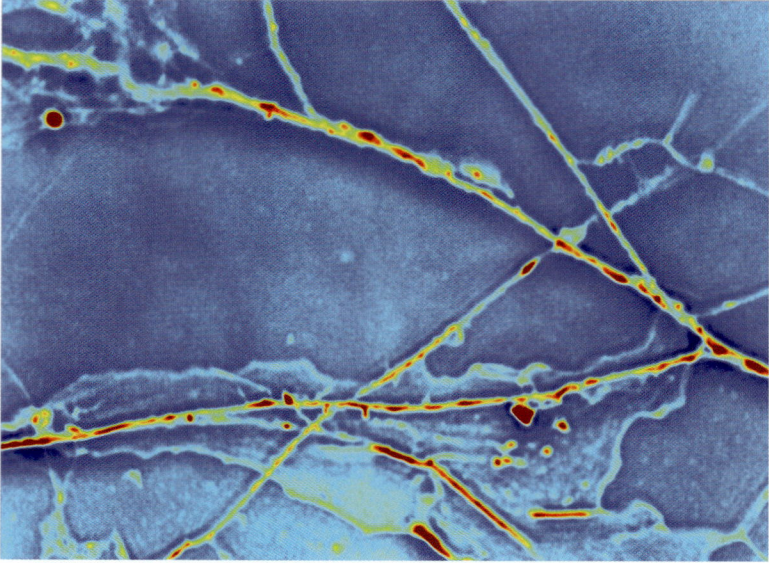

Plate 14. Neuron with nucleolus (acquired by SLIM).

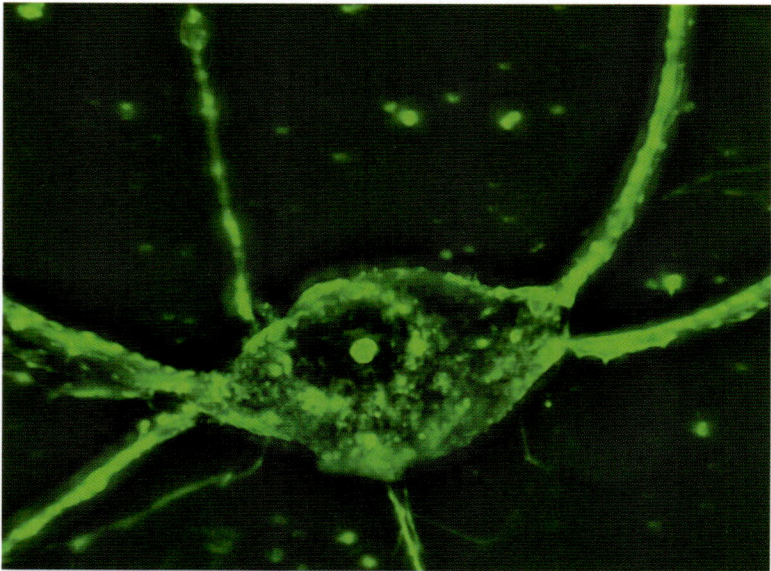

Plate 15. Fibroblasts (acquired by SLIM).

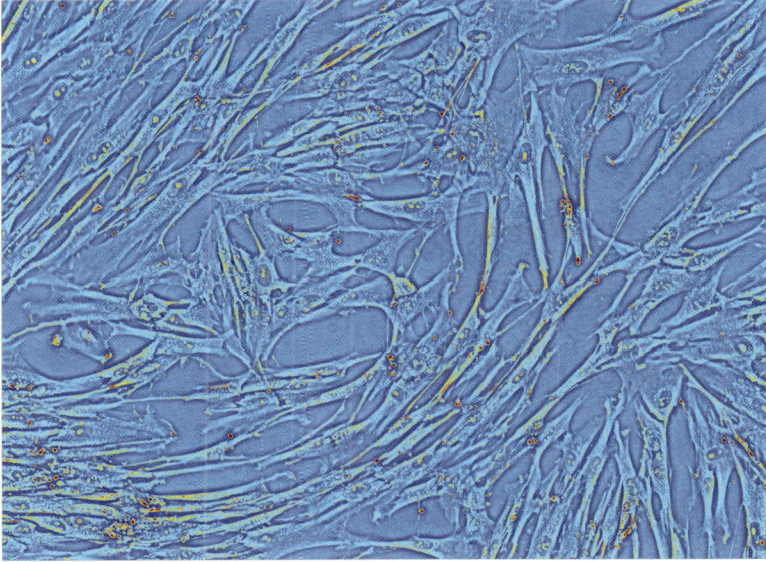

Plate 16. Chromosomes (acquired by SLIM).

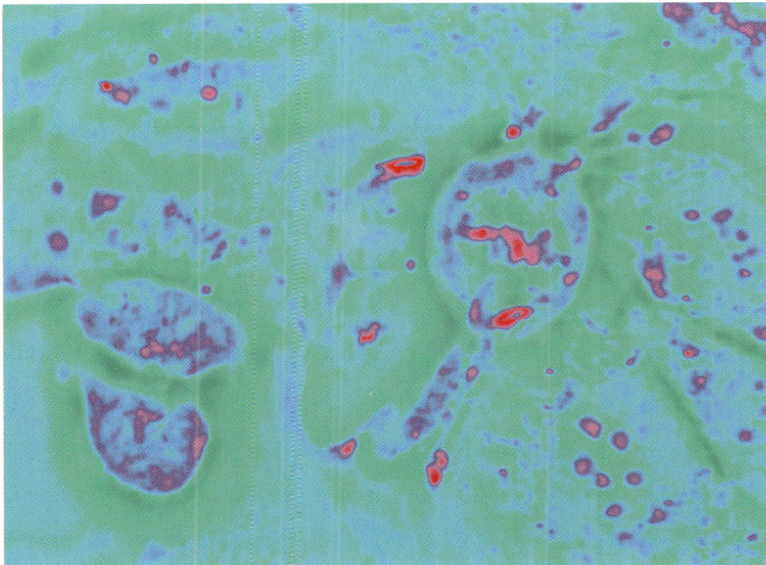

Plate 17. Prostate biopsy with blood vessel (acquired by SLIM).

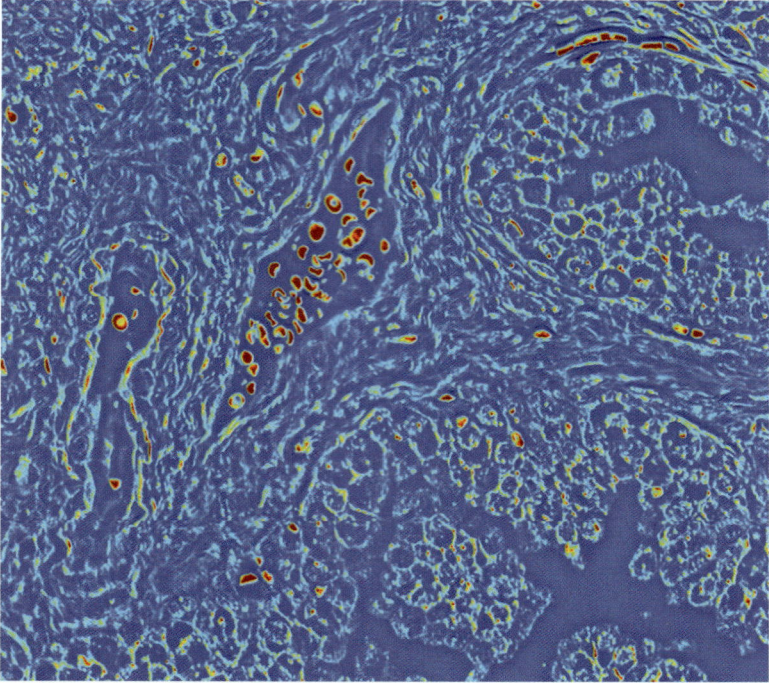

Plate 18. Red blood cells (acquired by SLIM).

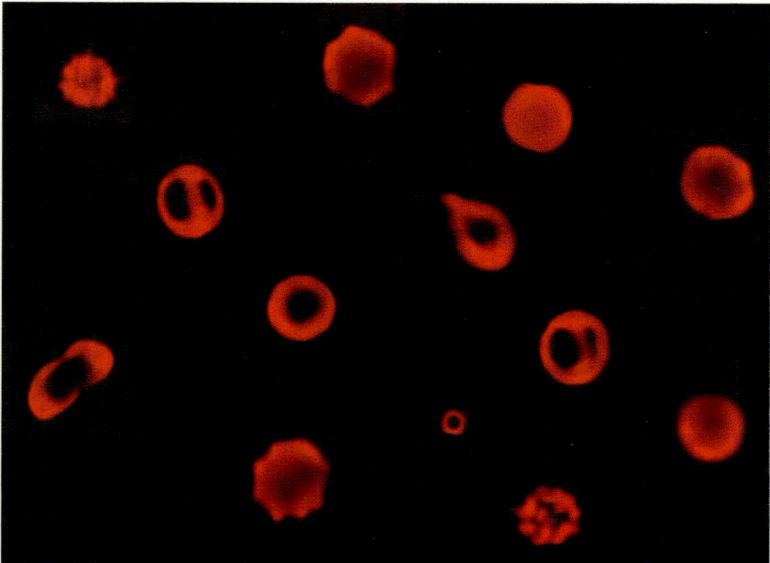

Plate 19. Cell cycle (acquired by SLIM).

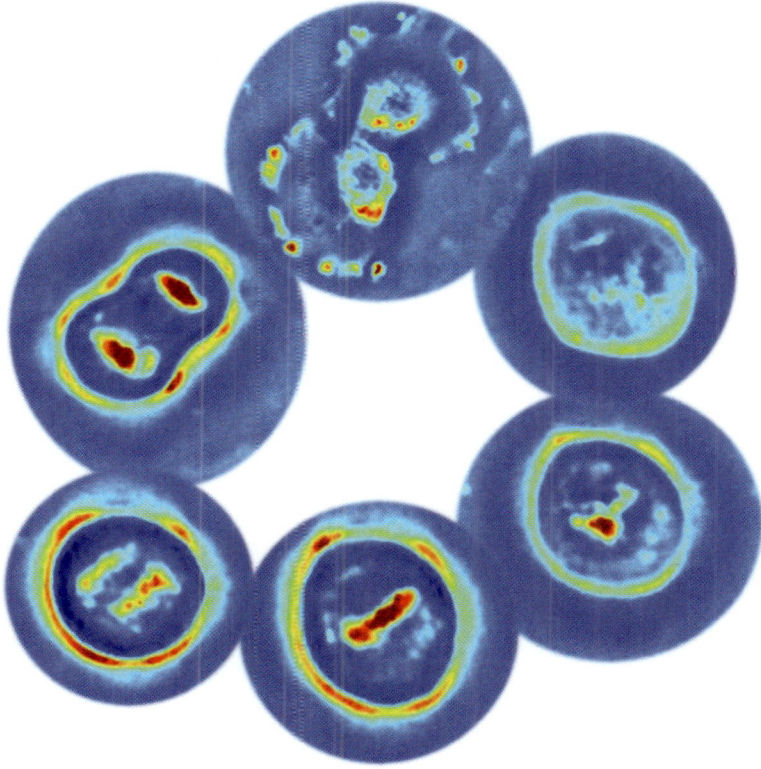

Plate 20. Blood smear (acquired by SLIM).

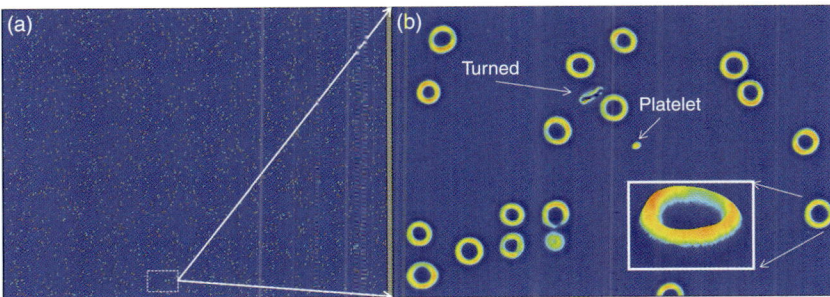

Index